The Science and Lore of the Plant Cell Wall

Biosynthesis, Structure and Function

Edited by
Takahisa Hayashi

BrownWalker Press
Boca Raton • 2006

The Science and Lore of the Plant Cell Wall:
Biosynthesis, Structure and Function

BrownWalker Press
Boca Raton , Florida
USA • 2006

ISBN: 1-58112-445-7 (paperback)
ISBN: 1-58112-446-5 (ebook)

BrownWalker.com

Contents

Cell Walls in Development

Wall Assembly and Loosening

Cell Wall Genomics, Proteomics and Glycomics

This volume is dedicated to the 2004 Anselme Payen Award Winner, Deborah P. Delmer. Dr. Delmer has received international recognition for her contributions to cellulose biosynthesis. She was also elected as a member of the National Academy of Sciences USA in April 2004. Her pioneering research from her challenging heart has stimulated our studies on plant cell wall.

The Anselme Payen Award is named for the distinguished 19th-century French scientist who identified and defined cellulose as the main fibrous component found universally in plant cell walls. The award is recognized internationally as the most significant honor in the science and technology of cellulose.

Takahisa Hayashi

Preface

My thanks go out to Takahisa Hayashi, the Editor of this book, for organizing such an interesting effort. Taka also asked ME to contribute an article, but, as I've moved on to a whole new life dealing with issues of agriculture in the developing world, I declined. I did so because of my feeling that the field of plant cell walls now has so many talented individuals, it was definitely time for me to step aside and let you all speak for yourselves. But I did agree to write this preface and reminisce very briefly about my own research with cell walls and with the topic of cellulose synthesis in particular..

I thought that one interesting way to "reminisce" might be to use my accumulated experiences in a different way and instead to ask the question: "If I were young and starting over again to work on plant cell walls and cellulose synthesis, where would I like to focus my efforts?" Obviously, we tend to focus on projects that suit our own particular set of skills, which in my case, are primarily as a biochemist and a Johnny-Come-Lately molecular biologist. But I'll cheat a little and just imagine I could learn all the new skills I would need in order for me to extend the range of possibilities I see emerging as potential topics for research.

Just to touch briefly on the non-cellulosic part of the wall, I am a bit hesitant to suggest new directions, knowing that others will have a much better perspective than I. But I can say that I think that, in terms of biosynthesis, the powers of both genomics and molecular genetics are beginning to reveal themselves as being huge assets. It is clear that, as for cellulose synthesis for so many years, it has proven exceedingly difficult to demonstrate synthesis of complex non-cellulosic polymers using isolated membrane preparations. One notable exception has been ability to synthesize galactomannans, but when it comes to pectins and other non-cellulosic polysaccharides, the complexity of the systems, the need for suitable primers and/or cooperative interactions among a large number of enzymes, and the low levels of such enzymes in membranes, has made this task exceedingly challenging. Clearly the ability to isolate mutants with altered polysaccharide composition provides one avenue for the identification of key genes encoding biosynthetic enzymes. Yet, possible redundancy of genes and/or inability to detect altered phenotypes for whatever reason can still limit this approach. This is where genomics can play an important role, and we are finally beginning to see how searching for important genes based upon their homology with other related glycosyltransferases is beginning to pay off for pathways involved in synthesis of polymers such as the pectins, and xyloglucan. And, when it comes to structure, well, I think it won't be long before we'll understand just how all those various structures that we lump as together as pectins really do associate with each other and perhaps also with other non-pectic polysaccharides and/or proteins. One hopes we will also finally get clarification of how expansins and XETs modulate extensibility, and maybe even one day we'll really understand that great white whale of plant biology—how does auxin really control cell elongation?

Discussion of these possibilities also helps me make one very important point—it is no longer possible just to be one kind of scientist as many of us used to be. Each group now needs to have members who are either skilled themselves in many fields, or find partners who can complement their own skills in order to really make progress. Certainly each group needs to be able to use model systems effectively because they offer ease of manipulation, complete genome sequences, and extensive mutant collections. Yet model systems also have limitations. I think most of us really do hope that our research will one day lead to greater benefits to agriculture. Thus, we hope that, at some point, the results of model systems can to be translated to enhancing our understanding and ultimate improvement of crops such as cotton, trees, and major cereal or legume crops. Model systems also may not be the best for hard biochemistry where large amounts of tissue may be required, and they may lack specialized tissues (like the cotton fiber for cellulose synthesis) where a particular process predominates at one particular stage of development in one distinct and easily-isolated cell type. In short, we need to be flexible and not wedded just to one approach for our entire careers. Certainly my own career took a turn for the better when I was able to make the leap beyond just biochemistry.

Turning to cellulose synthesis, I have to say that I am REALLY pleased to see how many new faces have joined the field and the new insights you have added to our understanding of this process. Through genomics and mutant studies, we now know that there are a number of *CesA* genes that indeed can be predicted to catalyze glucan chain elongation in some way, and that some of these distinct CESA proteins mysteriously seem to be required to work in combination to synthesize microfibrils. But why this is true still eludes us. It is clear that genetics can only suggest that this is true and may also provide indications of other new partner proteins that may be necessary for the process. But hard work involving good, serious biochemistry will still be important to provide a final confirmation of the structure of the rosette, the function of each CESA within the rosette, and the role of other accessory proteins. Questions also remain as to what distinguishes primary from secondary wall cellulose synthesis. Steryl glycosides seem to function as primers in vitro in membranes of cotton fibers engaged in secondary wall synthesis. Is this some *in vitro* artefact, is it unique to secondary wall synthesis *in vivo*, or is it a general phenomenon for all of cellulose synthesis? Is one CESA responsible for elongating primers while others may function in further elongation? What is different about the CESA's required for primary vs. secondary wall synthesis? Why do rosettes/ CESAs appear to have such short half lives? Only time will tell. Why should cellulases like Korrigan be needed? Perhaps to cleave the primers, to edit mistakes in growing microfibrils, or to facilitate chain termination? How do rosettes assemble? What are the domains that associate to form rosettes—e.g., do the zinc finger domains interact in a redox-dependent fashion? (In fact, if there were two topics I'm sure I'd address more seriously if I were to continue would be the general role of redox regulation and the role of protein turnover in the process of cell wall assembly.) Are there scaffolding proteins and/or other accessory proteins embedded within the rosettes? Do microtubules really interact directly with rosettes to guide their movement as suggested by some recent data? The answers to some of these questions are beginning and will continue to be answered with the additional help of the fantastic new tools for imaging that are improving at a very rapid pace. Will we one day really be able to see clearly the rosette moving within the membrane and show with such imaging techniques exactly what proteins are within the rosette and how its movement is guided? (Simon Turner's lab seems to be getting close to doing just that!) And what about proteins that may be involved but may not directly interact with the rosette, such as Korrigan or sucrose synthase? And, well, have we really solved the issue of how synthesis of cellulose and callose may or may not be connected? And how did these biosynthetic pathways evolve?

Finally, what may be the practical benefits that could arise from such work? One prediction is that work with enzymes like sucrose synthase may provide new insights into how carbon is partitioned in plant cells. Clearly the cell wall is a major sink for carbon, but, unless you are a ruminant, it's not exactly the world's best source of calories. And lignin is even less pleasant than cellulose—clearly of benefit to the plant, but with major impacts on digestibility for both humans and animals, and on the efficiency and safety of processing of plant tissues for production of paper, starch, and ethanol. So understanding how carbon may be partitioned between the digestible and non-digestible polymers may offer opportunities to alter the value of many plant products. As one who invariably seems to choose in the market the toughest old radishes, turnips, rutabagas, and carrots and the stringiest celery and green beans, I've also wondered if there may be opportunities to alter cellulose (and lignin) content in such veggies to enhance texture as well as digestibility. With suitable tissue-specific promoters and the right genes, it surely seems quite possible. Finally, I've been most gratified to see that some of our research may also prove valuable for the kind of work I'm now doing with the Rockefeller Foundation. Lodging of cereals is a problem throughout the world, not only for maize in the fields of commercial farmers, but also for crops such as rice, millet, sorghum and tef in the developing world. While the green revolution genes have played an important role in helping solve this problem, it now appears there may be opportunities to enhance resistance to lodging by over-expression of *CesA* genes in specific tissues.

So I leave this fascinating field just when the opportunities seem the greatest. But it makes leaving so much easier to see how many talented young (and older!) scientists have taken up the cause. Taka has asked

you all to speculate as you write the articles for this book. This is surely fitting since I've not ever been shy to speculate throughout my career. (How I remember Bruce Stone moaning "Oh Dear! Not another Delmer theory!!!"). But it's a valuable exercise even when we get it wrong, as it does challenge us and our colleagues to extend our way of thinking. Yet, as valuable as speculation can be, it is no substitute for good hard facts. And that is what's so gratifying now—you no longer have to do SO MUCH speculating, as the answers to many of the questions posed above are now well within your reach.
Enjoy the journey!

Deborah Delmer

Introduction

Plant cell walls are composed of complex carbohydrates, proteins, phenolic compounds, and inorganic ions, all of which play functional roles. Several models for plant cell walls help explain their structure, which consists of primary and secondary walls. The primary wall is the outer wall of a growing plant cell, and elongates and/or expands over the life of the cell. The secondary wall exists inside the primary wall in elongated and expanded cells. Structural analysis has primarily employed microscopy of walls, because "seeing is believing." Since cellulose microfibrils are skeletal and architectural components, their orientation determines the physical properties of plant cell walls. The fine structure of the wall has been determined by chemical analysis.

Cellulose (1,4-β-glucan) and callose (1,3-β-glucan) are synthesized in the plasma membrane, while other polysaccharides are synthesized in the Golgi. Although cellulose is the most abundant biopolymer on the earth, the details of its biosynthesis have been difficult to elucidate, despite the ongoing work of many scientists for many years. The success story of cellulose biosynthesis, by V. Bulone, serves as an excellent plenary chapter in this book.

Plant cell growth occurs with the loosening of the walls, which may be caused by several enzymatic actions. Many polysaccharides synthesized in the Golgi are exported to the cell wall, where hydrogen bonding occurs between xyloglucan and cellulose, together with chemical cross-linking between polymers. By measuring wall extensibility, tethers between xyloglucan and cellulose have been proven to play important roles. Xyloglucan metabolism is undoubtedly involved in the loosening.

Plant development is related to the morphological changes of cells and tissue, which is caused by structural changes of the walls: the generation of walls from grasses to seeds, induction of secondary wall formation, and cotton fiber development. The key action might be caused by XTHs and proline-rich wall proteins.

A recent genome project has revealed about 248 glycan hydrolases and 214 glycan transferase genes, including 10 cellulose-synthase (*CesA*) and 30 cellulose-synthase-like (*Csl*) genes in *Arabidopsis thaliana*. Genetic analysis of gene knockout mutants in *Arabidopsis thaliana* has revealed the function of these genes. Proteomics and glycomics also provide new insights into the studies on plant cell walls.

Woody plants make up 80% of total plant biomass, which is the carbon source for living organisms and also the biological sink of CO_2 on the Earth. Cell walls control plant cell growth and define the structure of plants, both in guiding the development of plants and in providing a role in their defense against pathogens. Other benefits of plant cell walls are their uses as food, various materials, and energy for human beings.

Takahisa Hayashi

Models of Plant Cell Walls

Kei'ichi Baba

The primary cell wall is composed of complex carbohydrates and proteins. Many models have been proposed to understand its structure, beginning with the first complete model by Keegstra *et al.* (1973), which was based on enzymatic and chemical analyses of the walls of growing sycamore cells (Figure 1). Their primary theory was that rhamnogalacturonan, arabinogalactan, xyloglucan, and hydroxyproline-rich proteins are interconnected by covalent bonds, and the connection between cellulose and xyloglucan is hydrogen bonds. The walls of growing plant cells were thought to act as a macromolecular complex. Based on this model, numerous other models were formed using additional findings (Figures 2-5). Figure 2 (Fry 1986) shows the interactions between the molecules in the cell walls, and Figure 3 shows the matrix polysaccharides arranged nearly parallel to the cellulose microfibrils.

The models can be divided into two approximate categories. One type stresses chemistry and the interactions and bonds between molecules, while the other attempts to express an artistic architectural construction of the molecules. Figure 4 (Roberts 1994) illustrates the most popular model of the latter type. It appears in many general biology textbooks; though the drawing is simple, it is a beautiful description of cell wall construction. Since some of the polysaccharides and the cell wall structure of some monocots are different from those of dicots and other monocots, they were drawn separately as Types I and II in Figure 5. Most dicots and noncommelinoid monocots have Type I walls, while commelinoid monocots have Type II walls.

A completely different model has been proposed for secondary walls in woody plants (*e.g.*, Esau 1977; Higuchi 1997). Figure 6 (Liese 1970) shows a gymnosperm tracheid and the alteration of microfibril orientation in the layered secondary wall. Later, the fiber wall of angiosperms was revealed to have the same structure as this model. However, bamboo has cell walls with more layers (Figure 7: Parameswaran and Liese 1981). The secondary wall is known to consist of cellulose microfibrils, hemicelluloses, and lignin. A model (Figure 8) has been proposed for their localizations, but no details of the molecular bonding level have been found. How they interact and construct the secondary wall remains unknown.

We can clarify the story by gathering the models together, focusing on each stage of wall material production, and so on. Because of limited space, only two figures for wall formation are shown here. Cellulose microfibrils are produced on the membrane along the cortical microtubules (Figure 9: Gunning and Steer 1996). New cellulose and xyloglucan meet just outside the membrane and immediately bond tightly to each other (Figure 10: Hayashi 1989).

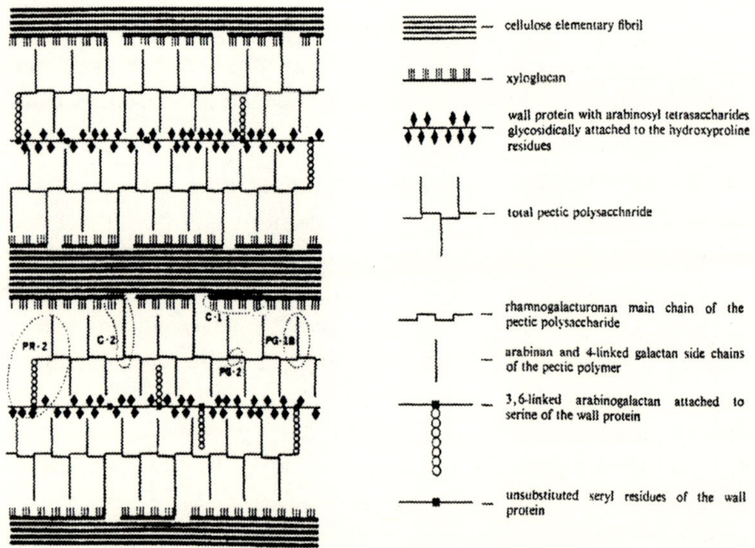

Fig. 1 Tentative structure of sycamore cell walls.
This model is not intended to be quantitative, but is instead an effort to present the wall components in approximately correct proportions. The distance between cellulose elementary fibrils is expanded to allow presentation of the interconnecting structure. The circled areas are representative wall fractions released by the degrading enzymes. They are fractions PG-1B and PG-2 released by endopolygalacturonase, fractions C-1 and C-2 released by endoglucanase, and fraction PR-2 released by pronase. The symbols shown at right represent the various components of the cell wall (Keegstra *et al.* 1973).

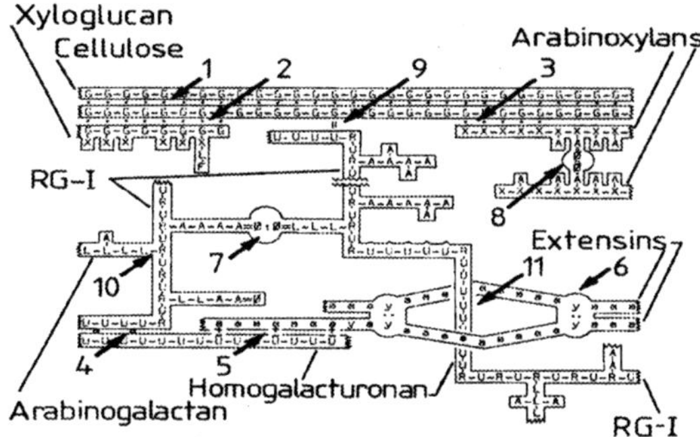

Fig. 2 Representative primary structures and possible cross-linking of wall polymers.

This is not a model of the plant cell wall, and no significance is placed on the chain length, orientation, conformation, or spacing of the molecules (Fry 1986).

(•) hydrogen bonds:
- **1. cellulose-cellulose**
- **2. xyloglucan-cellulose**
- **3. xylan-cellulose**

(o) calcium bridges:
- **4. homogalacturonan-homogalacturonan**

(±) other ionic bonds:
- **5. extensin-pectin**

(:) coupled phenols:
- **6. extensin-extensin**
- **7. pectin-pectin**
- **8. arabinoxylan-arabinoxylan**

(=) ester bonds:
- **9. pectin-cellulose**

(-) glycosidic bonds:
- **10. arabinogalactan-rhamnogalacturonan**

(<|>) entanglement:
- **11. pectin-in-extensin**

***A*, arabinose; *F*, fucose; *G*, glucose; *L*, galactose; *R*, rhamnose; *U*, galacturonic acid; *U^*, galacturonic acid methyl ester; *a*, amino acid other than tyrosine; *y*, tyrosine; *y:y*, isodityrosine; *ϕ*, ferulic acid; *ϕ:ϕ*, diferulic acid.**

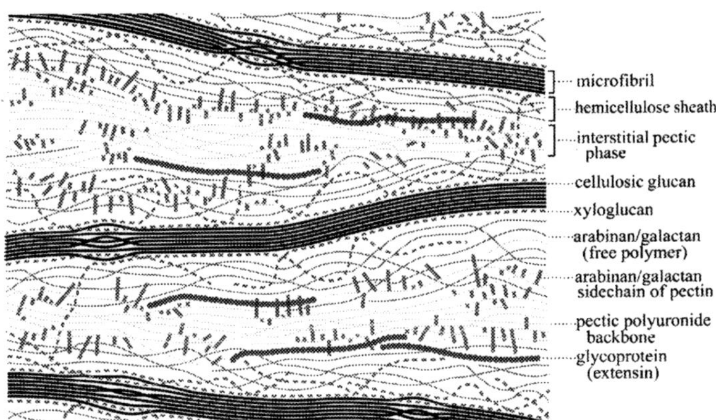

Fig. 3 Schematic of suggested polymer organization in pea primary walls, depicted in the plane of the cellulose microfibrils.

The xyloglucan and arabinogalactan layers form a hemicellulosic sheath ("cortex") around each microfibril, the non-crystalline portions of which contain intercalated xyloglucan chains. Interstices between the hemicellulose-coated microfibrils are occupied by pectin. Where microfibrils approach one another more closely, their hemicellulose sheaths may overlap, and some xyloglucan chains may extend from one microfibril to another. The differently labeled polymer types are depicted in the relative portions in which they occur in pea cell walls, according to this study and previous reports. Smooth portions of polyuronide backbones donate homogalacturonan, and side chain-bearing portions are rhamnogalacturonan blocks. The longer blocks represent lengthy runs of substituted galacturonosyl-rhamnose, such as RG-I. Except for extensin, minor components (*e.g.*, mannan and xylan) are not shown (Talbot and Ray 1992).

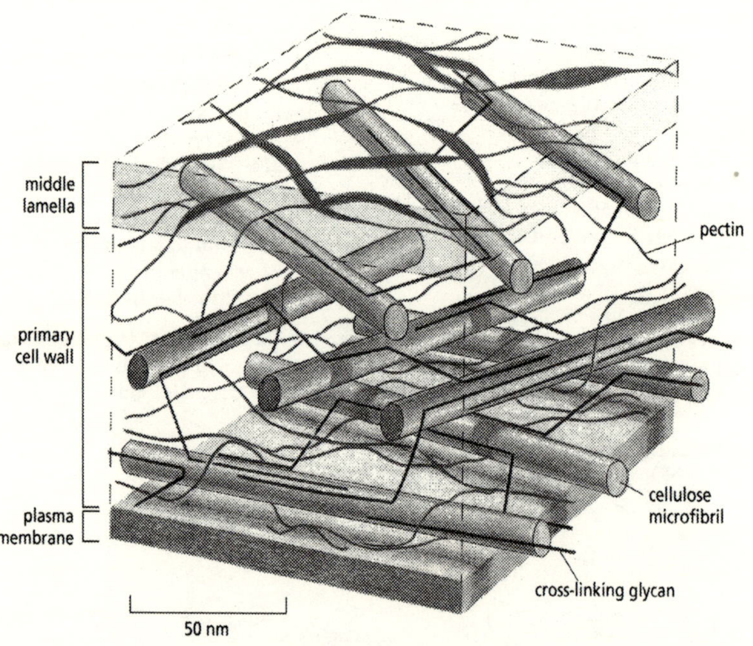

Fig. 4 Scale model of a portion of primary cell wall showing the two major polysaccharide networks.

The orthogonally arranged layers of cellulose microfibrils are cross-linked into a network by hydrogen-bonded hemicellulose. This network is coextensive with a network of pectin polysaccharides. The cellulose and hemicellulose network provides tensile strength, while the pectin network resists compression. Cellulose, hemicellulose, and pectin are typically present in roughly equal quantities in a primary cell wall. The middle lamella is pectin-rich and cements adjacent cells together (Roberts 1994).

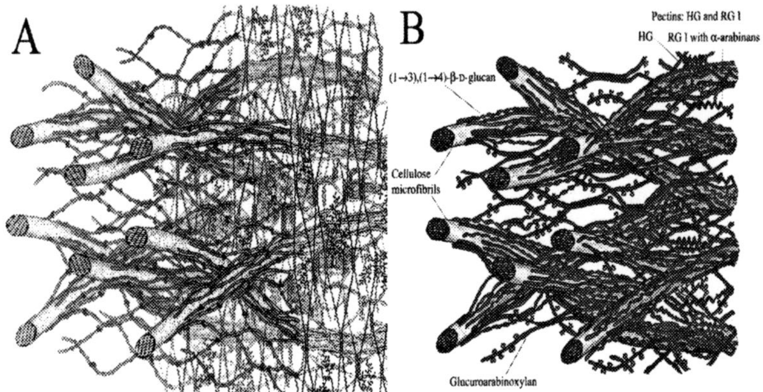

Fig. 5 (A) Architectural model of the Type I cell wall.
The Type I cell wall is a strong but dynamic network of cellulose, tethered by cross-linking xyloglucans and embedded in a gel of matrix pectins, which include simple and complex homogalacturonans (HG) and rhamnogalacturonan I (RG I). To the RG I backbone may be linked α-arabinans, β-galactans, and type I arbinogalactans. The wall is also residence for several structural proteins and hundreds of enzymes. (adapted from Carpita and Gibeaut 1993)

(B) Architectural model of the Type II cell wall.
The microfibrils coated with a dense layer of (1→3),(1→4)-β-glucan and relatively unsubstituted glucuronoarabinoxylan (GAX) are interlaced by primarily by GAX with greater degree of branching by single arabinofuranosyl units. Some highly-substituted GAX remains intercalated in the small amount of pectins that also are found in the primary wall. Unlike the Type I wall, a substantial portion of the non-cellulosic polymers are "wired on" the microfibrils by alkali-resistant phenolic linkages. (adapted from Buckeridge et al. 2004)

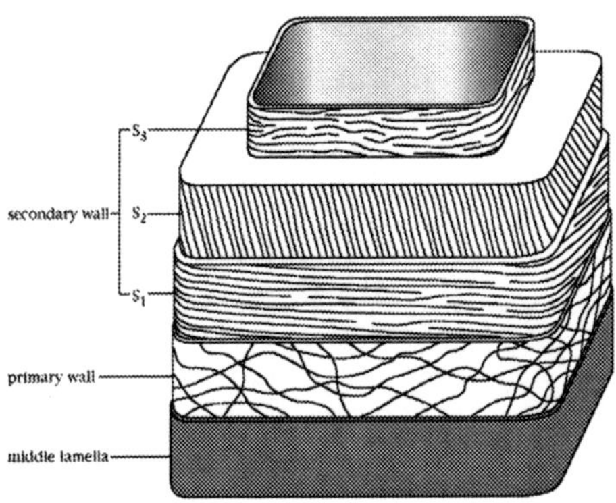

Fig. 6 Diagram of a piece of tracheid wall illustrating the layers and their microfibrillar organization.
S_1 to S_3 refers to the secondary wall. S_3 is sometimes interpreted as a tertiary wall layer (Adapted from Liese 1970).

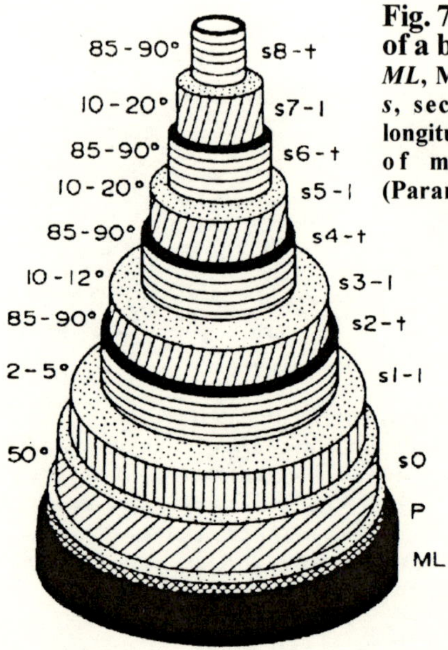

Fig. 7 Polylamellar wall structure of a bamboo culm fiber.
***ML*, Middle lamella; *P*, primary wall; *s*, secondary wall; *l* and *t* denote longitudinal and transverse orientations of microfibrils, respectively (Parameswaran and Liese 1981).**

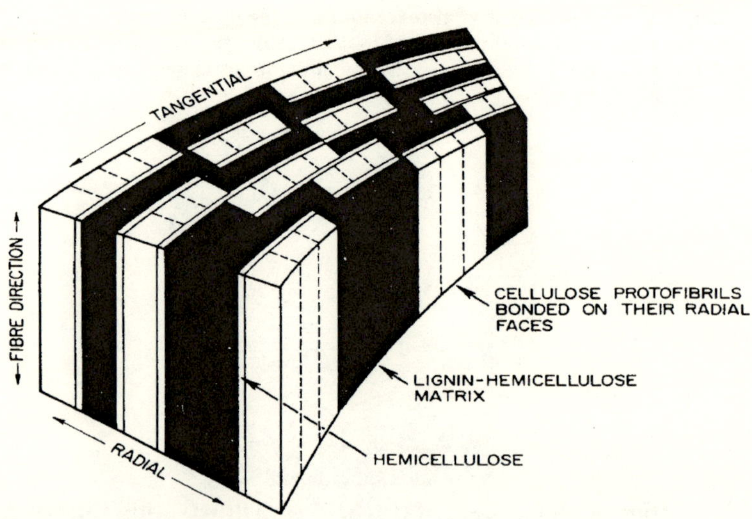

Fig. 8 Localization of and the relationships between cellulose, hemicelluloses, and lignin in the S_2 layer (Ruel *et al.* 1978).

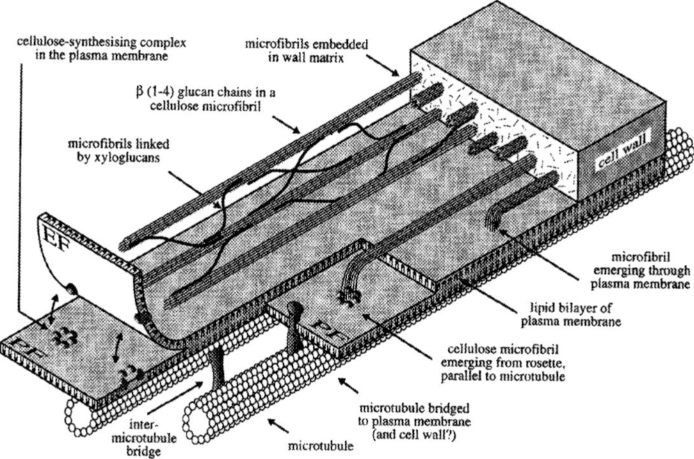

Fig. 9 Schematic diagram showing cellulose synthesis by the membrane synthase complex ("rosette") and its presumed guidance by the underlying microtubules in the cytoplasm (Gunning and Steer 1996).

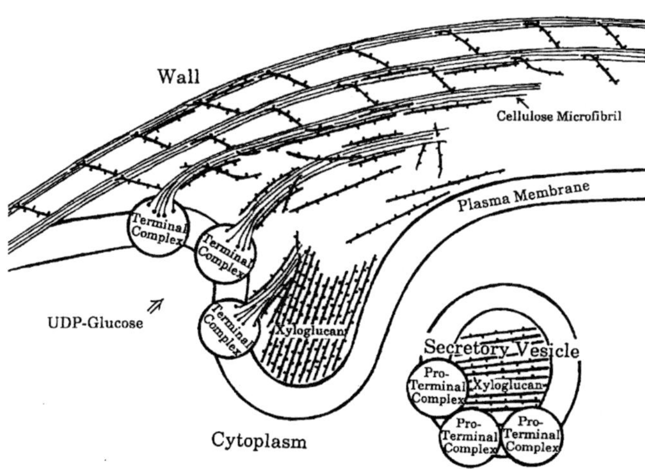

Fig. 10 Potential model of xyloglucan exocytosis and macromolecular organization.

As soon as the xyloglucan synthesized in Golgi is transported to extracellular sites, cellulose synthase (terminal complex) catalyzes cellulose synthesis at the cell surface, and the association of cellulose with xyloglucan takes place. (Hayashi 1989).

Acknowledgement
This work was supported in part by Program for Promotion of Basic Research Activities for Innovation Biosiences (PROBRAIN).

References

Buckeridge MS, Rayon C, Urbanowicz B, Tiné MAS, Carpita NC. (2004) Mixed-linkage (1→3),(1→4)-β-D-glucans of grasses.Cereal Chem 81: 115-127.

Carpita NC, Gibeaut DM (1993) Structural models of primary cell walls in flowering plants: consistency of molecular structure with the physical properties of the walls during growth. Plant J 3: 1-30.

Esau K (1977) Anatomy of Seed Plants 2nd ed. John Wiley & Sons, New York.

Fry SC (1986) Cross-linking of matrix polymers in the growing cell walls of angiosperms. Annu Rev Plant Physiol 37: 165-186.

Gunning BS, Steer MW (1996) Plant Cell Biology: Structure and function. Jones and Bartlett Publishers, Boston.

Hayashi T (1989) Xyloglucan in the primary cell wall. Annu Rev Plant Physiol Mol Biol 40: 139-168.

Higuchi T (1997) Biochemistry and molecular biology of wood. Springer, Berlin.Keegstra K, Talmadge KW, Bauer WD, Albersheim P (1973) The structure of plant cell walls. Plant Physiol 51: 188-196.

Liese W (1970) Elektronmikroskopie des Holzes. In: Handbuch der Mikroskopie in der Technik. Band V, Teil 1. Umschau, Frankfurt, pp 109-170.

Parameswaran N, Liese W (1981) The fine structure of bamboo. In: T Higuchi (ed) Bamboo production and utilization. Wood Res Inst, Kyoto Univ, 178-183.

Roberts K (1994) The plant cell wall. In: Molecular Biology of the Cell 3rd ed. Garland, New York, pp 1000-1009.

Ruel K, Barnoud F, Goring DAI (1978) Lamellation in S2 layer of softwood tracheid as demonstrated by scanning transmission electron microscopy. Wood Sci Technology 12: 287-291.

Taiz L, Zeiger E (2002) Plant Physiology 3rd ed. Sinauer, Sunderland.

Talbot LD, Ray PM (1992) Molecular size and separability features of pea cell wall polysaccharides. Plant Physiol 98: 357-368.

Imaging the Primary Cell Wall

Tobias I. Baskin

Introduction

As the son of an artist, I grew up surrounded by drawings, paintings, and sculpture. Perhaps for that reason, making images is central to my work as a scientist. The problem I study, organ morphogenesis, requires analysis of the cell wall. Fertile approaches stretch out in many directions (see the rest of this book) but I have been drawn to architectural problems, revealing how components are integrated into a whole. For structural appreciation, imaging is paramount.

For imaging the cell wall, my laboratory has developed or enhanced several methods. I will describe some of these here, and discuss advantages and limitations, without attempting to review cell wall imaging comprehensively. The first section is on immunocytochemistry at the light-microscope level. The second and third sections are on high-resolution imaging based on field-emission scanning electron microscopy. The first is convenient and allows many samples to be examined and a relatively large area of tissue to be viewed; the second and third are technically more demanding but resolve structure at virtually the level of macromolecules.

Butyl-methyl-methacrylate embedding for immunocytochemistry

Cell wall researchers will be familiar with the use of antibody probes to localize polysaccharide as well as protein epitopes in the cell wall (Knox 1997), but are probably less familiar with the use of butyl-methyl-methacrylate as an embedding matrix. I first encountered this resin while attempting to localize microtubules (Baskin et al. 1992). This methacrylate is easy to section, dry or wet; however, its primary advantage for immunocytochemistry is that following sectioning, the majority of the embedding matrix can be removed with a brief incubation in acetone, in contrast to nearly all other plastic embedments, which can be removed partially if at all only with harsh treatments. By removing the embedment, access to the antigen for the antibody is enlarged. Removability is shared by paraffin and wax but these preserve most samples poorly compared to plastic. At the electron-microscope level, adequate access for antibody to antigen is provided by various plastics that are sufficiently porous to allow antibodies to penetrate 10 nm or so into the section, thus sampling an appreciable proportion of the volume of an ultra-thin section (60 nm thickness). However, for the semi-thin sections (1 to 2 μm thick) typically used in light microscopy, a 10 nm penetration depth amounts to a negligible proportion of the section volume, and the greater penetration gained by removing the embedment becomes a significant advantage.

Butyl-methyl-methacrylate, like most methacrylates, polymerizes via a free radical-based mechanism. This is useful because it means that polymerization can be catalyzed by ultraviolet light, thus avoiding denaturation caused by high-temperature polymerization. However, early efforts to use this methacrylate mix were frustrated by oxidative damage to the sample that lowered antigenicity, damage presumably mediated by free radicals. I found that adding the free-radical scavenger dithiothreitol to the resin allowed polymerization but blocked the attack on the sample (Baskin et al. 1992). Subsequently, this resin has been used to localize tubulin and other antigens in a variety of samples (e.g., Herman et al. 1994; Stadler et al. 1995; Hoffman et al. 1998; Palmer et al. 2001).

In general, embedding in butyl-methyl-methacrylate is straightforward (Baskin and Wilson, 1997). However, the small size of arabidopsis roots (ca 0.15 mm diameter) makes them easy to lose while changing solutions. To retain the roots, I use a method that is not only convenient but also turns out to be beneficial for sample preservation. Originally, I encased each root in a small droplet of low-gelling-temperature agarose (Baskin et al. 1992), but this is messy and exposes the sample to heat, albeit briefly. Then, I modified a method from cryofixation where samples are mounted on a Formvar film (Baskin et al. 1996). A chemically fixed root tip is placed on a Formvar-coated wire loop, a second Formvar film secures the root tip on the loop. The Formvar films are readily permeated by solvents and small molecules. Between Formvar films, the thin arabidopsis root tip is prevented from bending or twisting. I call this "mechanical fixation" and beyond being convenient, it seems to enhance sample preservation.

Loops are made in advance and coated by casting Formvar rectangles (measuring a little more than the loop diameter on one side and a little more than twice the loop diameter on the other) and plunging the loop into the water over the rectangle so that the plane of the loop bisects the long axis of the rectangle. The Formvar rectangle wraps around the wire loop and the coated loop is removed at once from the water. Such loops remain stable for months. To secure a sample, the procedure is repeated: After the sample has been fixed and rinsed, a loop (already Formvar coated) is placed horizontally on a drop of water (or buffer) and the sample placed on the Formvar. Excess sample is trimmed if needed, and the loop (with sample) is plunged onto a new Formvar rectangle, thus encasing the sample between Formvar layers, held by the loop. Several loops can be placed in a vial and solutions exchanged without losing the sample. The loop is embedded with the sample, and removed during trimming. I use fine copper wire (36 gauge), which can be trimmed along with the block.

My colleagues and I have recently taken advantage of this methacrylate to characterize cell wall epitopes present in the *rhd1/reb1/cst2* mutant of arabidopsis (Andème-Onzighi et al. 2002). We found that selected arabinogalactan-protein epitopes were differentially expressed within the epidermis between root-hair forming cells (trichoblasts) and non-root-hair forming cells (atrichoblasts) and suggested that these proteins may be involved in repressing bulge formation in atrichoblasts (Figure 1).

Field-emission scanning electron microscopy for cell wall ultrastructure

The most common approach to cell wall ultrastructure is transmission

Fig. 1 Transverse sections of arabidopsis roots embedded in butyl-methyl-methacrylate and stained with anti-cell wall antibodies.

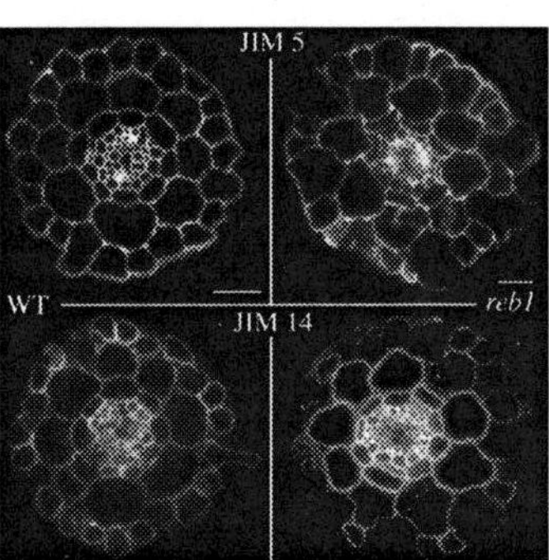

Top panels show JIM5, which recognizes pectin, bottom panels show JIM 14, which recognizes arabino-galactan epitopes. Left hand panels show wild type (Columbia), right hand panels show *reb1*. Note that JIM 5 stains more or less ubiquitously, as JIM 14 stains the wild type; however, JIM 14 staining in the trichoblasts of *reb1* appears specifically decreased. Bars = 25 µm (top), 15 µm (lower right). Figure modified from Andème-Onzighi et al. 2002).

electron microscopy, with samples embedded, sectioned, and stained with heavy metals. Such microscopy illuminates the ultrastructure of the wall, particularly to show its thickness and lamellation, and the sections can be used for immunocytochemistry; however, this approach suffers from the fact that the heavy metals do not stain cellulose or other polysaccharides reliably, and consequently images can be difficult to interpret (Emons 1988).

An alternative instrument for ultrastructure is field-emission scanning electron microscopy. The field-emission gun allows the microscope to scan the sample with a narrow diameter beam, with consequent improvement in resolution. As with all scanning electron microscopy, the image largely reflects the topography of the sample so it is useful for showing the structural arrangements of components. Additionally, sample preparation for field-emission scanning electron microscopy is faster and easier than for transmission electron microscopy, requiring dehydration and critical-point drying, but no embedding or ultra-thin sectioning.

A further difference is that for transmission electron microscopy the cell wall is usually viewed in cross-section whereas for field-emission scanning electron microscopy the wall is usually viewed from the surface. Surface views are particularly important for delineating architectural relations. Surface views can also been obtained for transmission electron microscopy by casting a metal-carbon replica from exposed cell walls (e.g., Iwata and Hogetsu 1989; McCann et al. 1990); however, replica making is technically difficult, and in field-emission scanning electron microscopy the sample itself is imaged rather than a replica.

Since the early development of the field-emission gun, wood anatomists have taken advantage of the instrument to examine the orientation of microfibrils at the innermost surface of tracheids and other cell types in wood (Hirakawa and Ishida 1981; Abe et al. 1994). Such cells are ideal for field-emission scanning electron microscopy because the cell walls are thick and strong and cell contents are absent or easily removed to expose the surface for imaging. In contrast, the use of field-emission scanning electron microscopy to image the primary cell wall has only been rather recently reported (Vesk et al. 1996; Crow and Murphy 2000).

An early and influential application of field-emission scanning electron microscopy to the primary cell wall was developed at Australian National University in Geoffrey Wasteneys' lab (Sugimoto et al. 2000). I happened to be there on sabbatical and was able to learn the procedure at first hand and to explore related methods. Again, the object of interest is the arabidopsis root and its small size requires heroic efforts. In Sugiomoto's method, roots are fixed, cryroprotected, frozen in liquid nitrogen, mounted on a cryo-ultramicrotome, and sectioned while frozen. About a dozen sections are discarded and the remaining portion of the root, now cut open, is warmed up, incubated

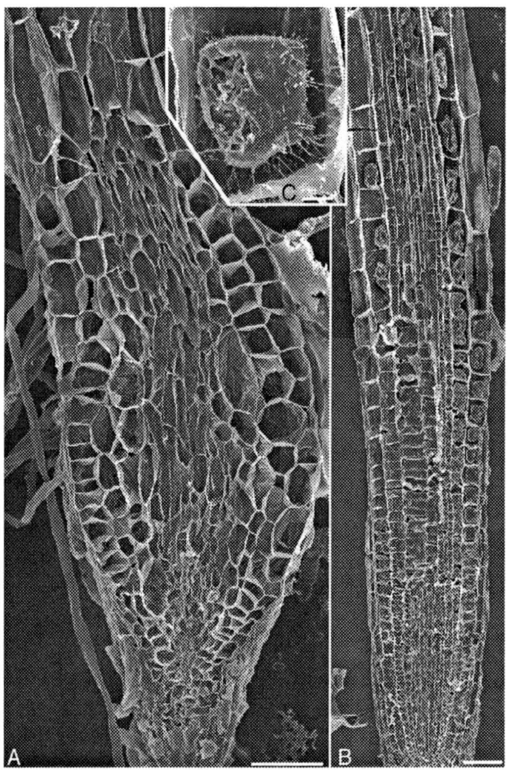

Fig. 2 Survey views of arabidopsis roots prepared for field-emission scanning electron microscopy.
A: Image of the *rsw4* mutant prepared by the cryo-ultramicrotme method. B: Image of wild type prepared by the polyethylene-glycol method. C: Higher magnification of an epidermal cell from B showing that many Hechtian strands are preserved. Bars = 60 µm (A); 30 µm (B); 3 µm (C).

in bleach to remove the cytoplasm, dehydrated, critically point dried, sputter-coated with platinum, and viewed. I tried out this method on radially swollen mutants (*rsw4* and *rsw7*; Wiedemeier et al. 2002) and an example is shown here (Figure 2A). Inside the emptied cells, the cell wall can be imaged at high resolution (not shown, but see Sugimoto et al. 2000).

While the method produces striking images, and has been used successfully by others (Zhong et al. 2002), it has some difficulties. The bleach, required to remove the cytoplasm, may also attack the cell wall. The pieces of root left after sectioning are minuscule and difficult to process for dehydration and sectioning. Finally, the cryo-ultramicrotome is a specialized and expensive piece of equipment. Therefore, I sought alternatives.

One alternative is butyl-methyl-methacrylate. The acetone treatment mentioned above while sufficient to give antibodies access nevertheless leaves an insoluble residue (ca. 10% of the original section mass) that obscures all but the grossest topography in the field-emission scanning electron microscope. Because an objective in my research is to assess the orientation of cellulose microfibrils, I examined methacrylate sections that were treated in acid mixtures commonly used to remove all organic matter other than cellulose, including a mixture of acetic and hydrochloric acids (Updegraff 1969), and a mixture of acetic acid and hydrogen peroxide (Wolters-Arts and Sassen 1991). Semi-thin sections (1.75 μm) were collected on coverslips, incubated in acetone to remove the majority of the methacrylate and then incubated in acid at 100 °C, usually for 1 h. Subsequently, sections were rinsed, dehydrated, critically point dried, sputter-coated and imaged.

The results were disappointing. Low magnification images (Figure 3A and B) resemble sections viewed in polarized light microscopy or treated with Calcofluor White and viewed in fluorescence microscopy, indicating that the overall cell pattern survived extraction. However, when cell walls present in face view were examined at high magnification, the microfibrils were invariably disorganized (Figure 3C). Each type of extraction gave similar results. Even though the sections are fixed to glass for the extraction, the hot acid presumably causes the cell wall to swell and allows some microfibrils to rearrange in the boiling reagent. Indeed, acid-extracted samples viewed in cross section in the transmission electron microscopy have swollen and distorted cell walls (Emons 1998).

As an alternative, I embedded roots in low molecular weight polyethylene glycol. Polyethylene glycol is water soluble allowing it to be fully extracted after sectioning. Additionally, I plasmolysed roots in sucrose prior to fixation and embedding so as to be able to expose

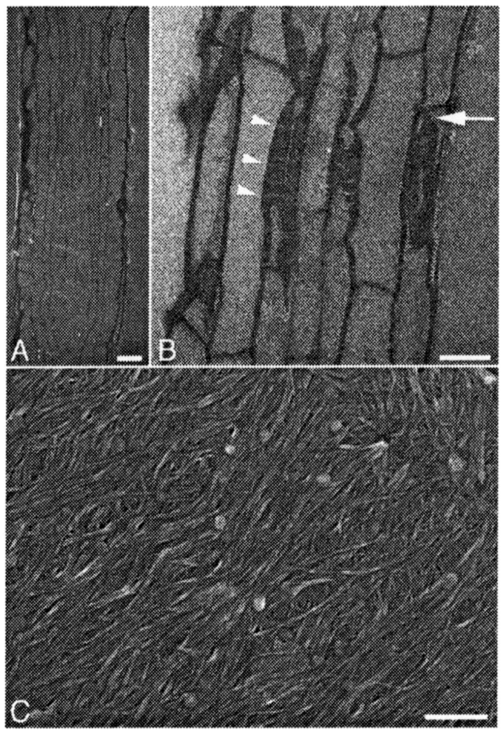

Fig. 3 Arabidopsis roots embedded in butyl-methyl-methacrylate, sectioned at 1.75 µm, acetone-treated to remove methacrylate, and treated with peroxide/actetic acid at 100 °C for 1h to remove everything except cellulose.

A: Survey view showing the section outlined by its remaining cell walls. The quiescent center is roughly at the bottom of the panel. B: View of elongation zone cells where some regions of cell wall are present in the plane of the section (dark areas). Faint transverse striations (white arrowheads) are probably pit fields. C: High magnification (region as indicated by arrow in B) showing microfibrils, which appear thick and significantly disorganized. Bars = 30 µm (A); 20 µm (B); 250 nm (C).

the cell wall surface for imaging without bleach. After infiltration and embedding, sections were collected onto coverslips, incubated in water to remove the embedment, and then prepared for field-emission scanning electron microscopy (dehydrated, critical-point dried, sputter coated). An example of a wild-type root is shown (Figure 2B). The method is sufficiently gentle to retain many Hechtian strands linking the cell wall to the protoplasts (Figure 2C). While the cell wall can be imaged in large cells in regions not covered by the protoplast, in small cells, the whole wall is obscured. Therefore, sections were incubated in salt and dilute protease, which removed protoplasts completely.

As with cryo-ultramicrotomy, polyethylene-glycol embedding allowed the cell wall surface to be imaged at high resolution (Figure 4A and B). Some tissues had distinct cell wall textures, for example the root cap cell walls microfibrils appear thick and stiff (Figure 4C). By incubating sections in various reagents it was possible to examine their effects. Bleach treatment changed the appearance of the cell wall, making the microfibrils narrower and less encrusted (Figure 4D). Removing pectin with calcium chelation changed the appearance modestly (Figure 4E), perhaps straightening microfibrils, whereas removing pectin with a pectolyase (which presumably removes more pectin than chelation does) caused microfibrils to appear to merge (Figure 4F). Others have also seen this merging in pectin-depleted samples (Sugimoto et al. 2000; Pagant et al. 2002), but the phenomenon has yet to be explained.

Although polyethylene-glycol embedding succeeded, it was difficult. The blocks are so soft that sectioning them was only possible at about 16 °C, provided that the relative humidity was less than around 70%; even then, many sections were lost. Furthermore, sectioning radially swollen mutants disrupted the integrity of the tissue, perhaps because their cell walls are weak. I tried polyethylene glycols only up to a molecular weight of 1450. Larger glycols have been used in transmission electron microscopy to provide "resinless embedding" (Penman and Penman 1997), and these should prove easier to section. On the other hand, the larger the molecular weight the more difficult to remove completely with water. Nevertheless, high molecular weight glycols might give better results for field-emission scanning electron microscopy views of arabidopsis cell walls.

Imaging cell walls in organs with large cells

Arabidopsis is an object of intense scrutiny, nevertheless other plants continue to be useful subjects for research and techniques for imaging cell wells in such species should be developed. In working on maize roots, I found a reliable way to prepare cell walls for field-emission scanning electron microscopy imaging that, when compared to the difficulties described above for arabidopsis, is amazingly simple. All

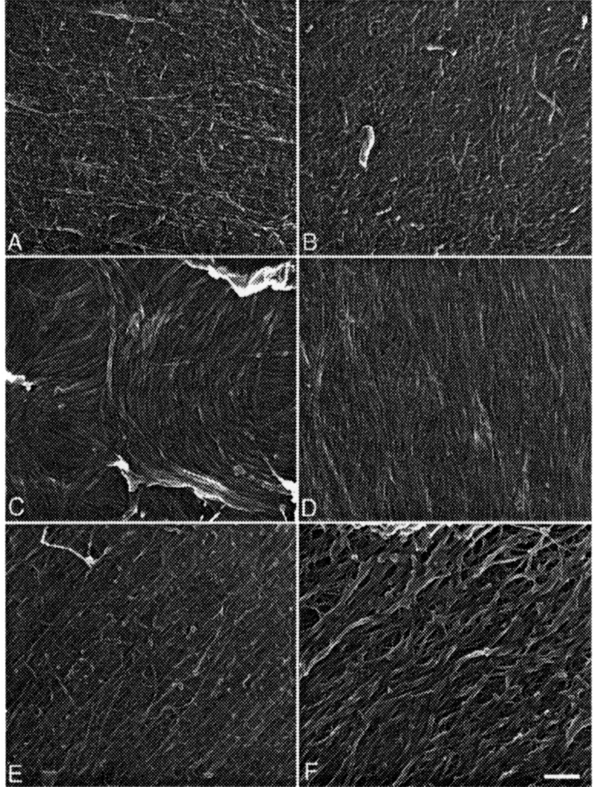

Fig. 4 High magnification field-emission scanning electron micrographs of polyethylene-glycol embedded arabidopsis roots.

A, B: Epidermal cells in the growth zone with different textures. C: Central columella cell. D – F: Sections treated as described after removing the polyethylene glycol and before dehydration. Images are from the growth zone. D: Bleach (0.5 % for 8 min). E: CDTA (50 mM for 16 h). F: Pectolyase (0.1 % for 1 h; Seishin Y-23). Bar = 200 nm.

that is required is to section the root, without fixation, in dilute buffer. Sectioning without fixation is accomplished readily for an organ the size of a maize root on a Vibratome, an instrument that vibrates a razor blade through a sample, producing sections on the order of 100 µm thick. However, if a Vibratome is not available, bisecting the organ by hand suffices. The sectioned material is incubated in buffer for a short period (10 to 60 min) and then fixed, dehydrated, critically point dried, sputtered with platinum, and viewed in the field-emission scanning electron microscopy (Figure 5).

Sectioning before fixation ejects cytoplasm and the buffer incubation washes out any residual material (Figure 5A). The method works similarly for cucumber and tobacco hypocotyls and inflorescence

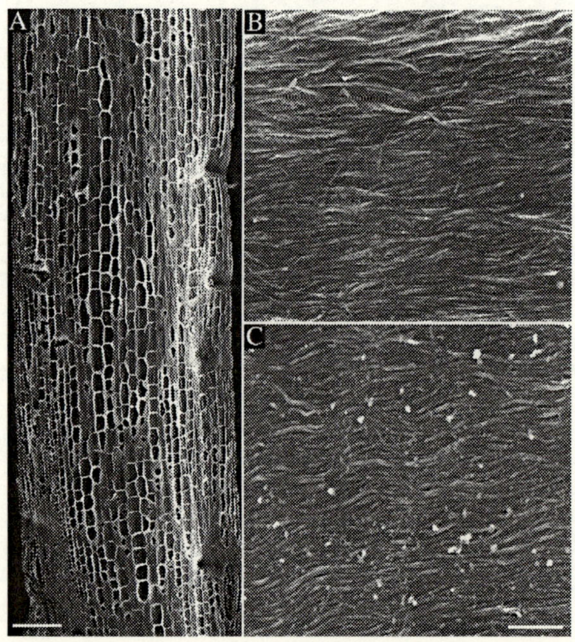

Fig. 5 Field-emission scanning electron micrographs from cutting sections prior to fixation.

A: survey view of a cucumber hypocotyl indicating the widespread absence of cytoplasm (no bleach required). B, C: High magnification images of cortical parenchyma from a maize root (B) and a cucumber hypocotyl (C). Samples cut on Vibratome (maize) or free-hand (cucumber). Bars = 300 µm (A); 250 nm (B, C).

stems of arabidopsis; however, when arabidopsis roots are cut with a razor blade under water, the plasma membrane is retained tenaciously inside the cells, precluding this approach. In images from larger organs, the microfibrillar texture as well as cross-links of various kinds are clear (Figure 5B and C). A consistent feature of cell walls prepared in this way, as well as of arabidopsis roots prepared as above, is that the microfibrillar texture appears as a single lamina with little or no visibility of underlying layers. Additionally, the diameters of the microfibrils vary continuously between roughly 8 and 40 nm. Given that the diameter of the crystalline cellulose microfibril, presumably as synthesized by a single rosette, is 3 to 5 nm (Davies and Harris 2003), it appears as though the cellulose microfibrils are sheathed in matrix polysaccharides and proteins to make structures that are thicker but still highly fibrillar.

These images depict cell walls that presumably remain close to their native composition, with the only material lost being that extracted by a low ionic-strength wash before fixation or by ethanol or liquid CO_2 after fixation. However it is an open question to what extent the architecture remains native, because dehydration and critical point drying themselves could rearrange cell wall components, particularly those architectural features that depend on non-covalent bonds. To resolve this question, it will be necessary to use the capability of the field-emission scanning electron microscopy to image frozen samples or to turn to other techniques such as atomic force microscopy so that the ultrastructure of the cell wall can be imaged in a hydrated state.

In conclusion, images may inspire artists to create for our enjoyment and scientists to experiment for our enlightenment.

Acknowledgements

I thank Geoffrey Wasteneys and Keiko Sugimoto for introducing me to the joys of field-emission scanning electron microscopy, and Jan Judy-March for stalwart technical assistance. Field-emission scanning electron microscopy was carried out at the Electron Microscopy Unit of the Australian National University and the Core Facility for Electron Microscopy of the University of Missouri (Columbia). Work in my lab on the cell wall is supported by a grant from the United States Department of Energy (award No. 03ER15421), which does not constitute endorsement by that Department of views expressed herein.

References

Abe H, Ohtani J, Fukazawa K (1994) A scanning electron microscopic study of changes in microtubule distributions during secondary wall formation in tracheids. IAWA J 15: 185-189.

Andème-Onzighi C, Sivaguru M, Judy-March J, Baskin TI, Driouich A (2002) The reb1-1 mutation of Arabidopsis alters the morphology of trichoblasts, the expression of arabinogalactan-proteins and the organization of cortical microtubules. Planta 215: 949-958.

Baskin TI, Busby CH, Fowke LC, Sammut M, Gubler F (1992) Improvements in immunostaining samples embedded in methacrylate: Localization of microtubules and other antigens throughout developing organs in plants of diverse taxa. Planta 187: 405-413.

Baskin TI, Miller DD, Vos JW, Wilson JE, Hepler PK (1996) Cryofixing single cells and multicellular specimens enhances structure and immunocytochemistry for light microscopy. J Microsc 182: 149-161.

Baskin TI, Wilson JE (1997) Inhibitors of protein kinases and phosphatases alter root morphology and disorganize cortical microtubules. Plant Physiol 113: 493-502.

Crow E, Murphy RJ (2000) Microfibril orientation in differentiating and maturing fibre and parenchyma cell walls in culms of bamboo (Phyllostachys viridiglaucescens (Carr.) Riv. & Riv.). Bot. J. Linnean Soc. 134: 339-359.

Davies LM, Harris PJ (2003). Atomic force microscopy of microfibrils in primary cell walls. Planta 217: 283-289.

Emons AMC (1988) Methods for visualizing cell wall texture. Acta Bot Neerl 37: 31-38.

Herman EM, Li X, Su RT, Larsen P, Hsu H, Sze H (1994) Vacuolar-type H^+-ATPases are associated with the endoplasmic reticulum and provacuoles of root tip cells. Plant Physiol 106: 1313-1324.

Hirakawa Y, Ishida S (1981) A SEM study on the layer structure of secondary wall of differentiating tracheids in conifers. Research Bulletins of the College Experiment Forests, Hokkaido Univ 38: 55-71.

Hoffman JC, Vaughn KC, Mullins JM (1998) Fluorescence microscopy of etched methacrylate sections improves the study of mitosis in plant cells. Microsc Res Tech 40: 369-76.

Iwata K, Hogetsu T (1989) Orientation of wall microfbrils in Avena coleoptiles and mesocotyls and in Pisum epicotyls. Plant Cell Physiol. 30: 749-757.

Knox JP (1997) The use of antibodies to study the architecture and developmental regulation of plant cell walls. Int Rev Cytol 171: 79-120.

McCann MC, Wells B, Roberts K (1990) Direct visualization of cross-links in the primary plant cell wall. J Cell Sci 96: 323-334.

Pagant S, Bichet A, Sugimoto K, Lerouxel O, Desprez T, McCann M, Lerouge P, Vernhettes S, Hofte H (2002) KOBITO1 encodes a novel plasma membrane protein necessary for normal synthesis of cellulose during cell expansion in Arabidopsis. Plant Cell 14: 2001-2013.

Palmer JH, Harper JDI, Marc J (2001) Control of brittleness in butyl-methylmethacrylate resin embedding mixtures to facilitate their use in immunofluorescence microscopy. Cytobios 104: 145-156.

Penman J, Penman S(1997) Resinless section electron microscopy reveals the yeast cytoskeleton. Proc Natl Acad Sci USA 94: 3732-3735.

Stadler R, Brandner J, Schulz A, Gahrtz M, Sauer N (1995) Phloem loading by the PmSUC2 sucrose carrier from Plantago major occurs into companion cells. Plant Cell 7: 1545-1554.

Sugimoto K, Williamson RE, Wasteneys GO (2000) New techniques enable comparative analysis of microtubule orientation, wall texture, and growth rate in intact roots of arabidopsis. Plant Physiol 124: 1493-1506.

Updegraff DM (1969) Semimicro determination of cellulose in biological materials. Anal Biochem 32: 420-424.

Vesk PA, Vesk M, Gunning BES (1996) Field emission scanning electron microscopy of microtubule arrays in higher plant cells. Protoplasma 195:168-182.

Wolters-Arts AMC, Sassen MMA (1991) Deposition ond reorientation of cellulose microfibrils in elongating cells of petunia stylar tissue. Planta 185: 179-189.

Wiedemeier AMD, Judy-March JE, Hocart CH, Wasteneys GO, Williamson RE, Baskin TI (2002) Mutant alleles of arabidopsis RADIALLY SWOLLEN 4 and RSW7 reduce growth anisotropy without altering the transverse orientation of cortical microtubules or cellulose microfibrils. Development 129: 4821-4830.

Zhong R, Burk DH, Morrison WH 3rd, Ye ZH (2002) A kinesin-like protein is essential for oriented deposition of cellulose microfibrils and cell wall strength. Plant Cell 14: 3101-3117.

Excursions in Cell-Wall Biophysics

Mike C. Jarvis

There are various experimental ways in which the mechanical function of the cell wall can be approached: stress-strain experiments, NMR and vibrational spectroscopy, electron microscopy. This chapter however is mainly concerned with the underlying concepts: how cell walls deform under stresses of different kinds, especially turgor stress during growth, and what internal rearrangements are needed for deformation to take place. It follows a wandering train of thought through three decades of research, my own and other people's, arriving at the conclusion that we do not really know how cell walls function.

The non-woody, primary-walled parts of plants are phenomenally high-performance materials, if one considers their stiffness and toughness in relation to the metabolic investment in cell-wall polymer biomass. A lot of plant can be kept in shape by very little polymer in the primary cell wall. This chapter is about how cell walls carry loads, a question that has preoccupied me since the 1980s.

When I first came to study plant cell walls in the early 1970s, though, I was driven by a slightly different technical challenge: how to find out the structure of an insoluble polymer network when the methods available only worked on solutions. Whereas Peter Albersheim and his group concentrated on releasing soluble polymer fragments by controlled degradation with enzymes (Keegstra et al. 1974), I tried to bring chemical solubilisation methods under better control (Jarvis et al. 1981). Seeing the limitations of this approach, I later went over to spectroscopic methods that could be used directly to look at polymer structure in insoluble materials. The key inspiration for this came from the little-known work of Morikawa et al. (1978) on polarised infrared (FTIR) spectroscopy, although initially my group focussed on solid-state NMR and left it to Maureen McCann and her co-workers (McCann et al. 1992) to develop FTIR microscopy into a routine technique for cell-wall analysis.

By the time when I was becoming more involved in spectroscopic methods, I had also become interested in how cell walls fulfil their load-bearing function and I was aware of the potential of spectroscopy for elucidating this – polarised FTIR to probe polymer orientation under load (Morikawa et al. 1978) and NMR relaxation experiments to probe the local rigidity of specific parts of the structure (Irwin et al. 1984).

At that time I was particularly interested in the contribution of pectic gels to the mechanical properties of the primary cell wall (Jarvis 1984), seeing fascinating possibilities in the physical chemistry of polyelectrolyte gels and their responsiveness to the ionic environment (e.g. Tanaka et al. 1982). But in retrospect this was a pity, because I missed out on the heyday of research on the geometry of cellulose networks by Preston (1974), Roland (Roland et al 1982) and Green (Green and Lang 1981) – I knew it was going on, but didn't take as much interest as it deserved.

As a cautionary tale on the consequences of such selective reading, let me describe an experiment that I did then on the effects of calcium ions on pectic cohesion in muro (Jarvis et al. 1984). I isolated long strands of cell walls from celery collenchyma and built a rather sensitive extensiometer to study the kinetics of their extension under tensile stress, submerged in solutions that buffered the calcium content of their pectins at different percentages of saturation. I thought that at low calcium levels I might observe extension that was linear with time, as in growth, but the results were inconclusive. With hindsight, Preston's work (Preston 1974) should have shown me that the collenchyma system is completely unsuitable for studying the role of pectin or any other matrix component in tension because, unlike most growing tissues, its microfibril orientation is more or less uniformly axial. In the absence of enzyme activity, all that one observes is the tensile properties of cellulose plus some loss of intercellular adhesion at very low calcium levels. It was only much later that Julian Vincent advised me how to probe the matrix properties of a composite with axially oriented fibres by making it into a torsional pendulum. This experiment revealed a completely different set of properties, including temperature dependence that implied mixed entropy-driven (rubber-type) and enthalpy-driven elasticity of the matrix, and an apparent phase transition around 40°C.

We can take two messages from this. First, some consideration of polymer orientations is needed to interpret the mechanical behaviour of intact cell walls, and often to design the experiments in which their mechanical behaviour is measured. So the fairly large amount of information that has been built up on the amount – not the direction – of thermal motion of cell-wall polymers in situ using NMR relaxation experiments (Foster et al. 1996; Ha et al. 1997; Renard and Jarvis 1999), only starts to make sense at the macromolecular level when we also know in some detail how microfibrils are oriented and how the 'matrix' polymers run between them.

Secondly, experimental systems that mimic living plants deforming under external mechanical stress are quite difficult to design. It is still more difficult to simulate the delicately tailored, time-linear, turgor-driven deformation that we know as growth. Growth requires the activity of enzymes acting on the cell wall and is usually, although perhaps not necessarily, accompanied by the deposition of new cell-wall material. One-dimensional stretching experiments ignore the three-dimensional nature of growth (Proseus et al. 1999), although for certain elongating tissues this may be an approximation that is just about acceptable (Hejnowicz and Sievers 1996). Cell walls almost certainly deform in completely different ways when stressed on timescales of milliseconds (Kawamura et al. 1995), seconds (Jarvis et al. 1984) and hours (Cosgrove 1993). Growth takes hours, and the circumstances where shorter-term measurements seem to correlate with growth have never quite been understood (Peters et al. 2001). Biophysical confusion has been exacerbated by confusion over terminology, often adopted from materials science but uncritically applied to living organisms that don't obey the same rules as classical materials. There is no classical material that grows.

The growth of plants is, in a sense, simply the growth of their cell walls. Whether the development of plant form is considered to operate at the cell scale or the organ scale, cell walls must expand in the directions in which the plant will grow (Green and Lang, 1981). The fused pair of primary walls between two cells may, for example, elongate simultaneously with cell division until these walls lie between two long files of cells; even then, expansion of the cell walls is a prerequisite for growth. Moreover it is often possible to assume that growth is limited by the capacity of the cell walls to expand under enzymic influence, and is controlled in direction by the anisotropy of this capacity for expansion. These assumptions may not always hold (e.g. Koch et al. 2004) but they direct our attention to the cell-wall expansion process as a central factor in growth.

What actually happens, then, inside the cell wall when a plant cell expands? I have thought about this question over a long period but I don't think we really know. Here are some ideas distilled out of this uncertainty, owing a considerable debt to Preston (1974) and Roland et al. (1982).

The behaviour of synthetic composites or of wood (Preston 1974) makes it clear that fibre orientation has a major effect on tensile rigidity, even though synthetic composites are rather poor models for primary cell walls (Ha et al. 1997). The relationship between fibre orientation and tensile modulus is quite complex (Preston 1974). To a first approximation, for one small domain with the cell wall, the resistance to expansion in any one direction depends on the cosine of the angle α between that direction and the local orientation of the microfibrils, i.e. the projection of the microfibril axis on the direction of stress. For this purpose 'small' means small enough for variation in microfibril orientation within the domain to be minor and random. The dependence on cos α may not be linear, indeed terms in $\cos^2\alpha$ and $\cos^4\alpha$ were included by Preston (1974).

Cos α expresses three things - (a) the distribution of load between microfibrils and matrix, (b) the extent to which the microfibrils are being stretched rather than pulled apart sideways, and (c) the extent to which the matrix between the microfibrils is loaded in shear or in tension. Note that we do not assume that anisotropy in cell-wall properties all comes from the microfibril orientation (Wiedemeier et al. 2002). Probably most of it does but matrix polymers too can be oriented (Morikawa et al. 1978).

For elongation growth, α is the angle between the microfibrils and the cell axis. Figure 1 shows this and also how cos α may vary across the thickness of a simplified, model elongating cell wall, as progressively older microfibrils turn towards the cell axis and thin out. Polylamellate or helicoidal cell walls would of course behave somewhat differently (Roland et al. 1982).

So for a newly deposited layer of the cell wall with α close to 90°, it seems that the matrix provides most of the resistance to elongation

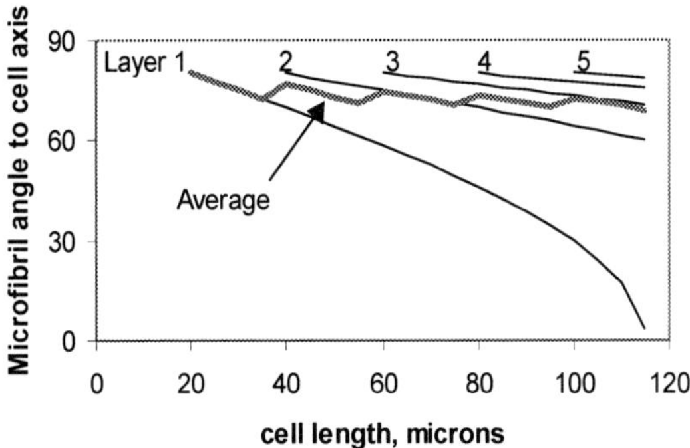

Fig. 1 Modelled changes in microfibril orientation as a cell, initially 20 x 20 μm, elongates to 120 μm in length without change in diameter. **Successive layers of microfibrils are added at the inner face every 20 μm of cell elongation. Each layer thins out with elongation so layers 1-3 are likely to have merged when the cell length reaches 100 μm, and the average microfibril angle does not vary greatly because it is dominated by the more abundant inner microfibrils (Preston 1974). It is assumed that it is easier for the microfibrils to separate than to slide parallel to one another, but parallel sliding is the only mechanism possible once the microfibrils are parallel to the cell axis.**

just by restraining lateral separation of the microfibrils. Of course what we call the 'matrix' consists of diverse polymers located and oriented between the microfibrils in specific ways that are only hazily understood. Xyloglucan cross-links normal to the microfibrils would restrain their separation well (McCann et al. 1990). Simply embedding the microfibrils in a network of non-cellulosic polymers, like pectins, would also be effective if the network were stiff enough. Elucidating how this 'matrix' stretches between microfibrils, with enzymic assistance, is a key objective if we wish to understand how cell walls grow.

Older microfibrils further out in the primary wall, with α closer to 0°, must either slide parallel to one another or stretch. If not, the cell would twist or contract in width when it elongates. The irreversible deformation of wood (Keckes et al. 2003) illustrates what might happen.

It is not clear how the control of cell-wall elongation is distributed between the inner and outer layers of the primary wall, and hence which of the above mechanisms is limiting. Perhaps older microfibrils mediate the eventual cessation of growth (Roland et al 1982). It is clear, however, that all layers of the cell wall must elongate somehow. So do the microfibrils that are closer to axial orientation stretch, or do they slide in parallel?

Stretching of the microfibrils is usually, and probably correctly, discounted. It is quite difficult to see how parallel sliding of microfibrils might work, because they are very long – just how long, we do not really know. Certainly it would not require much shear strength in the matrix to restrain such parallel sliding. But if microfibrils cannot be stretched, then some of them must somehow slide in this way, and the way in which the non-cellulosic polysaccharides are arranged to constrain this sliding - the 'shear strength' of the matrix - is relevant to growth (Roland et al., 1982).

Perhaps terms like 'shear strength' force us too far into a materials scientist's way of thinking, towards an image of the matrix as a kind of amorphous goo. Polymer bridges between microfibrils exist (McCann et al. 1990) and look as if they would be configured to restrain the microfibrils from both separating and sliding, but how these bridging chains deform or detach under loads of any kind has never yet been observed. At any rate, separating the internal deformation of the cell wall into these 'stretching' and 'sliding' modes gives us an agenda for experiments towards the mechanism of growth, even if the technical details are still obscure.

References

Cosgrove DJ (1993) Wall extensibility – its nature, measurement and relationship to plant-cell growth. New Phytol 124: 1-23

Foster TJ, Ablett S, McCann MC, Gidley MJ (1996) Mobility-resolved C-13-NMR spectroscopy of primary plant cell walls. Biopolymers 39: 51-66

Green PB, Lang JM (1981) Toward a biophysical theory of organogenesis – birefringence observations on regenerating leaves in the succulent, *Graptopetalum paraguayense* E. Walther. Planta 151: 413-426

Ha MA, Apperley DC, Jarvis MC (1997) Molecular rigidity in dry and hydrated onion cell walls. Plant Physiol 115: 593-598

Hejnowicz Z, Sievers A (1996) Tissue stresses in organs of herbaceous plants .3. Elastic properties of the tissues of sunflower hypocotyl and origin of tissue stresses J Exp Bot 47: 519-528.

Irwin PL, Pfeffer PE, Gerasimowicz WV, Pressey R, Sams CE (1984) Ripening-related perturbations in apple cell-wall nuclear-spin dynamics. Phytochemistry 23: 2239-2242

Jarvis MC (1984) Structure and properties of pectin gels in plant cell walls. Plant Cell Environ 7: 153-164

Jarvis MC , Logan AS, Duncan HJ (1984) Tensile characteristics of collenchyma cell walls at different calcium contents. Physiol Plant 61: 81-86

Jarvis MC, Threlfall DR, Friend J (1981) The polysaccharide structure of potato cell walls – chemical fractionation. Planta 152: 93-100

Kawamura Y, Hoson T, Kamisaka S, Yamamoto R (1995) Formulation of pre-extension in a practical stress-relaxation measurement of the plant cell wall. Biorheology 32: 611-620.

Keckes J, Burgert I, Fruhmann K, Muller M, Kolln K, Hamilton M, Burghammer M, Roth SV, Stanzl-Tschegg S, Fratzl P (2003) Cell-wall recovery after irreversible deformation of wood. Nature Materials 2: 810-814.

Keegstra K, Talmadge KT, Bauer WD, Albersheim P (1973) The structure of plant cell walls. III. A model of the walls of suspension-cultured sycamore cells based on the interconnections of the macromolecular components. Plant Physiol 51: 188-196.

Koch GW, Sillett SC, Jennings GM, Davis SD (2004) The limits to tree height. Nature 428: 851-854.

McCann MC, Hammouri M, Wilson R, Belton P. Roberts K (1992) Fourier-transform infrared microspectroscopy is a new way to look at plant-cell walls. Plant Physiol 100: 1940-1947.

McCann MC, Wells B, Roberts K (1990) Direct visualization of cross-links in the primary plant-cell wall. J. Cell Sci. 96: 323-334

Morikawa H, Hayashi R, Senda M (1978) Infrared analysis of pea stem cell walls and oriented structure of matrix polysaccharides in them. Plant Cell Physiol 19: 1151-1159.

Peters WS, Farm MS, Kopf AJ (2001) Does growth correlate with turgor-induced elastic strain in stems? A re-evaluation of de Vries' classical experiments. Plant Physiol 125: 2173-2179.

Preston RD (1974) The physical biology of plant cell walls. Chapman and Hall, London.

Proseus TE, Ortega JKE, Boyer JS (1999) Separating growth from elastic deformation during cell enlargement. Plant Physiol 119: 775-784.

Renard CMCG, Jarvis MC (1999) A CP-MAS ^{13}C NMR study of the polysaccharides in sugar beet cell walls. Plant Physiol 119: 1315-1322

Roland J-C, Reis D, Mosiniak M, Vian B (1982) Cell-wall texture along the growth gradient of the Mung Bean hypocotyl - ordered assembly and dissipative processes. J. Cell Sci. 56: 303-313

Tanaka T, Nishio I, Sun ST, Ueno-Nishio S (1982) Collapse of gels in an electric field. Science 218: 467-469

Wiedemeier AMD, Judy-March JE, Hocart CH, Wasteneys GO, Williamson RE, Baskin TI (2002) Mutant alleles of Arabidopsis RADIALLY SWOLLEN 4 and 7 reduce growth anisotropy without altering the transverse orientation of cortical microtubules or cellulose microfibrils. Development 129: 4821-4830

From Cell Wall Architecture to Wall Modeling: A Systems Biology Approach '*avant la lettre*'

Anne Mie C. Emons

Cellulose microfibrils are mesmerizing polymers, critically important in plant life and invaluable for industrial use. It is hard to imagine a world without cellulose. Many researchers have asked, "Why did you choose the *Equisetum hyemale* root hair to study the creation of its wall architecture?" In the polarized microscope, its young cell wall is isotropic, indicating that the wall texture is random or helicoidal. If helicoidal, it would be an excellent system to prove the hypothesis that cortical microtubules do indeed orient cellulose microfibrils. We began working with *Equisetum* in the early 1980's, before *Arabidopsis* became the model system of choice. A helicoidal texture is a construction built of one microfibril-thick lamellae in which the microfibril orientation shifts in every lamella with a constant angle. The cortical microtubules should behave the same if they are orienting the nascent cellulose microfibrils directly, while microtubule drugs should alter the orientation of microfibril deposition.

The cell wall texture of *E. hyemale* root hairs, as well as other *Equisetum* species, turned out to be helicoidal. However, the two sets of polymers at the two sides of the plasma membrane, microtubules and microfibrils, are not parallel to each other either in these cells or in other root hairs with helicoidal texture. This phenomenon was also observed in various other secondary walls, most notably algae. In addition, the texture of the root hairs remained helicoidal when the microtubules were depolymerized. Thus, two questions remained: how are cortical microtubules organized, and what organizes the cellulose microfibrils? In 2005, the first question has only just begun to be addressed.

A possible mechanism that would provide a solution to the question of cellulose microfibril organization is the geometrical model for microfibril orientation, which I discuss in this article. Since work on mutants by Geoff Wasteneys's group has indicated that the general idea of cortical microtubules orienting nascent cellulose microfibrils is highly unlikely, the geometrical model is steadily gaining importance, as it can explain all identified wall textures and predict texture if a number of cell parameters is known. It can also be verified or invalidated, as a working 'Systems Biology' model should do.

History

When I began my PhD studies in 1980, my thesis supervisor, Prof. André Sassen of the University of Nijmegen, wanted to prove the multi-net growth theory, formulated by Roelofsen from cell wall data obtained with an electron microscope. Unfortunately, this work had ended in 1968 after Roelofsen died in a skiing accident. The multi-net growth theory states that cellulose microfibrils are deposited transverse to the orientation of cell elongation, a finding repeatedly observed, and that existing cellulose microfibrils change their orientation to a longitudinal direction during cell elongation (Roelofsen 1958), which has been difficult to proof experimentally. Additional indications that they do reorient came from the work on styles of *Petunia*, by Wolters-Arts and Sassen in 1991.

Meetings

Cell Walls '81 in Göttingen was my first scientific congress and the second cell wall meeting, organized by David Robinson. The first had been organized in Nijmegen by André Sassen in 1978, when Alan Wardrop was on sabbatical in his lab. In Göttingen, Sassen reported my work on the cell wall of *E. hyemale* root hairs. The cell wall meeting in Fribourg in 1984 was tumultuous, during which I gave my first international presentation and told the audience that in *E. hyemale* root hairs, the cortical microtubules and nascent cellulose microfibrils are not in the same orientation. I was nearly physically attacked by the audience, but Russell Jones came to my rescue and pointed out that "data are data." At this meeting, Jan Kijne predicted that legume–*Rhizobium* interactions would become one of the most important fields in plant biology, and indeed, cytoskeletal reorganization during this interaction is one of my current research topics. Debby Delmer reported biochemical research indicating that *in vitro,* 1,3-β-glucan polymers (callose) were formed instead of 1,4-β-glucan (cellulose), while a presentation from a laboratory in Jerusalem stated that they had achieved high rates of *in vitro* synthesis of 1,4-β-glucan in a cell free system from *Phaseolus aureus*. Later, this proved to be incorrect. The next Cell Wall Meeting was organized by the Roland group, in Paris in 1986. In Paris, I had only 10 minutes to present my geometrical cell wall model, but people became very enthusiastic about it. The 1992 meeting in Nijmegen was organized by the Sassen group; as Sassen was retiring that year, it was co-organized by myself from Wageningen University. I did attend the next meeting in Norwich, but by then had become more interested in the cytoskeleton and was not actively working on the cell wall.

At the first Gordon Conference on the Plant Cell Wall in 1997, I was invited to present the geometrical cell wall model, which by now had been published and matured. At this meeting, Debby Delmer presented studies of glucosyltransferases in cotton. The breakthrough in the field of cellulose microfibril-synthases is from Richard Williamson's group, published in *Science* (Arioli *et al.* 1998), which had been preceded by a paper from Delmer's group in which the authors showed that higher plants contain homologs of the bacterial celA genes encoding the catalytic subunit of cellulose microfibril-synthase (Pear *et al.* 1996). The labs of Malcolm Brown and Takaoh Itoh proved that the rosettes are indeed cellulose microfibril synthases (Kimura *et al.* 1999). Rosettes were found for the first time in higher plants by Mueller and Brown in the plasma membrane of cellulose microfibril-depositing cells (Mueller and Brown 1978), and were observed in the Golgi apparatus by Haigler and Brown (1986). The finding that the plasma membrane of root hairs of *E. hyemale* contains rosettes (Emons 1985) was important in the development of the geometrical model. It showed that helicoidal walls are not a special case in which bulk material is deposited in the exoplasm between the plasma membrane and the cell wall, which would allow microfibrils to self-organize like a liquid crystal, as was proven possible for cellulose/hemi-cellulose mixtures *in vitro* (Vian *et al.* 1994).

The geometrical model

What is the geometrical model? What is the idea behind it, and how did I arrive at the model? In many intercalary growing cells, cortical microtubules and nascent cellulose microfibrils are both transverse to the direction of elongation. In non-elongating cells, chiefly studied in algae (Itoh 1989) and the non-elongating tubes of root hairs with non-axial cellulose microfibril orientation, the two polymers do not run parallel to each other (reviewed in Emons *et al.* 1992). Therefore, I needed a model to explain the ordering mechanism of cellulose microfibrils. I began with the simplest question: What types of textures are possible, if microfibrils 'grow' straight in the absence of obstruction? If this is the rule, could a helicoidal wall texture be formed?

Cellulose microfibrils are stiff, crystalline rods. Malcolm Brown has shown that cellulose microfibrils can tilt and move *Acetobacter* bacteria when they spin them out from non-moving complexes buried inside the plasma membrane. The cells rotate along their longitudinal axes when they spin the cellulose ribbons, caused by the force of crystallization. Microfibril stiffness, therefore, is a characteristic that should be taken into account when formulating theories regarding mechanisms for its orientation. Another physical feature that must influence microfibril ordering is the number of microfibrils being deposited at any moment in a cell, both locally and simultaneously, and hence the number of active cellulose microfibril synthases making microfibrils at one time.

The geometrical model for cellulose microfibril ordering during deposition takes into account the characteristics of cellulose microfibril stiffness and the number of active cellulose synthases (Emons 1994; Emons and Kieft 1994; Emons and Mulder 1998, 2000 and 2002; Mulder and Emons 2001; Emons *et al.* 2002; Mulder *et al.* 2004). The model proposes that the default mechanism determining the orientation of deposited microfibrils, in the absence of other influences, is geometrical in origin. Based on the observation that microfibrils always appear approximately evenly spaced in close-packed lamellae, and that their average distance of separation does not depend on their orientation with respect to the cell axis, led me to posit the geometrical close packing rule (Emons 1994).

$$\sin\alpha = \frac{Nd}{2\pi R}$$

This formula relates the microfibril winding angle α to the number of microfibrils being deposited by cellulose synthase complexes (N), the distance d between them, and the radius R of the cell. This explicit mathematical rule is the cornerstone of a dynamic developmental model. The model rests on the assumption that new active cellulose synthases insert into the plasma membrane, presumably during exocytosis when Golgi membranes fuse with the plasma membrane, or else are activated within moving, localized regions along the cell.

We call these regions the cellulose synthase activation domains. The rate at which new synthases become active is regulated and under cellular control, and microtubules may well play an as-yet unknown role in this process, as we discussed previously (Emons 1994; Emons and Mulder 1998, 2000; Emons *et al.* 2002). Once activated in the plasma membrane, the cellulose synthases move forward, propelled by the forces generated by the microfibril crystallization process. In time, their angle of motion with respect to the cell axis continuously adapts to the changing number of nearby cellulose synthases, in order to satisfy the constraints of close geometrical packing. The deposited microfibrils follow the tracks of the cellulose synthases and, as such, constitute a "recording" of their motion (Figure 1). The final component of the model is that cellulose synthases have a finite active lifetime. We have shown that by varying the parameters of this fully predictive mathematical model, several known cell wall textures can be reproduced: the axial, helical, helicoidal, and crossed polylamellate wall textures (Figure 2) (Emons *et al.* 2001, 2002). We recently published the random wall texture (Mulder *et al.* 2004) and are currently working on the transverse texture, as well as on the banded pattern seen in xylem cells.

This formula was deduced from the data. Root hairs are tip-growing cells; therefore, the wall deposited at the tip is a primary cell wall. A secondary cell wall, which is a wall deposited in a non-expanding cell, or part of a cell (definition agreed upon in a discussion session at the first cell wall meeting in 1978), is deposited in the tube of the hair against this primary cell wall (Figure 3). Consequently, in a growing root hair, we have the unusual situation that a secondary wall, *i.e.* a wall texture unchanged by cell expansion forces, is deposited in a cell that still elongates at one side (Emons and Wolters-Arts 1983). Therefore, fortunately for the researcher, the cell wall shows its deposition history not only in the deposited lamellae perpendicular to the wall at any site seen in sections, but also along the length of the root hair at the inner surface of the wall in surface preparations (Figure 3). Thus, sequentially from root hair base to root hair tip, all stages of wall development are represented in a single root hair. Since the cell elongates at the tip while the secondary wall is laid down, transmission electron micrographs reveal the deposition history of the microfibrils. This makes root hairs ideal for the study of cell wall texture formation, as well as the relationship with cortical microtubules, independently from the function the latter have in cell elongation. Of course, this becomes especially interesting in root hairs with non-uniform microfibril deposition angles, such as in helicoidal wall textures. In the pre-computer era of the 1980's, I used a protractor to measure the angles that microfibrils make with the long axis of root hairs on room-long transmission electron micrographs of surface preparations made for cellulose microfibril analysis. Since the micrographs could only be interpreted as that all the different orientations were deposited simultaneously along the hair (I saw few cellulose microfibril ends, and cellulose microfibrils were known to be long), I realized that microfibrils must wind around the plasma membrane of the root hair while gradually assuming the different orientations (Figure 3).

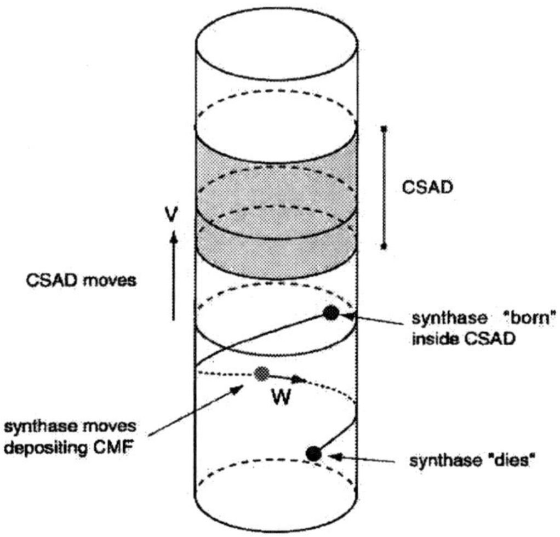

Fig. 1 The cellulose synthase life cycle.

After insertion into the plasma membrane within a cellulose synthase activation domain (CSAD: located between the circles at the time of deposition), the cellulose synthase moves with an average speed W within the plasma membrane, leaving a cellulose microfibril (CMF) in its wake. The direction of motion—and hence the angle the deposited microfibril makes with the cell axis—is determined by the local density of the synthases. The cellulose synthase becomes inactive after a characteristic lifetime, which determines the length of the microfibrils. The CSAD itself, shown here in gray, moves with a velocity V in the direction opposite to that of the cellulose synthases.

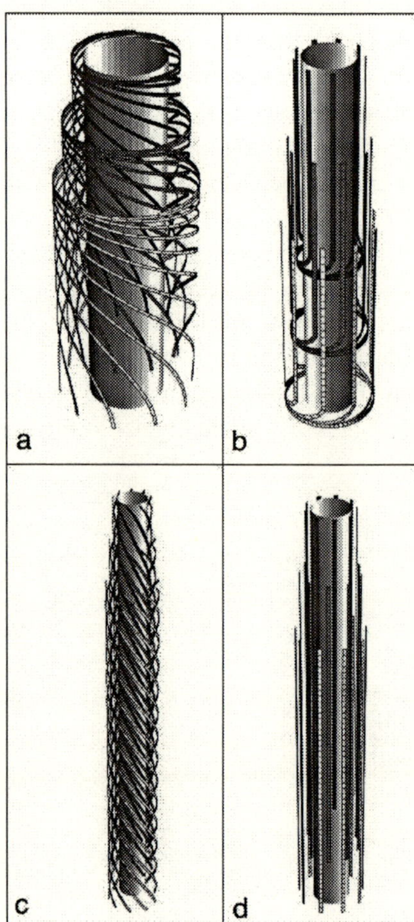

Fig. 2 Different cell wall textures as produced by the geometric model.

The ribbons shown represent the tracks of microfibrils obtained from the explicit solutions to the microfibril evolution equation. a) The helicoidal texture in which the angle of orientation between subsequent lamellae changes by a constant amount. b) A crossed polylamellate texture with alternate lamellae with transverse and axial oriented microfibrils. c) A helical texture. d) A axial texture in which the microfibrils have an almost constant winding angle.

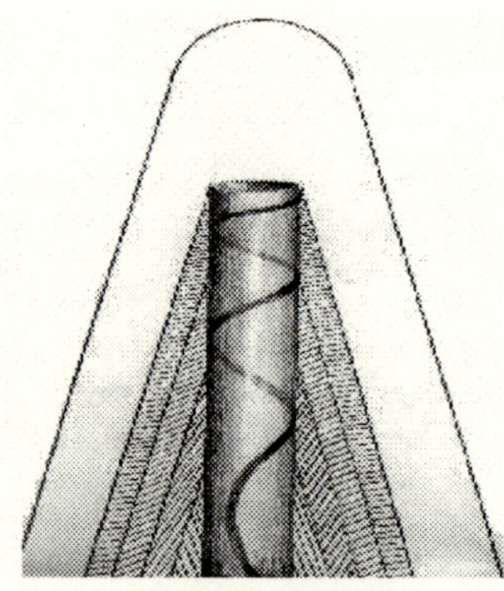

Fig. 3 Drawing, not to scale, of the cellulose microfibril deposition orientation at the plasma membrane of the secondary wall in the root hair tube.

The cellulose microfibril deposition orientation, i.e. the path along which the cellulose synthase complexes travel, changes as the number of cellulose microfibril synthases active inside the plasma membrane changes.

But how could such a winding procedure give rise to a regular, helicoidal cell wall? Since the distance (d) between individual microfibrils did not depend on their angle with the cell's long axis, and the inner root hair diameter (2R) was the same for all orientations, the only variable determining microfibril angle (α) could be N, the number of active cellulose synthases. If that number increased from 1 to N_{max}, α would gradually change from transverse to longitudinal. This produces only half the arc of a helicoid. To produce the second half and change the microfibril angle from longitudinal to transverse to the cell's long axis, N had to decrease gradually, which it could do by itself if the cellulose synthases have a finite lifetime. Thus, in order to make a helicoidal texture, synthase insertion/activation had to stop as soon as the longitudinal microfibril orientation was reached. In other words, there should be an insertion/activation domain in the plasma membrane, an area where the rosettes enter and become active; this is the aforementioned cellulose synthase activation domain.

Since I had accepted an assistant professorship at Wageningen University to work on the cell biology of somatic embryogenesis of *Zea mays*, it took eight years for the geometrical model to be published (Emons 1994). In this paper, I explained the orientation determination using drawings. My technical assistant, Henk Kieft, and I could show (with the help of simple computer simulations) that the model was able to produce all kinds of cell wall textures (Emons and Kieft 1994). The computer calculated microfibril orientation in a stepwise manner. Fortunately, I met theoretical physicist Bela Mulder, who was already interested in polymer organization and became fascinated by biological polymers in living cells. His partial differential equations (for full mathematical elaboration, see Mulder and Emons 2001) are fed into a computer, which then produces simulations as seen in Figure 2. Our studies have thus come full-circle. From 2-D electron microscopy puzzle pieces, cell biologists could study structures and see the 3-D structure as a whole. From partial differential equations for the synthase density N (z,t), which can be solved as a function of the position z along the cell and time t, theoretical physicists can envisage the texture as it is formed. These computer simulations demonstrate that the actual textures can be produced by these formulae.

The geometrical model provides a conceptual framework for the alignment mechanism of cellulose microfibrils, which unites examples where cortical microtubules are and are not parallel to nascent cellulose microfibrils, and in which they do not directly move or channel individual cellulose synthases, but may be involved in their insertion or activation inside the plasma membrane. The basic theory is as follows: by default, cellulose microfibrils "grow" straight unless obstructed, and align depending primarily on the number of cellulose synthases that are simultaneously active at any position in the plasma membrane. The geometrical model does not itself rule out that cortical microtubules bind to the plasma membrane so tightly that cellulose synthase movement is obstructed, which could be the case in elongating cells in which both polymers are often found in line with each other and transverse to the direction of cell elongation. However, recent work by Geoff Wasteneys's group shows that this is unlikely (Sugimoto *et al.* 2003; Himmelspach *et al.* 2003).

Why are the cortical microtubules important, and where could they enter the model? The relationship between cortical microtubule orientation and cell elongation perpendicular to the cortical microtubules is undisputed. There must be a cellular structure determining which wall facets should elongate, *i.e.* into the plasma membrane of which expandable wall facets the Golgi vesicles containing the cell wall matrix should be inserted, and cellulose synthases inserted/activated. The geometrical model certainly does not rule out the idea that cortical microtubules are part of the mechanism regulating the sites of exocytosis, and the amount of cellulose synthase insertion/activation in the plasma membrane. If cellulose synthase insertion occurs via exocytosis, then by inference from our knowledge of tip growing cells, cellulose synthase insertion would require the actin cytoskeleton (Miller *et al.* 1999; Ketelaar *et al.* 2002) and microtubules (Sieberer *et al.* 2002; Ketelaar *et al.* 2003), as well as calcium ion gradients at those sites (de Ruijter *et al.* 1998). Indeed, if the Golgi vesicles contain the cellulose synthases, as suggested by the work of Haigler and Brown (1986), the microtubules could determine the sites of exocytosis. This will soon be more easily confirmed by GFP-cellulose microfibril-synthase research.

Acknowledgements

I thank André Sassen, who introduced me to the field of cell wall studies; Mieke Wolters, who was my first collaborator; Michel Ebskamp, Genetwister Wageningen; Joost van Opheusden, Biometris, Wageningen University; Kim Boutilier, Plant Research International; Wageningen University and Research Center; Fabiana Diotallevi, FOM-Institute AMOLF; and Jan. Schel and the other members of the new cell wall group now in my laboratory: Tiny Franssen-Verheijen, Adriaan van Aelst, Miriam Akkerman, Carolina Cifuentes, and Peter Twumasi. Most of all, I thank Bela Mulder, who listens patiently to my explanations of the biological complexity of plant cells and my ideas about how these cells structure their cellulose microfibrils, and makes elegant, predictive simulations from differential equations.

References

Arioli T, Peng LC, Betzner AS, Burn J, Wittke W, Herth W, Camilleri C, Höfte H, Plazinski J, Birch R, Cork A, Glover J, Redmond J, Williamson RE. (1998) Molecular analysis of cellulose biosynthesis in *Arabidopsis*. Science 279: 717-720.

De Ruijter NCA, Rook MB, Bisseling T, Emons AMC (1998) Lipochito-oligosaccharides reinitiate root hair tip growth in Vicia sativa with high calcium and spectrin-like antigen at the tip.1998. Plant J 13: 341-350.

Emons AMC (1994) Winding threads around plant cells: a geometrical model for microfibril deposition. Plant Cell Environment 17: 3-14.

Emons AMC (1985) Plasma-membrane rosettes in root hairs of *Equisetum hyemale*. Planta 163:350-359.

Emons AMC, Derksen J, Sassen MMA (1992) Do microtubules orient plant cell wall microfibrils? Physiol Plant 84: 486-493.

Emons AMC, Kieft H (1994) Winding threads around plant cells: applications of the geometrical model for microfibril deposition. Protoplasma 180: 59-69.

Emons AMC, Mulder BM (1998) The making of the architecture of the plant cell wall: How cells exploit geometry. Proc Natl Acad Sci USA 95: 7215-7219.

Emons AMC, Mulder BM (2000) How the deposition of cellulose microfibrils build cell wall architecture. Trends in Plant Science 35-40.

Emons AMC, Schel JHN, Mulder BM (2002) The geometrical model for microfibril deposition and the influence of the cell wall matrix. Plant Biol 4: 22-26.

Emons AMC, Wolter-Arts AMC (1983) Cortical microtubules and microfibril deposition in the cell wall of root hairs of *Equisetum hyemale*. Protoplasma 117: 68-81.

Haigler CH, Brown RM (1986) Transport of rosettes from the Golgi apparatus to the plasma membrane in isolated mesophyll cells of *Zinnia elegans* during differentation to tracheary elements in suspension culture. Protoplasma 134: 111-120.

Himmelspach R, Williamson RE, Wasteneys GO (2003) Cellulose microfibril alignment recovers from DCB-induced disruption despite microtubule disorganization. The Plant J 36: 565-575.

Ketelaar T, Faivre-Moskalenko C, Esseling JJ, de Ruijter NCA, Grierson CS, Dogterom M, Emons AMC (2002) Positioning of nuclei in *Arabidopsis* root hairs: an actin-regulated process of tip growth. The Plant Cell 14: 2941-2955.

Ketelaar T, de Ruijter NCA, Emons AMC (2003) Unstable f-actin specifies the area and microtubule direction of cell expansion in *Arabidopsis* root hairs. The Plant Cell 15: 285-292.

Kimura S, Laosinchai W, Itoh T, Cui XJ, Linder CR, Brown RM (1999) Immunogold labeling of rosette terminal cellulose-synthesizing complexes in the vascular plant *Vigna angularis*. The Plant Cell 11: 2075-2085.

Miller DD, de Ruijter NCA, Bisseling T, Emons AMC (1999) The role of actin in root hair morphogenesis: studies with lipochito-oligosaccharide as a growth stimulator and cytochalasin as an actin perturbing drug. Plant J 17: 141-154.

Mueller SC, Brown RM (1978) Characterization of presumptive cellulose synthesizing system in corn root-cells. J Cell Biol 79: A237-A237.

Mulder BM, Emons AMC (2001) A dynamical model for plant cell wall architecture formation. J Math Biol 42: 261-289.

Pear JR, Kawagoe Y, Schreckengost WE, Delmer DP, Stalker DM (1996) Higher plants contain homologs of the bacterial celA genes encoding the catalytic subunit of cellulose synthase. P Natl Acad Sci USA 93: 12637-12642.

Roelofsen PA (1958) Cell-wall structure as related to surface growth. Acta Bot Neerl 7: 77-89.

Sugimoto K, Himmelspach R, Williamson RE, Wasteneys GO (2003) Mutation or drug-dependent microtubule disruption causes radial swelling without altering parallel cellulose microfibril deposition in *Arabidopsis* root cells. The Plant Cell 15: 1414-1429.

Sieberer B, Timmers ACJ, Lhuissier FGP, Emons AMC (2002) Endoplasmic Microtubules configure the Subapical Cytoplasm and Are Required for Fast Growth of *Medicago truncatula* Root Hairs. Plant Phys 130, 977-988

Vian B, Reis D, Darzens D (1994) Cholesteric-like crystal analogs in glucuronoxylan-rich cell wall composites. Experimental approach of a cellular re-assembly from native cellulose suspension. Protoplasma 180: 70-81

Wolters-Arts AMC, Sassen MMA (1991) Deposition and reorientation of cellulose microfibrils in elongating cells of *Petunia* stylar tissue. Planta 185: 179-189.

Implications of Emergence, Degeneracy and Redundancy for the Modeling of the Plant Cell Wall

Marcos S. Buckeridge

A wall composed of a mixture of polymers, mainly carbohydrates, proteins and secondary metabolites, surrounds every plant cell. These polymers interact through a mixture of covalent and non-covalent linkages to form a "supramolecular complex" that is thought to be a multifunctional structure responsible for controlling the mechanical properties of the cell. The wall is, in fact, embedded in a solution containing several different independent components such as ions, free sugars, enzymes and secondary metabolites, to name but a few.

In order to elucidate plant cell wall structure and functions, wall scientists have followed some key steps mainly during the second half of the 20th Century. The first step was to set up methods for extraction and determination of composition and structure of wall components. It was found that the main compounds are carbohydrate polymers. This step was followed by the development of methods to determine the structure (linkage types) and physico-chemical features of these polymers. This work culminated in the proposition of a succession of static models trying to picture a general assembly for the polymers in the wall (Keegstra et al. 1973, McCann and Roberts 1991 and Carpita and Gibeaut 1993). Likewise, many aspects of the biochemistry of the wall were developed, culminating in the discovery of several hydrolases (Fry 2004) that have been shown to correlate with physiological events in plants, which are controlled by plant hormones. With the development of cell and molecular biology, part of the wall research community invested in finding and correlating wall related genes (e.g. Holland et al. 2000; Lima et al. 2001) and also succeeded to find probes (enzymes and monoclonal antibodies) capable to localise wall components in the cell (Willats et al. 2001).

Presently, we are living the "omics" era and as a consequence of the use of these spectacular tools, the view of the wall is becoming progressively reduced. The tendency seems to be going inside the genome and scrutinising cell mechanisms in order to understand in detail wall production dynamics, from gene to polysaccharide. However, a number of researchers have focused on the consequences of wall properties at the organism level (e.g. Freshour et al. 2000; Wojtaszek 2001) and even less has been discussed about the importance of the wall for the plant as a whole or even at the ecophysiological level (Lambers et al. 1998).

This reductionist view imposed by actual circumstances, makes us to call it the *cell* wall so that its properties are thought usually as strictly related to the cell it "belongs". Cells are cemented to each other by a slightly different and apparently independent kind of supramolecular complex (the middle lamella), so that there is a possibility of redundancy of cell types, forming tissues. In this sense, the middle lamella is a sort of "wall of the walls" and probably one of the most important components that gives structure to the tissue. This raises the question whether we are missing some of the emergent properties that rise at higher levels of organisation of the wall.

In this chapter it will be proposed that the wall is not only a "cell wall", but it can also be viewed as the wall of a tissue, an organ or the entire plant. If one accepts this point of view, one has also to accept that the wall has emergent properties, acquiring different new properties at different levels of complexity. This is to say that when a group of cells form similar walls or a cell mosaic with certain features (e.g. the vascular system), it brings to existence a property that was not present at the cell level. New emergent properties become apparent also at the evolutionary level mainly because of two other properties: **degeneracy** (two different structures can have the same function or one structure can have more than one function) and **redundancy** (multiplication of a given structure and related mechanism that have the same function in the system) (Edelman and Gally 2001).

The reductionism embedded in static wall models

There are two types of modelling, static and dynamic (Holland 1998). The main feature of a static model is that details are suppressed so that only the more evident points of reference are shown. For the wall, to say that it is composed of cellulose, hemicelluloses and pectins can be termed modelling whereas as knowledge increases, more details are added, including the dynamics of how the polymers are synthesised, interact and are degraded. Once these aspects are unveiled, the new stage of modelling turns to a more dynamic approach and answers to questions about the meaning, for the plant, of the presence of a wall with a given structure, in a given part of the organism, at a given moment requires understanding properties of the wall that the static models do not have. These are emergent properties that appear as a result of the redundancy of cell types that form plant tissues and when different plant organs integrate.

Another feature of the current models of the wall is that the common approach is a sort of inside-to-outside. A typical view is that any change or even turnover of the wall is coordinated exclusively by the genome of the cell that a given wall involves. It is as if every cell would have its particular wall and events would be coordinated by the interaction genome-transcriptome-proteome-metabolome, essentially at the cellular level.

This view is not wrong or less important, but it has been historically contaminated with reductionism mainly because we needed to determine the wall composition and the functions of its components and also because most cell wall scientists were chemists, physicists, biochemists, molecular biologists and cell biologists. This is important, since the more reduced levels of complexity are those that first evolved in nature and bare the basic mechanisms on which more complex levels built up.

Wall properties are a result of the interaction between inside (cellular events) and the outside (environmental events). In other words, a given wall is a result of the expression of genes encoded by the genome as well as from the interaction of the plant with its environment.

Thus, the wall can be viewed as a supramolecular polymer that separates all cells from each other so that a liquid crystal inner environment is maintained as a *continuum spacer* among virtually all cells. It is as if the cells are all suspended in an extensive extracellular matrix through which cells can communicate either directly by plasmodesmata or indirectly using apoplastic transport through the wall. The former is used for short distances and the second for medium distances since the vascular system can be used for long distance communication and nutrient transport.

The importance of biological degeneracy for dynamically modelling the wall

McCann and Roberts (1991) coined the term architecture of the cell wall in a reference to the existence of three different independent domains in the wall. This added a certain dynamic element to the view of the wall and led many researchers to think that every domain could have different properties. Although this implied that the three domains might interact and this sole interaction would have emergent properties, most researchers preferred to follow the analytical route and study the properties of the parts (i.e. the domains). Carpita and Gibeaut (1993) showed that at the level of superior plants, the wall structure is degenerated, i.e. that different structures (e.g. arabinoxylan and xyloglucan) may play similar functions. Thus, the "architectural" model points out that the domains are changing during evolution, i.e. monocots changed their hemicelluloses and decreased the pectin domain. We have now analysed 11 species of Pteridophytes (arranged phylogenetically) and found that a third type of wall appears to exist since in this ancient group of plants, mannans are the principal hemicellulose (GB Silva, MA Tiné, J Prado and MS Buckeridge, unpublished results). All this is adding more dynamics to the architectural view of the wall at the evolutionary level; however, the emergent properties above the cellular level are still missing. A possible mechanism involved in the changes in the wall during evolution has been proposed by Buckeridge et al. (2000). It was proposed that degeneracy (i.e. the same structure having different functions) made possible the appearance and intensification of certain functions, so that galactomannans and xyloglucans seem to have been transferred from a structural role in the primary wall to a storage function in seeds.

Building a dynamic integrated model of the wall that contemplates its functioning as entities of tissues, organs or the whole plant will not be possible if we look only for physical and chemical contacts of polymers of the walls at higher levels of complexity. We may take into consideration the influence that the wall has, for example, at the level of the apoplast of a tissue, in the functioning of an organ and even in the ecophysiological behaviour of a plant. The existence of a certain type of wall bearing redundancy and degeneracy is probably what affords the appearance of emergent properties that are characteristic of an organ and its functions.

The evolutionary transition leaf-to-cotyledon will be taken as an example of integration among walls in different tissues and the environment. Let's first subdivide the leaf organ into the component tissues that present similar types of walls, forming a redundant-degenerated wall system: the palisade, the parenchyma; the epidermis (including stomata) and the vascular system. The latter tissue is a special case in which there is a higher level of degeneracy, but one single property: long distance transport. The epidermis is degenerated in the sense that the simplest form will have epidermal cells and stomata. A recent work by Jones et al. (2003) brought evidences of the direct involvement of a wall component and the functioning of stomata. Some species posses, in their leaves, a subepidermal layer of cells that seem to be associated with early ecological succession stages in tropical forests (Boeger and Wisniewski 2003). The ecophysiological function of this layer, as well as the composition and architectural wall model of the cells in this tissue is virtually unknown. The function of the palisade is to perform photosynthesis. In this case the form of the cell is crucial and it is well known (from the procedure of making protoplasts) that their wall gives the form of these cells. In this case of foliar sclereids, cell form is important also in controlling light propagation within the leaf tissue (Karabourniotis 1998). The parenchyma cells are, on one hand, a means to produce free space to store CO_2 in a gaseous form and on the other hand to function as a storage layer, where starch can accumulate transiently and then be degraded to produce sucrose, that will be transported to other tissues and organs. This storage function, which is coupled to the diurnal cycle of events of photosynthesis, is thought to have been transferred to cotyledons so that seed plants use this adaptation for a long term storage, including the wall in the case of xyloglucan-storing cotyledons (Buckeridge et al. 2000). A clear morphological gradient is seen in cotyledons of legumes (Smith 1981). The vascular system is based on extreme adaptations of walls. The phloem is a complex tissue with the walls being one of the principal functional components. It forms the framework for symplastic continuity in leaf tissue [see for example Ayre et al. (2003)]. The xylem and fibres are formed of dead cells whose function (physical support, transport of water and ions) can be almost strictly attributed to the wall. Acting together and synchronically, these tissues form the above level of complexity (the leaf), which has its own emergent properties thanks to the redundancy and degeneracy of most components (among them the wall). If for a moment we look at the wall as being from a tissue and not from a single cell, we can immediately appreciate the emergent properties of the wall at the above levels of complexity. Although the communication among these tissues is thought to occur within the organ, it is known that leaf tissue may keep long distance communication, via auxin, with cotyledon-storage tissue [for example Santos et al. (2004)].

As a consequence of redundancy, degeneracy and subsequent emergent properties, when we perform an in vitro fractionation of a homogenate of leaves of one species and determine the polymers of which it is composed and the constituent glycosidic linkages, we are destroying the organisation of the tissue and consequently loosing the map of correlations of the functional components of the system. Thanks to the development of probe tools such as antibodies, we can now extend the view of the wall to the whole plant (see for example Trethewey and Harris 2002). We can partially view this map by using specific antibodies to actually see where a given component is and how it changes in time, but we will need other tools capable to evaluate simultaneously wall properties of tissues and organs in order to probe the emergent properties of the wall as an integrated structure in the tissue/organs it belongs. The difficult task will be to find how to integrate the static models into progressively more dynamic ones, so that we understand the communication among different levels of complexity as well as the natural history of the extracellular matrix of plants.

Acknowledgements

Although the ideas presented here are my own, some colleagues contributed with challenging discussions about the biology of walls that were fundamental for the assembly of these ideas. They are: Grant Reid, Gustavo Maia Souza, Henrique Santos, Joss Rose, Marcia Braga, Marcos Aidar, Marco Tiné, Maureen McCann, Michael Hahn and Nick Carpita.

References

Ayre BG, Kelle F, Turgeon R (2003) Symplastic continuity between companion cells and the translocation stream: long-distance transport is controlled by retention and retrieval mechanisms in the phloem. Plant Physiol 131: 1518-1528.

Boeger MRT, Wisniewske C (2003) Comparação da morfologia foliar de espécies arbóreas de três estádios sucessionais distintos de floresta ombrófila densa (Floresta Atlântica) do Sul do Brasil. Rev Bras Bot 26:61-72 (http://www.scielo.br/scielo.php/script_sci_serial/pid_0100-8404/lng_en/nrm_iso).

Buckeridge MS, Santos HP, Tiné MAS (2000) Mobilisation of storage cell wall polysaccharides in seeds. Plant Physiol Bioch 38: 141-156.

Carpita NC, Gibeaut DM (1993) Structural models of primary cell walls in flowering plants: consistency of molecular structure with the physical properties of the walls during growth. Plant J 3: 1-30.

Edelman GM, Gally JA (2001) Degeneracy anc omplexity in biological systems. Proc Natl Acad Sci USA 98: 13763-13768.

Freshour G, Bonin CP, Reiter WD, Albersheim P, Darvill A, Hahn MG (2000) Distribution of fucose-containing xyloglucans in cell walls of the *mur1* mutant of Arabidopsis. Plant Physiol 131: 1602-1612.

Fry S (2004) Primary cell wall metabolism: tracking the careers of wall polymers in living plant cells. New Phytol 161: 641-675.

Holland, JH (1995) Hidden order: how adaptation builds complexity, Reading, Mass: Addison-Wesley.

Holland N, Holland D, Helentjaris T. Dhugga KS, Xoconostle-Cazares B, Delmer DP (2000) A comparative analysis of the plant cellulose synthase (*CesA*) gene family. Plant Physiol 123: 1313-1323.

Jones L, Milne JL, Ashford D, McQueen-Mason SJ (2003) Cell wall arabinan is essential for guard cell function. Proc Natl Acad Sci USA 100: 11783-11788.

Karabourniotis G (1998) Light-guiding function of foliar sclereids in the evergreen scleropholl *Pillyrea latifolia*: a quantitative approach. J Exp Bot 49: 739-746.

Keegstra K, Talmadge K, Baurer WD, Albersheim P (1973) Structure of the plant cell walls. Plant Physiol 51: 188-196.

Lambers H, Chappin III FS, Pons TL (1998) Plant Physiological Ecology. Springer-Verlag, New York.

Lima DU, Santos HP, Tiné MA, Molle FD, Buckeridge MS (2001) Patterns of expression of cell wall related genes in sugar cane. Gen Mol Biol 24: 191-198.

McCann MC, Roberts K (1991) Architecture of the primary cell wall. In: The Cytoskeletal Basis of Plant Growth and Form, CW Lloyd (ed) London: Academic Press, pp 109-129.

Smith DL (1981) Cotyledons of Leguminosae. In Advances in Legume Systematics, RM Polhill and P Raven (eds) Part 2, pp 927-940. Royal Botanic Gardens Kew, Richmond.

Santos HP, Purgato E, Mercier H, Buckeridge MS (2004) The control of storage xyloglucan mobilization in cotyledons of *Hymenaea courbaril* L. Plant Physiol 135: 287-299.

Trethewey JAK, Harris PJ (2002). Location of (1-3) and (1-3;1-4)-beta-glucans in vegetative cell wall of barley (*Hordeum vulgare*) using immunogold labelling. New Phytol 154: 347-358.

Willats SGT, McCartney L, Mackie W, Konx P (2001) Pectin: cell biology and prospects for functional analysis. Plant Mol Biol 47: 9-27.

Wojtaszek P (2001) Organismal view of the plant and a plant cell. Acta Biochimica Polonica. 48: 443-451.

Chemistry in the Determination of Cell Wall Structure

Andrew J. Mort

The structures of the various polysaccharide types in plant cell walls have been studied for many years and the general structure of each has been determined. However, details of the structures, which may vary from species to species or cell type to cell type, are largely unexplored. I and my associates over the years have devised various chemical methods to explore these structural details.

I consider my most important contributions to cell wall research, so far, to be the application of previously known chemical reactions in new ways to the study of plant cell wall structure.

Deprotection of chemically synthesized peptide turns into protein deglycosylation and oligosaccharide generation

My first project as a graduate student in the laboratory of Derek Lamport was to test out ways to remove sugars from glycopeptides generated from the structural cell wall protein, extensin. Extensin is heavily glycosylated and insolubilized into many plant cell walls. Removal of the sugars would greatly facilitate proteolysis of the protein, purification of peptides generated from it, and sequencing. I also thought that cleavage of all of the sugar linkages in a cell wall should release the protein in a soluble form. Roy Vagelos (an expert on acyl carrier protein who was visiting) suggested that treatment of the peptides with 60 % aqueous HF might work to remove the sugars, since his group had been using it to cleave the phosphate that links the pantothenoic acid to a serine hydroxyl in acyl carrier protein (Prescott et al. 1969). Fortunately though, I did not try this procedure because I couldn't think of an easy way to remove the HF solution from the peptides after the treatment. In fact, 48% aqueous HF is now used fairly frequently to dephosphorylate lipopolysaccharides without cleaving glycosyl linkages (Mort and Pierce 2002). The suggestion did, however, lead me to read about deprotection of chemically synthesized peptides. Anhydrous liquid HF seemed to be the reagent of choice for complete removal of protecting groups from amino, hydroxyl, and sulfhydryl groups. Because of the aggressive nature of anhydrous HF, I did not want to try it. An alternative approach was to dissolve the peptide in trifluoroacetic acid and then bubble HBr gas through the solution. This reaction allowed me to remove galactose residues from serine hydroxyl groups without cleaving peptide bonds. Since anhydrous HF was used much more frequently for peptide deprotection, and a commercially available apparatus would make its use relatively safe and convenient, we switched to using HF. We have used HF in the commercial apparatus (with some modifications (Mort 1983)) for over 30 years now with only a few minor leakages of HF into the fume hood.

By using the fairly well characterized glycoproteins fetuin and porcine submaxillary mucin, we found that HF at 0 °C cleaved the glycosidic linkages of all neutral sugars without cleaving any peptide bonds, so it could be used to remove most of the sugars from any glycoprotein (Mort and Lamport 1977). The sugars that were not removed from the protein were those amino sugars linked directly to the protein. In the case of N-linked oligosaccharides, the chitobiose unit remained attached to the asparagines. For O-linked oligosaccharides, just a single N-acetyl-galactosamine remained linked to the hydroxyl groups of serine and threonine residues. If we did the HF treatment at room temperature (approx. 23 °C), the O-linked amino sugar linkages were also cleaved but not the N-glycosidic link between N-acetyl-glucosamine linked directly to the asparagine. Differences in resistance to aqueous acid hydrolysis of various glycosidic linkages were well known, but they were often not high enough to achieve high yields of selective hydrolysis. The difference in susceptibility of amino sugar linkages and neutral sugar linkages at 0 °C in HF seemed very distinct. HF or reagents with similar properties (trifluoromethane sulfonic acid (Edge et al. 1981) and HF/pyridine) have been used many times now to deglycosylate glycoproteins.

My expectation that HF treatment of plant cell walls would release extensin in a soluble form proved to be wrong. Essentially all of the extensin remained insoluble. This observation has been used as evidence for extensive covalent crosslinking between extensins (Fry 1986). Unfortunately, the nature of these crosslinks is still unclear, although they may involve oligomerization of certain tyrosine residues in the protein mediated by peroxidases (Brady et al. 1996).

In subsequent work in the laboratory of Dietz Bauer, I was characterizing the structure of the extracellular polysaccharide of *Bradyrhizobium japonicum*. This polysaccharide specifically binds soybean seed lectin, perhaps indicating an involvement in host specificity between host and symbiont. At the same time (in the late 1970s), there was a lot of interest in biomass conversion for production of liquid fuels, especially ethanol. Realizing that HF could give almost a 100 % yield of monosaccharides from all of the carbohydrate constituents, including cellulose, in plant cell walls and that it might be possible to work out a system to use the HF over and over again since it boils at 18 °C, I convinced the Charles F. Kettering Foundation to give me a small grant to study HF conversion of biomass. The conversion studies worked out well, but the recycling of the HF was a problem. This project meant that I had access to an HF apparatus and could modify it to allow much more careful control over the reaction conditions (Mort 1983). If I took the bacterial polysaccharide and treated it in HF at -23 °C (the temperature of a dry ice/carbon tetrachloride bath), I cleaved the linkages between the α-linked neutral sugars but not those of β-linked glucose or the α-linked galacturonic acid. The polysaccharide was a repeating pentasaccharide and the treatment gave me an almost 100 % yield of a trisaccharide from the polysaccharide's backbone (Mort and Bauer 1982). If I did the treatment at -40 °C, only the linkage of the branched glucose residue was cleaved to any great extent, which gave a pretty good yield of the entire repeat unit (Mort et al. 1983).

As an assistant professor at Oklahoma State University, I pursued the specificity of HF at low temperatures generating, along with a Soviet group at the Zelinski Institute in Moscow, a table of susceptibilities vs. temperature for a wide range of sugar linkages (Knirel et al. 1989; Mort et al. 1989).

Treatment of cell walls with HF at -23 °C does just as one would predict: the cellulose remains intact, xyloglucans lose their xylose, and the various regions of pectins remain polymeric or are severely degraded, depending on their structure. Homogalacturonans remain intact including their methylesterification; xylogalacturonans lose their xylose residues; the sidechains of rhamnogalacturonans are completely degraded to monomers and the backbone is converted into disaccharides of rhamnose and galacturonic acid. When we investigated these disaccharides by NMR spectroscopy, we found that many of them were acetylated (Komalavilas and Mort 1989) but none were methyl esterified. This was the first direct evidence that most of the acetyl groups but none of the methyl esters in pectin are in the RG region.

Cleaving methyl ethers becomes specific degradation of uronic acids

As alluded to above, in the late 1970s I was working on the hypothesis that the extracellular polysaccharide of *B. japonicum* was involved in host recognition via lectin binding. There was a striking correlation of loss of lectin binding ability with a predominance of the 4-O methyl ether of galactose rather than galactose in the polysaccharide. It has been found (Khuong-Huu et al. 1971) that treating glycosides of methyl ethers of sugars with lithium metal in an amine solvent leads to removal of the methyl ether without cleavage of the glycoside. We reasoned that we should be able to convert the polysaccharide that did not bind lectin into one which did by using this treatment. We found that in addition to cleaving the methyl ethers, lithium in ethylenediamine destroyed uronic acids. In the case of the *B. japonicum* polysaccharide, the treatment caused it to fall apart into tetrasaccharides, which were used to help determine the structure of the polysaccharide.

Since the RG region of pectin is made up of a repeating rhamnose-galacturonosyl disaccharide backbone with sidechains of neutral sugars on many of the rhamnosyl residues, a treatment of pectins with lithium in ethylenediamine can be used to release the sidechains for purification and characterization. Lau et al. (1987) were very successful in characterizing many sidechains of RG I from sycamore this way. Unfortunately, it seems that arabinofuranosyl linkages are not entirely resistant to the treatment.

Sometimes more is better: more reagent plus more buffering capacity gives quantitative conversion of esterified galacturonic acid to galactose and allows determination of degrees and patterns of methylesterification

The degree and pattern of methyl esterification of pectins is probably very important for their function. Traditionally, the degree of esterification of pectins has been determined by titration of the acid groups before and after saponification or by quantitation of methanol and galacturonic acid by colorimetric methods. Both of these methods require mg or higher amounts of sample. It is well known that esters, but not carboxylic acids, can be reduced to alcohols in organic solvents with sodium borohydride, or better yet lithium borohydride. However, working with polysaccharides, especially pectins, is much easier in aqueous systems for solubility reasons. Sodium borohydride is unstable in aqueous solutions at neutral pH and its breakdown leads to a rapid increase in pH. Lithium borohydride tends to catch on fire as it is added to water. Methyl esters are quite labile to alkaline pHs. We reasoned that if we could quantitatively reduce esterified GalA residues to galactose despite these difficulties, we could estimate the percent of GalA that was esterified in a pectin sample by determining the sugar composition of the pectin before and after reduction of the esterified residues to galactose. After much experimentation, we found that the reduction could be consistently made to go to completion without premature saponification of the esters if the pectin sample, or cell wall sample, was dissolved in, or equilibrated in, ice cold 1 M imidazole buffer at pH 7 and a 100-fold molar excess of $NaBH_4$ was added all at once with rapid stirring to keep the solution from bubbling out of the tube. Addition of small amounts of butanol as an anti-foamant was useful, especially if the reaction were being carried out on a mg to gm scale in a beaker. Once quantitative reaction was achieved and methods to remove all of the reagents were worked out, we succeeded in getting reliable measures of the degree of esterification of pectins using 10-100 μg samples, using GLC for the sugar analysis (Maness et al. 1990).

Once converted to galactose, previously esterified GalA residues are permanently identified and converted into a distinct chemical entity. We have made use of this to investigate patterns of methyl esterification in pectins (Mort et al. 1993). For example, if we take a commercial apple pectin, convert its methyl esterified GalA residues to galactose, and treat it in HF containing 1% water at -15 °C, the pectin is fragmented at every galactosyl linkage. An analysis of the products by ion exchange chromatography or capillary zone electrophoresis (Chen and Mort 1996) showed the relative frequency of occurrence of lengths of uninterrupted stretches of non-esterified

GalA residues. One can predict statistically what these relative frequencies should be if the esters are randomly distributed. If the frequencies don't fit the prediction, the pattern is not random. By this criterion apple pectin from Sigma is randomly esterified, but esterification in a 50% esterified section of pectin extracted from cotton suspension culture cell walls is far from random. Wiethöelter and co-workers (Wiethöelter et al. 2003) have used this approach to show distinct patterns of methyl esterification in isogenic lines of wheat resistant or susceptible to stem rust.

What next?

What I really want to do is find out how the different regions of pectins are linked together and why there is so much structural complexity in pectins.

A major impediment is still the lack of specific ways to cleave polysaccharides. Many microorganisms use cell walls as a carbon source, so they must be able to break them down. Our present goal is to clone and express a representative of every known cell wall polysaccharide-degrading enzyme coded for in the genome of *Aspergillus nidulans.* The whole genome of this fungus has been sequenced, so it is possible to deduce the sequences of the enzymes. We, Chris Somerville's group and mine, are now expressing the enzymes in the yeast *Pichia pastoris* and characterizing their modes of action. Very soon we will have a selection of highly specific enzymes directed against almost all sugar linkages in plant cell walls. The combination of chemical and enzymic degradations should allow us to generate fragments representing junctions between the various regions of pectins, which after HPLC purification, can be characterized by mass spectrometry and NMR spectroscopy.

Acknowledgements

Most of the work described here was supported by the Department of Energy. I certainly appreciate the guidance and freedom provided by Derek Lamport and Dietz Bauer during my early years in research, and from that time, too, Debbie Delmer has been an inspiration to me. We worked in adjacent labs for a few years at the Plant Research Laboratory at Michigan State University. She has been a long-time mentor and friend.

References

Brady JD, Sadler IH, Fry SC (1996) Di-isodityrosine, a novel tetrameric derivative of tyrosine in plant cell wall proteins: A new potential cross-link. Biochem J 315: 323-327.

Chen EMW, Mort AJ (1996) Nature of sites hydrolyzable by endopolygalacturonase in partially esterified homogalacturonans. Carbohydr Polym 29: 129-136.

Edge ASB, Faltynek CR, Hof L, Reichert LE, Jr., Weber P (1981) Deglycosylation of glycoproteins by trifluoromethanesulfonic acid. Anal Biochem 118: 131-137.

Fry SC (1986) Cross-linking of matrix polymers in the growing cell walls of angiosperms. Annu Rev Plant Physiol 37: 165-186.

Khuong-Huu Q, Monneret C, Kabore I, Goutarel R (1971) Steroid alkaloids. CXXIII. New method of O-demethylation of methoxy sugars. Demethylation of cymarose and amino glycosteroids. Tetrahedron Lett 22: 1935-1938.

Knirel YA, Vinogradov EV, Mort AJ (1989) Application of anhydrous hydrogen-fluoride for the structural-analysis of polysaccharides. Adv Carbohydr Chem Biochem 47: 167-202.

Komalavilas P, Mort AJ (1989) The acetylation at O-3 of galacturonic acid in the rhamnose-rich region of pectins. Carbohydr Res 189: 261-272.

Lau JM, McNeil M, Darvill AG, Albersheim P (1987) Treatment of rhamnogalacturonan I with lithium in ethylenediamine. Carbohydr Res 168: 245-274.

Maness NO, Ryan JD, Mort AJ (1990) Determination of the degree of methyl esterification of pectins in small samples by selective reduction of esterified galacturonic acid to galactose. Anal Biochem 185: 346-352.

Mort AJ (1983) An apparatus for safe and convenient handling of anhydrous liquid HF at controlled temperatures and reactions times. Application to the generation of oligosaccharides from polysaccharides. Carbohydr Res 122: 315-321.

Mort AJ, Bauer WD (1982) Application of two new methods for cleavage of polysaccharides into specific oligosaccharide fragments. Structure of the capsular and extracellular polysaccharides of *Rhizobium japonicum* that bind soybean lectin. J Biol Chem 257: 1870-1875.

Mort AJ, Komalavilas P, Rorrer GL, Lamport DTA (1989) Anhydrous hydrogen fluoride and cell-wall analysis, In: Linskens HF, Jackson JF (eds) Plant Fibers. Springer-Verlag, Berlin, 10: pp 37-69.

Mort AJ, Lamport DTA (1977) Anhydrous hydrogen fluoride deglycosylates glycoproteins. Anal Biochem 82: 289-309.

Mort AJ, Pierce ML (2002) Preparation of carbohydrates for analysis by modern chromatography and electrophoresis, In: El Rassi Z (ed) Carbohydrate Analysis by Modern Chromatography and Electrophoresis. Elsevier, Amsterdam, 66: pp 3-38.

Mort AJ, Qiu F, Maness NO (1993) Determination of the pattern of methyl esterification in pectins. Distribution of contiguous non-esterified residues. Carbohydr Res 247: 21-35.

Mort AJ, Utille JP, Torri G, Perlin A (1983) High selectivity in the partial degradation of an extracellular polysaccharide of *R. japonicum* with liquid hydrogen fluoride. An N.M.R. spectroscopic study. Carbohydr Res 121: 221-232.

Prescott DJ, Elovson J, Vagelos PR (1969) Acyl carrier protein XI. The specificity of acyl carrier protein synthetase. J Biol Chem 244: 4517-4521.

Wiethöelter N, Graessner B, Mierau M, Mort AJ, Moerschbacher BM (2003) Differences in the methyl ester distribution of homogalacturonans from near-isogenic wheat lines resistant and susceptible to the wheat stem rust fungus. Mol Plant-Microbe Interact 16: 945-952.

Rhamnogalacturonan II-Borate Complex in Plant Cell Walls

Tadashi Ishii

Chemical structure of RG-II

Rhamnogalacturonan II (RG-II) belongs to a group of pectic polysaccharides referred to as substituted homogalacturonans (O'Neill et al. 2004). These polysaccharides all have a backbone composed of 1,4-linked á-galactosyluronic acid residues but differ in the types of oligosaccharide side chains attached to the backbone (O'Neill et al. 2004). RG-II is a quantitatively minor (0.1 - 5.0 % w/w) component of the primary wall. Nevertheless, RG-II is present in the primary walls of all vascular plants that have been studied. Perhaps more remarkable is the recent demonstration that the glycosyl sequence of RG-II is conserved in lycopodiophytes, pteridophytes, gymnosperms, and angiosperms (Matsunaga et al. 2004). Thus, the machinery responsible for the biosynthesis of RG-II is likely to have been in existence for at least 400 million years (O'Neill et al. 2004). Bryophytes apparently contain small amounts of RG-II (Matsunaga et al. 2004).

Albersheim and co-workers pioneered the use of homogeneous endoglycanases to selectively release polysaccharides from primary walls (Albersheim 1975). This group was also one of the first to appreciate the complexity of the primary wall and introduced plant scientist to the notion that cell walls and their components may have other functions in addition to their structural roles.

During studies to isolate and characterize the polysaccharides released by endopolygalacturonase treatment of sycamore cell walls, Darvill et al. (1978) identified a low molecular mass (5 - 10 kDa), structurally complex, yet quantitatively minor rhamnose-containing anionic polysaccharides. Endopolygalacturonase treatment also solubilized a high molecular mass (>100 kDa) rhamnose-containing pectic polysaccharide (O'Neill et al. 1990). This larger polysaccharide was named rhamnogalacturonan I (RG-I) because it was the first to elute from an Agarose 5M size-exclusion chromatography column. RG-I has a backbone composed of the repeating disaccharide -4)-α-Gal$_p$A-

1, 2-α-Rha$_p$-(1- (O'Neill et al. 1990). Arabinan and galactan side chains are linked to *O*-4 of some of the Rha residues (O'Neill et al. 1990). To distinguish the low molecular mass pectic polysaccharide from RG-I, the second rhamnose-containing polysaccharide was named rhamnogalacturonan II in large part because it eluted after RG-I on the size-exclusion chromatography column. Despite the similarity in trivial names it must be emphasized that RG-I and RG-II are not related structurally since they differ in the nature of their backbones and in the types and structures of the oligosaccharides that are linked to the backbones.

In a series of studies extending over thirteen years, Albersheim's group has shown that endopolygalacturonase-released RG-II is a low molecular mass (5-10 kDa) polysaccharide that contains 12 different glycosyl residues linked together by more than 20 different glycosyl linkages. RG-IIcontains the rarely observed sugars apiose, L-aceric acid (3-*C*-carboxy-5-deoxy-L-xylose), 2-*O*-methyl fucose, 2-*O*-methyl xylose, L-galactose, 2-keto-3-deoxy-D-*lyxo*-heptulosaric acid and 2-keto-3-deoxy-D-*manno*-octulosonic acid. 3-*O*-methyl rhamnose is present in RG-II synthesized by some but not all pteriolophytes and lycophytes (Matsunaga et al. 2004). All RG-IIs contain galacturonic acid, glucuronic acid, rhamnose, galactose, arabinose, and fucose (Figure 1).

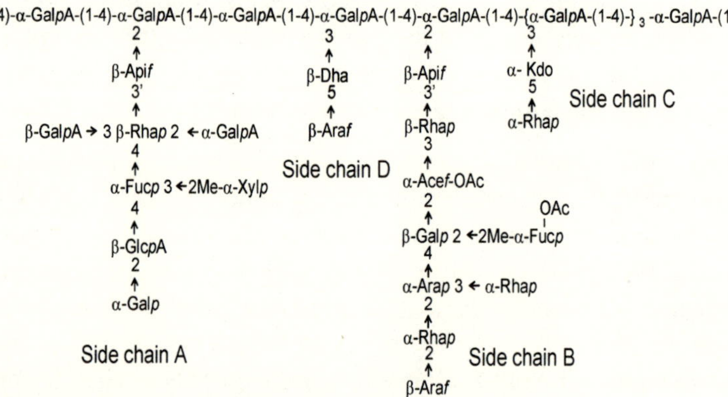

Fig. 1 Structure of rhamnogalacturonan II.
The location of side chains is tentative.

Most, if not all, of the early studies of the primary structure of RG-II were performed by Albersheim and co-workers. This group has developed numerous chemical methods to selectively fragment RG-II and employed GC, GC/MS, NMR spectroscopy, and mass spectrometry (MS) to characterize the released oligosaccharides. In two papers published in 1983 Spellman et al. described the detailed characterization of an unusual sugar component of RG-II that was named aceric acid (O'Neill et al. 1990) and an aceric acid-containing heptasaccharide (side chain B). Aceric acid is the only naturally occurring branched deoxy acidic sugar and to date has only been found in RG-II. Subsequently, the primary structure of side chain B has been refined using ^{1}H NMR spectroscopy together with electrospray-ionization mass spectrometry (Whitcome et al. 1995). A second side chain (side chain A) proved to be somewhat more challenging to structurally characterize since it is not released by treating RG-II with acid nor are there endoglycanases available that release the intact side chain. Nevertheless, the primary sequence of side chain A was deduced by Melton et al. (1986) and Stevenson et al. (1988) using the glycosyl sequencing procedure that had been pioneered in the Albersheim laboratory. A disaccharide containing 2-keto-3-deoxy-D-*manno*-octulosonic acid (side chain C) and a disaccharide containing 2-keto-3-deoxy-D-*lyxo*-heptulosaric acid (side chain D) were identified by York et al. and Stevenson et al., respectively. The identification of 2-keto-3-deoxy-D-*manno*-octulosonic acid and 2-keto-3-deoxy-D-*lyxo*-heptulosaric acid as components of sycamore RG-II was unexpected since these glycoses are typically components of bacterial polysaccharides. However, studies of other plant cell walls have firmly established that 2-keto-3-deoxy-D-*manno*-octulosonic acid and 2-keto-3-deoxy-D-*lyxo*-heptulosaric acid are present in all RG-IIs that have been isolated.

One of the greatest challenges with RG-II has been to establish which sugar in each of the four oligosaccharide side chains is itself directly linked to the backbone and how the four side chains are distributed along the homogalacturonan backbone. Again, most of the studies in this area emanated from the Albersheim group and utilized chemical methods to selectively fragment RG-II. The linkage position of side chains A, B, C, and D on the backbone was determined by Stevenson et al. and Puvanesarajah et al. Side chains A and B are attached to *O*-

2 of the backbone and disaccharide side chains C and D are attached to *O*-3. Progress has also been made in determining the locations of side chains B, C, and D on the backbone with respect to each other by using enzymically-generated oligosaccharides (Vidal et al. 2001) and by NMR spectroscopic analysis of the RG-II monomer (Perez et al. 2003) (Figure 1, Ridley et al. 2001).

I worked in Albersheim's laboratory as a postdoc twice. I remember that a graduate student and a postdoc always worked on RG-II. During my first stay a graduate student Tom Stevenson determined the structure of side chain D and also obtained strong evidence for the points of attachment of all four of the side chains. A problem that faced Stevenson during his graduate research was that determining the glycosyl-linkage compostions of RG-II become difficult. For some reason the Hakomori procedure that was used to per-*O*-methylate polysaccharides had ceased to work. After numerous modifications to the procedure Stevenson accidentaly found that adding a small amount (-5 μL) of water to the RG-II allowed the methylation reaction to go to completion. This was considered "odd" since we had all been taught that Hakomori methylation required a dry sample - indeed all samples were rigorously vacuum dried before the reaction. No one has actually determined why water is required for the methylation of RG-II although it is possible that "dry" RG-II forms a colloidal suspension and not a true solution in dimethyl sulfoxide - the solvent for the methylation reaction. Adding a trace amount of water (or glycerol) is still recommended for the effective methylation of RG-II. During my second stay in the Albersheim's lab, Whitcome et al. refined the structure of side chain B and determined the location of the two *O*-acetyl groups. At the same time Puvanesarajah was applying the Smith (peridoate oxidation) degradation and several of Aspinalls chemical degradation methods to demonstrate that the apisoyl and 2-keto-3-deoxy-D-*lyxo*-heptulosaric acid residues are directly linked to the backbone. There was obviously a lot of time and effort still being expended on RG-II. I asked Alan, "Why are you working RG-II for such a long time? He said, "RG-II is complicated. We would like to elucidate its structure. I do believe RG-II should have some physiologically important functions."

Isolation and function of RG-II borate cross-linked dimer

A function for RG-II remained elusive for many years despite our reasonably detailed knowledge of the glycosyl sequence of this polysaccharide. There were many discussions in the Albersheim's lab as to why plants synthesize a polysaccharide that contains such a large number of different glycoses and why the structure was so highly conserved in angiosperms. This led to speculations that RG-II was a signaling molecule rather than a structural component of the wall. There was an interesting speculation, although it is one that has not been substantiated by experimental evidence. Several of the individuals (Willie York and Tom Stevenson) who worked in Albersheim's lab had observed that RG-II had some unusual chromatographic properties - its molecular weight appeared to increase or decrease depending on the nature of the treatment. There were suggestions that RG-II molecules may interact with themselves to form a dimer, or a trimer, or a multimer, or that the affects were simply artifacts. Our ability to examine these phenomena was impeded by the limitations of the chromatographic procedures available at the time. These methods typically required large amounts (>25 mg) of sample, had runs times of up to 24 hours, and did not fully resolve all the components. Moreover, numerous fractions had to be collected and each one analyzed for glycoses using colorimetric assays. Nowaday we can analyze a small amount of RG-II (10 μg) by using HPLC with a Superdex 75 column. Thus, the unusual chromatographic properties of RG-II were "put on the back burner" but still remained the subject of much lab speculation. In hindsight it is clear that the chromatographic affects were real but an explanation would not be forthcoming until the work of Matoh and colleges almost 10 years later which established that RG-II existed in primary walls as a borate cross-linked dimer and the studies of O'Neill et al. that showed that the monomer and dimer were readily interconverted *in vitro*.

Studies with RG-II illustrates what has become all to common problem in modern research. The vast amount of data that is and will continue to be published makes it a daunting task to "keep up" with the literature, particularly in fields that are not directly related to one's favorite topic. For example, plant nutritionist had known since the 1920s that boron (B) was essential for plant growth and there were several reports (e.g. Kochi and Kumazawa 1976, Matoh 1997) that boron is associated with cell wall formation and that boron-deficiency leads to the formation of abnormal primary walls. Boron has also been implicated in numerous other cellular processes, although compelling evidence for these roles is lacking (Brown et al. 2002). A comprehensive review by Loomis and Durst (1992) described studies showing that borate can form a stable diol ester with furanoses that have *cis* hydroxyl groups. Apiose is one such sugar and these authors suggested that borate esterification of wall-bound apiose results in the formation of acid-labile cross-links that regulate wall expansion. The suggestion that the acid lability of borate-apiose cross-links regulated auxin-induced wall creep may not be correct. However, the notion that borate may have profound influences on plant cell wall organization and function seems to have, at least until recently, gone largely unnoticed by the cell wall community in general. Matoh et al. (1993) were the first to report evidence for the existence of a pectin borate complex that is released by pectinase treatment of radish root walls. Subsequently they provided compelling evidence that the pectin fraction is RG-II and that a borate diol diester cross-links two RG-II molecules to form a dimer (dRG-II-B) (Kobayashi et al. 1996) (Figure 2). dRG-II-B was identified from sugar beet (Ishii and Matsunaga 1996), sycamore (O'Neill et al. 1996), bean, red pine, bamboo shoot and red wine (see O'Neill et al. 2004). Ishii et al. (1999) demonstrated that apiosyl residue of side chain A is the borate esterification site without regardless of plant species and the dimer formed *in muro* and *in vitro*. The formation of a borate ester cross-link between two apiosyl residues results in a boron atom that is chiral. Thus, two diasteroisomers may form. The existene of these diasteroisomers has been confirmed by ^{13}C NMR spectroscopic analysis of the products formed when methyl ß-D-apioside reacts with borate (Ishii and Ono 1999).

The interconversion of the RG-II dimmer and monomer occurred *in vitro*. For example, the dimmer is converted to the monomer and boric acid at pH 1. The RG-II dimer is formed *in vitro* between pH 3 and 4 by reacting the monomer with boric acid. RG-II dimmer formation is much more rapid in the presence of divalent cations (Sr^{2+}, Ba^{2+}, and Pb^{2+}) with ionic radii > 1.1 angstrom (refer to O'Neill et al. 2004).

You may be surprised to learn that red wine contains between 100 - 150 mg of RG-II. Wine has become the material of choice in many studies where large amounts of RG-II are required. The finding that

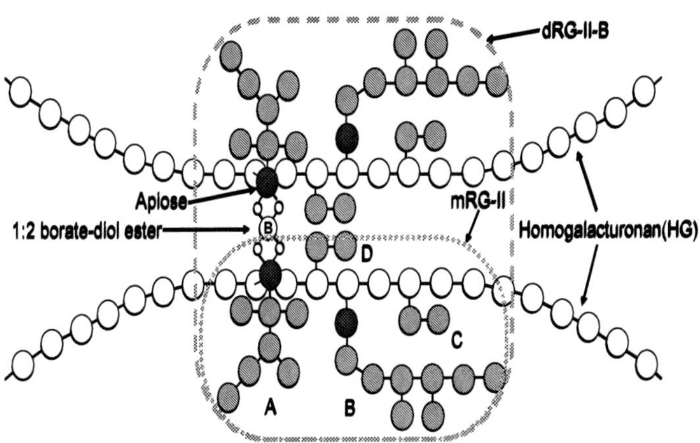

Fig. 2 The model of pectin cross-linked by dRG-II-B formation. **mRG-II: monomeric RG-II. dRG-II-B: dimeric RG-II borate. A borate diol diester cross-links two mRG-II molecules to form dRG-II-B. The borate forms a 1:2 complex with an apiosyl residue in side chain A of each mRG-II.**

RG-II exists predominantly as a dimer that is covalently cross-linked by a borate diester implied that a function of boron is to cross-link pectin chains and to stabilize three-dimensional pectin network *in muro*. Evidence to support this notion has been obtained by studies of boron-deficient plants and with mutant plants that synthesize RG-II with altered structure (O'Neill et al. 2004).

When I started my own pectin projects after returning from the Albersheim lab, Dr. Kouchi told me that boron should be associated with pectin and advised me how to grow boron-deficient plants. I then met Dr. Matsunaga who has a strong background in analytical chemistry and had been studying boron in plant tissues using inductively coupled plasma mass spectrometry and ^{11}B NMR spectroscopy. I grew boron-deficient pumpkin, separated boron-containing carbohydrates from the cell walls and analyzed sugar composition. Dr. Matsunaga analyzed B content by inductively coupled plasma mass spectrometry-MS and characterized the boron-containing fraction by ^{11}B NMR spectroscopy. We showed that dimer formation of RG-II controls the cell wall thickness (Ishii et al. 2001). Since I got acquainted with him, the pectin projects rapidly progressed.

My research on RG-II has benefited greatly from numerous collaborations with individual at National Agricultural Research Center for Kyushu Okinawa, National Food Research, and Complex Carbohydrate Research Center, the University of Georgia. Through my cell wall projects, I have realized that if we meet a friend who understands what we want to do, the half is almost done.

Acknowledgements

I thank Malcolm A. O'Neill (Complex Carbohydrate Research Center, University of Georgia, Athens, GA, USA) for his critical reading of the manuscript. This work was supported in part by Program for Promotion of Basic Research Activities for Innovation Biosiences (PROBRAIN).

References

The detailed references are cited in review articles by O'Neill et al. 1990 and O'Neill et al. 2004.

Albersheim P (1975) The primary cell wall of plants. Scientific American 232: 80-95.

Brown PH, Brellaloui N, Wimmer MA, Bassil ES, Ruiz J, Hu H, Pfeffer H. Dannel F. Römheld V (2002) Boron in plant biology. Plant Biol. 4: 205-223.

Darvill AG, McNeil M, Albersheim P (1978) Structure of plant cell walls VIII. A new pectic polysaccharide. Plant Physiol 62:418-422.

Ishii T, Matsunaga T, Pellerin P, O'Neill MA, Darvill A, Albersheim P (1999) The plant cell wall polysaccharide rhamnogalacturonan II self-assembles into a covalently cross-linked dimer. J Biol Chem 274: 13098-13104.

Ishii T, Matsunaga T, Hayashi N (2001) Formation of rhamnogalacturonan II-borate dimer in pectin determines cell wall thichkness of pumpkin tissue. Plant Physiol 126: 1698-1705.

Kobayashi M, Matoh T, Azuma J (1996) Two chains of rhamnogalacturonan II are cross-linked by borate-diol ester bonds in higher plant cell walls. Plant Physial 110:1017-1020.

Kouchi H , Kumazawa K (1976) Anatomical responses of root tips to boron deficiency III. Effect of boron deficiency on sub-cellular structure of root tips, particularly on morphology of cell wall and its related organelles. Soil Sci. Plant Nutr. 22:53-71

Looms WD, Durst RW (1992) Chemistry and biology of boron. Bio Factors 3:229-239.

Matoh T, Ishigaki K, Ohno K, Azuma J. (1993) Isolation and characterization of a boron-polysaccoride complex from radish root. Plant Cell Physiol 34:639-642.

Matsunaga T, Ishii T, Matsumoto S, Higuchi M, Darvill A, Albersheim P, O'Neill MA (2004) Occurrence of the primary cell wall polysaccharide rhamnogalacturonan II in pterdophytes, lycophytes, and bryophytes. Implications of the evolution of vascular plants. Plant Physiol 134: 339-351.

O'Neill M, Albersheim P, Darvill A. (1990) The pectic polysaccharides of primary cell walls. In Dey PM (ed), Methods in Plant Biochemistry, Vol.2:Carbohydrates, Academic, London: pp.415-441.

Ridley B, O'Neill MA, Mohnen D (2001) Pectin: structure, biosynthesis, and oligogalacturonide-related signaling. Phytochemistry 57: 929-967.

O'Neill MA, Ishii T. Albersheim P, Darvill AG (2004) Rhamnogalacturonan II: Structure and function of a borate cross-linked cell wall pectic polysaccharide. Ann Rev Plant Biol 55:109-139.

Screening for Cell-Wall Phenotypes by Infrared Spectroscopy

Maureen McCann and Nick Carpita

The plant cell wall is both figuratively and literally the farthest structure from the nucleus. The vast majority of the individual constituents of this complex and dynamic amalgam of polysaccharide, protein, and phenylpropanoids are themselves the products of catalysis by thousands of gene products that we are just beginning to discover and characterize. Unlike proteins and nucleic acids, the sequencing of polysaccharides is far from straightforward, and no direct relationship exists between the structure of the products, the enzymes that synthesize them, or, by extension, the genes that encode those enzymes. In part this explains why, even up to the mid-1980s, after decades of careful carbohydrate chemistry to elucidate the unit structures of cell wall polysaccharides, the plant cell wall community had yet to identify cell wall-related genes. Even then, the first cell wall-related gene to be characterized encoded a cell wall protein rather than a synthase or transferase to build a polysaccharide (Chen and Varner 1985). In the early 1990s, map-based cloning of mutated genes began to yield a few of the enzymes involved in nucleotide-sugar interconversion pathways (Reiter et al. 1993). Over a decade passed between the first report of the cloning of a plant gene and the discovery by Debby Delmer and her colleagues of the first cellulose synthase gene (Pear et al. 1996). Coincident with the completion of the Arabidopsis genome sequence (Arabidopsis Genome Initiative 2000), this hallmark report opened a door of discovery into genes encoding many more potential polysaccharide synthases and glycosyl transferases.

Debby Delmer spent her academic life in pursuit of the catalytic mechanism for the synthesis of the world's most abundant biopolymer, cellulose, and much has been learned as a result of her efforts (Delmer 1977, 1987, 1999). But Debby would be the first to urge us to consider that the dynamic wall is much more than cellulose. Many classes of genes function in the construction of a plant cell wall, comprising genes involved in substrate generation, backbone polymerization and decoration, secretion and targeting, assembly into a wall, and disassembly during developing and differentiation (http://cellwall.genomics.purdue.edu). As the Arabidopsis and rice gene families are calculated, almost one-half of the genomes consist of genes for which we have no idea of function. An estimated 10% of the known genes of the genome encode cell-wall related products, which means that as many as 2500 genes could function in the generation and dynamics of a cell wall (Carpita et al. 2001).

One way to deduce the function of potential cell-wall relevant genes is to characterize mutant plants defective in a specific gene for changes to their wall composition and architecture. However, finding cell wall mutants is not always straightforward. Cell wall architecture transcends its component parts, and when one of the components is altered or missing, the cell tweaks the biogenesis pathways to accommodate the change without altering normal growth and development. Many mutants in cell wall-related genes, now available in collections of insertional mutants (see http://signal.salk.edu), have no visible phenotypes when compared to wild type. Many others are altered chemically in ways that are hard to predict. Nearly two years elapsed between the time that Delmer and her colleagues discovered a plant cellulose synthase (*CesA*) gene and Richard Williamson's laboratory identified a temperature-sensitive *CesA* mutant with a phenotype of reduced cellulose content (Arioli et al. 1998). This mutant was discovered serendipitously in a screen for radial swelling of roots, a screen that had been designed to uncover cytoskeletal genes involved in regulating the direction of growth. Other cell wall mutants were found in genetic screens for growth and developmental phenotypes, for example, by looking at hypocotyl growth (Fagard et al. 2000) or at xylem phenotypes (Turner and Somerville 1997). The cell wall community was hungry for mutants and rapid ways to detect them directly in mutagenized populations of plants. Wolf-Dieter Reiter and colleagues in the laboratory of Chris Somerville (Reiter et al. 1993) initiated a program to identify mutants in Arabidopsis and identified eleven mutants where particular cell wall monosaccharides were either over- or under-represented based on comparison to a wild-type population (for review, see Rieter et al. 1997). This was a labor-intensive screen involving analysis of alditol acetates derived from acid hydrolysis of cell walls. These mutants remain among the best characterized in cell wall research even today.

In the mid-1990s, Tatiana Gorshkova from the Russian Academy of Science in Kazan visited the Carpita laboratory and brought flax (*Linum usitatissimum*) as a model system to study secondary wall formation in developing phloem fibers (Gorshkova et al. 1996). Flax possesses a very small genome, in the size-range of that of Arabidopsis, and is a model system for finding secondary wall biosynthetic genes. In a pilot experiment, Carpita chemically mutagenized a population of flax seeds and grew the M1 seeds in his garden. The M2 seeds were collected from this self-pollinating species, and planted in a greenhouse under controlled conditions to look for homozygous recessive mutants. The intention was to follow the strategy of cell-wall monosaccharide analysis that Reiter had shown to be successful in Arabidopsis and look for mutants that expressly affected secondary wall formation. However, a visit by McCann to the Carpita lab in 1994 changed the strategy drastically.

In the early 1990's McCann, working in the laboratory of Keith Roberts at the John Innes Centre, teamed with Reg Wilson and Kate Kemsley, at the Institute of Food Research, to use Fourier transform infrared (FTIR) spectroscopy as a way of looking at isolated cell walls and the individual molecules they comprise (McCann et al. 1992). FTIR spectroscopy had been used for many years as a research tool to detect molecular bonds and functional groups in samples, and together with nuclear-magnetic resonance spectroscopy and mass spectrometry, could provide all the information needed to deduce the chemical structure of complex organic molecules. Underivatized materials could be probed by FTIR to quickly identify methyl, carboxylate, carbonyl ester, hydroxyl, and amine groups. Analyses of plant cell walls presented some new challenges because the molecular bonds (for example, C-H or C-O bonds) were common to different wall components and had overlapping frequencies of absorption. Nevertheless, a region called the "carbohydrate fingerprint" region (1200 to 800 cm^{-1} within a useful spectrum, for plant cell walls, of 1800 and 800 cm^{-1}) gave spectral features that were distinctive for a particular polymer or mixture of polymers. However, the spectrum of a cell wall is not merely the sum of the spectra of its component molecules (McCann et al. 1992). The infrared spectrum also contains information that reflects the local environment of molecular bonds, such as their hydration state, the conformation of the molecule, and interactions with other molecules (McCann et al. 1997, 2001). Therefore, the spectrum is a fingerprint that is characteristic of the architecture as well as the composition of the wall. The difference between two spectra becomes more obvious if one is digitally subtracted from the other (Figure 1). More importantly, McCann and colleagues showed that cell walls isolated from cell cultures of different species could be distinguished from each other solely on the basis of their characteristic infrared spectra

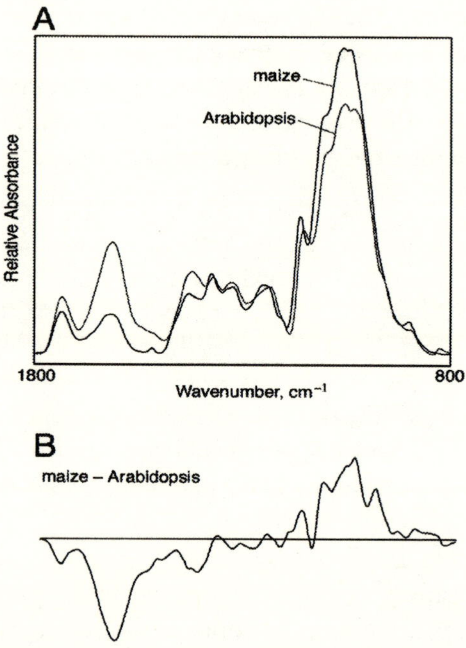

Fig. 1 Isolated cell walls prepared from maize coleoptile and young Arabidopsis seedlings can be distinguished from each other on the basis of their Fourier transform infrared spectra. **A. Area-normalized spectra from cell walls of fully expanded Arabidopsis leaves and maize coleoptiles at the maximum rate of elongation. B. Digital subtraction of the Arabidopsis spectrum from the maize spectrum to generate a difference spectrum reveals that the maize walls are relatively enriched in components absorbing in the carbohydrate fingerprint region, while the Arabidopsis walls are relatively enriched in acidic and esterified pectins.**

(Séné et al. 1994), and they speculated that walls of different composition and architecture could be chemically "fingerprinted".

During this time, Estie Shedletzky in Debby Delmer's lab published two very interesting reports (Shedletzky et al. 1990, 1992) on how tomato, barley and other cells in liquid culture became 'resistant' to grow in the presence of a potent and specific inhibitor of cellulose synthesis called dichlobenil, or DCB. The big surprise was that cellulose synthase did not become resistant to the herbicide, as was expected, but the cells adapted by producing more cross-linked pectin (for tomato cells) or more cross-linked polyphenolics (for barley cells) to grow in the near absence of cellulose! These results were confirmed by FTIR microspectroscopy in a collaboration with Debby Delmer (Wells et al. 1994). The infrared beam was diverted from passing through a sample in the sample compartment in the body of the spectrometer to passing through a small area of a sample mounted on the stage of an infrared microscope. In the infrared microscope, beams of visible and infrared light are collinear, and an area of the specimen delimited by double-bladed apertures is imaged and then infrared data collected (McCann et al. 1992, 1997). Because individual cells could be analyzed, FTIR imaging revealed a few cells that seemed to produce a significant amount of cellulose, and this was shown also by staining with the cellulose-binding fluorescence dye, Calcofluor White. Just how these few cells continued to make cellulose is something that has not been explained to this day.

Hearing of the plan to screen the mutagenized flax population for cell wall alterations, McCann suggested that FTIR spectroscopy might prove to be a more rapid screen for many different kinds of cell wall mutants, not solely those affecting neutral sugar content. What a fruitful and timely suggestion this was. Although mutagenized Arabidopsis populations were commercially available then, the mutagenized flax population was just about to be harvested. Because the Refractive Index beam could penetrate an intact leaf, then leaves harvested from the flax M2 population had only to be cleared of pigmentation by extraction with methanol, washed with water and freeze-dried. Then during a summer at John Innes Centre, spectra were taken from thousands of leaf samples. Statistical tools such as Principal Components Analysis and Linear Discriminant Analysis were applied to the spectral data in order to select potential mutants, rather than attempting to do this by eyeballing thousands of similar-looking spectra (Kemsley 1996). Selection of potential mutants by these chemometric analyses applied to FTIR spectra was validated by subsequent analysis of monosaccharide composition (Chen et al. 1998). While a majority of the selections showed a distinct monosaccharide sugar distribution, another group were obviously of altered FTIR 'spectrotype' but showed no difference in monosaccharide content from wild-type. For example, some apparent mutants had altered absorbances of esters or amide peaks.

Although the flax seemed a suitable model for cell wall genomics at the time, the sequence of the Arabidopsis genome would soon be complete. With an entire collection of cell-wall related genes at hand, flax genomics would have to wait while we redirected our efforts toward Arabidopsis mutants that could be identified by FTIR spectroscopy. The first step was to acquire funding for what would be a major research effort. McCann, with colleagues Roberts and Wilson, obtained UK funding from the BBSRC, and Carpita obtained funding for a pilot project from NSF. The aim of both projects was to demonstrate the validity of this approach to characterize cell wall phenotypes in a high throughput manner, such that populations of 20 or 30 plantlets of each homozygous mutant line could be screened. The NSF project also screened for maize mutants as outliers of segregating populations. Initially, the thinking was that we would apply Refractive Index spectroscopy in a forward screen, relying on map-based cloning to identify the mutated gene responsible for the phenotype. However, a new resource became available to the plant science community: a collection of transfer DNA insertion mutants developed by Joe Ecker, who had the vision of generating a mutant in every single gene of Arabidopsis (Alonso et al. 2003). Suddenly, it made more sense to take a reverse genetics approach to establish spectroscopic signatures, "spectrotypes", for mutants that we knew were in cell wall-related genes, and use these to categorize mutants in genes of unknown function. This would allow mutants to be targeted for specific chemical assays in an efficient way. The pilot projects were a success – we could deliver the high throughput that we had proposed and we could use spectrotypes to give us some information about the molecular basis for the cell wall phenotype. Now we needed a lot more money! We began to assemble a team of experts: Sara Patterson and Tony Bleecker at Wisconsin University would generate homozygous lines of Arabidopsis mutants from seed stocks provided from the SALK institute, while Karen Koch and Don McCarty would send as close to an equivalent in maize from University of Florida. Wolf Dieter-Reiter, who had generated many of the original cell wall mutants, would provide much-needed cell wall bioinformatics. Steve Thomas, from National Renewable Energy Lab in Colorado, and Wilfred Vermerris, at Purdue, are interested in the applications of the modification in ligno-cellulosics in animal nutrition and ethanol production and apply near-infrared spectroscopy in assay of field-grown mutagenized maize. Because the formation of the cell wall is integrally linked to cytoskeletal dynamics, Chris Staiger, at Purdue, also joined the team. NSF agreed to fund our giant project in September 2002. To date, we have collected spectrotypes for most of the mutants for which we know that there is an alteration in cell wall composition and architecture. Now we are beginning to work our way through collections of cell-wall-related gene families. We sample spectra from isolated cell walls of two-weeks-old Arabidopsis plantlets and so we expect that only some of the mutants in gene family members will show a spectrotype – many genes in gene families are restricted in their temporal or spatial expression patterns. Soon we will begin to screen families of genes of unknown function to discover new proteins and enzymes, for which we have as yet no other assay, that contribute to building the highly complex structure of the cell wall.

References

Alonso JM, Stepanova AN, Leisse TJ, Kim CJ, Chen HM, Shinn P et al. (2003) Genome-wide insertional mutagenesis of *Arabidopsis thaliana*. Science 301: 653-657.

Arabidopsis Genome Initiative (2000) Analysis of the genome sequence of the flowering plant *Arabidopsis thaliana*. Nature 408: 796-815.

Arioli T, Peng LC, Betzner AS, Burn J, Wittke W, Herth W, Camilleri C, Hofte H, Plazinski J, Birch R, Cork A, Glover J, Redmond J, Williamson RE (1998) Molecular analysis of cellulose biosynthesis in Arabidopsis. Science 279: 717-720.

Carpita N, Tierney M, Campbell M (2001) Molecular biology of the plant cell wall: searching for the genes that define structure, architecture and dynamics. Plant Mol Biol 47: 1-5.

Chen J, Varner JE (1985) Isolation and characterization of cDNA clones for carrot extensin and a proline-rich 33-kDa protein. Proc Natl Acad Sci USA 82: 4399-4403.

Chen LM, Carpita NC, Reiter WD, Wilson RH, Jeffries C, McCann MC (1998) A rapid method to screen for cell wall mutants using discriminant analysis of Fourier transform infrared spectra. Plant J 16: 385-392.

Delmer DP (1977) Biosynthesis of cellulose and other plant cell wall polysaccharides. Rec Adv Phytochem 11: 105-153.

Delmer DP (1987) Cellulose biosynthesis. Annu Rev Plant Physiol Plant Mol Biol 38: 259-290.

Delmer DP (1999) Cellulose biosynthesis: exciting times for a difficult field of study. Annu Rev Plant Physiol Plant Mol Biol 50: 245–276.

Fagard M, Höfte H, Vernhettes S (2000) Cell wall mutants. Plant Physiol Biochem 38: 15-25.

Gorshkova TA, Wyatt SE, Salnikov VV, Gibeaut DM, Ibragimov MR, Lozovaya VV, Carpita NC (1996) Cell-wall polysaccharides of developing flax plants. Plant Physiol 110: 721-729.

Kemsley EK (1998) Discriminant Analysis of Spectroscopic Data. John Wiley and Sons, Chichester, UK.

McCann MC, Hammouri M, Wilson R, Belton P, Roberts K (1992) Fourier transform infrared microspectroscopy is a new way to look at plant cell walls. Plant Physiol 100: 1940-1947.

McCann MC, Chen L, Roberts K, Kemsley EK, Sene C, Carpita NC, Stacey NJ, Wilson RH (1997) Infrared microspectroscopy: Sampling heterogeneity in plant cell wall composition and architecture. Physiol Plant 100: 729-738.

McCann MC, Bush M, Milioni D, Sado P, Stacey NJ, Catchpole G, Defernez M, Carpita NC, Hofte H, Ulvskov P, Wilson RH, Roberts K (2001) Approaches to understanding the functional architecture of the plant cell wall. Phytochemistry 57: 811-821.

Pear JR, Kawagoe Y, Schreckengost WE, Delmer DP, Stalker DM (1996) Higher plants contain homologs of the bacterial *celA* genes encoding the catalytic subunit of cellulose synthase. Proc Natl Acad Sci USA 93:12637-12642.

Reiter W-D, Chapple CCS, Somerville CR (1993) Altered growth and cell walls in a fucose-deficient mutant of Arabidopsis. Science 261: 1032-1035.

Reiter W-D, Chapple C, Somerville CR (1997) Mutants of *Arabidopsis thaliana* with altered cell wall polysaccharide composition. Plant J 12: 335-345.

Séné CFB, McCann MC, Wilson RH, Grinter R (1994) Fourier-transform Raman and Fourier-transform infrared spectroscopy - an investigation of 5 higher plant cell walls and their components. Plant Physiol 106: 1623-1631.

Shedletsky E, Shmuel M, Delmer DP, Lamport DTA (1990) Adaptation and growth of tomato cells on the herbicide 2,6-dichlorobenzonitrile leads to production of unique cell walls virtually lacking a cellulose-xyloglucan network. Plant Physiol 94: 980-987.

Shedletsky E, Shmuel M, Trainin T, Kalman S, Delmer D (1992) Cell wall structure in cells adapted to growth on the cellulose synthesis inhibitor 2,6-dichlorobenzonitrile – a comparison between two dicotyledonous plants and a gramineous monocot. Plant Physiol 100: 120-130.

Turner SR, Somerville CR (1997) Collapsed xylem phenotype of Arabidopsis identifies mutants deficient in cellulose deposition in the secondary cell wall. Plant Cell 9: 689-701.

Wells B, McCann MC, Shedletzky E, Delmer D, Roberts K (1994) Structural features of cell walls from tomato cells adapted to grow on the herbicide 2,6-dichlorobenzonitrile. J Micros-Oxford 173: 155-164.

The Biochemistry of Plant Polysaccharide Biosynthesis: Four Decades and Counting

Kenneth Keegstra

The primary goal of this chapter is to describe an interesting anecdote regarding the scientific contributions of Professor Deborah (Debby) Delmer. At the same time, I hope to provide a brief historical perspective on polysaccharide biosynthesis in plants, especially on the biochemical investigations of this process. This is not intended to be a comprehensive review and I apologize to those scientists whose contributions are not mentioned here.

Cell wall polysaccharides constitute a major proportion of the dry matter in most plant tissues, depending, of course, upon the species and the tissue under consideration. In addition to the quantitative importance of cell wall polysaccharides as a major form of biomass, they also have enormous fundamental significance for the biology of plants. Cell walls play a critical role in defining the structure of plants, in guiding the development of plants and in providing a role in plant defense against pathogens. Finally, plant cell wall polysaccharides have great practical significance, including important roles in the materials industry, the food industry and many cell wall polysaccharides have important health benefits as dietary fiber. Given the abundance and importance of plant cell wall polysaccharides, it is amazing and disappointing that we still understand so little about their biosynthesis. One reason for this deficiency is the difficulty of studying the biochemistry of cell wall polysaccharide biosynthesis. Even today, most of the current understanding of cell wall biosynthesis comes from the application of molecular genetics and genomics while many important biochemical questions remain unanswered. Before addressing some of these issues, I want to go back nearly 40 years to a time when Debby Delmer began working on plant cell walls.

The sucrose synthase saga

My familiarity with the story begins in the fall of 1967, when I arrived in Boulder to begin graduate work with Peter Albersheim in the Chemistry Department at the University of Colorado. While I decided to work on projects related to the structure of cell walls (Keegstra et al. 1973), others in the Albersheim laboratory were studying polysaccharide biosynthesis (for example, see Villemez et al. 1968). In order to understand the rationale for the experiments described below, it is necessary to briefly summarize the situation at that time. Very little was understood about the biosynthesis of plant polysaccharides. For example, in a 1965 review on this topic, Nordin and Kirkwood (1965) began a section entitled *"Synthesis of other polysaccharides by enzymes from higher plants"* with the following statement: "The amount of information on the synthesis of polysaccharides other than starch and cellulose by enzymes derived from higher plants is very small." However, about this time, important advances were reported in bacterial systems. Specifically, it had been shown that lipid-linked sugars and oligosaccharides were important intermediates between sugar nucleotides and the final polymers found in the peptidoglycan of bacterial cell walls (Higashi et al. 1967) and the lipopolysaccharides found on the bacterial cell surface (Wright et al. 1965). Thus, there was great interest in investigating the possibility that lipid intermediates were involved in the biosynthesis of plant cell wall polysaccharides.

One effort to investigate polysaccharide biosynthesis was being pursued by Bill Grimes, a graduate student in the Albersheim laboratory. While I was not directly involved in this project, I observed these events first hand. In addition, because Bill and I became good friends and we did lots of hiking together in the mountains around Boulder, I learned many details about his research project. During the preparation of this chapter I checked with Bill to make sure that my recollections agreed with his. Bill observed that when extracts of etiolated mung beans seedlings were incubated with radiolabeled UDP-glucose, a product was formed that did not migrate during borate electrophoresis (Assay III in Grimes et al. 1970). This assay was used because most free sugars and small oligosaccharides formed charged borate complexes and migrated away from the origin during electrophoresis, while other products of interest did not migrate and remained at the origin. One of the ideas under consideration was that the non-mobile product was a lipid intermediate or some other type of interesting intermediate in polysaccharide biosynthesis. However, careful analysis of the product and the acceptor requirements for this reaction revealed that the product was sucrose, the acceptor was fructose and that the enzyme activity under investigation was sucrose synthase (Grimes et al. 1970). Thus a project aimed at studying polysaccharide biosynthesis, with a hope of possibly identifying lipid intermediates, resulted in the identification of sucrose synthase.

Toward the end of Bill's time in the lab, Debby Delmer arrived to begin postdoctoral studies in the Albersheim laboratory. Because Bill was ready to leave for his postdoctoral position at MIT with Phillips Robbins (one of the discoverers of lipid intermediates in bacterial polysaccharide biosynthesis), Debby took over the project and began working on sucrose synthase. She investigated its distribution in plant tissues and provided evidence that this reversible enzyme was most likely involved in the biosynthesis of sugar nucleotides from sucrose (Delmer and Albersheim 1970). Debby continued to work on sucrose synthase after she left the Albersheim lab, purifying the enzyme and performing a detailed characterization (Delmer 1972). Later, when Debby moved to the Plant Research Laboratory she began her long quest to understand cellulose biosynthesis (Delmer and Stone 1988; Delmer and Amor 1995, Delmer 1999; Doblin et al. 2002).

It was almost 25 years later that Debby and her colleagues found a direct connection between her early studies on sucrose synthase and her life-long investigations of cellulose biosynthesis. They observed that in developing cotton fibers, at least half of the sucrose synthase was tightly associated with the plasma membrane (Amor et al. 1995). They postulated that sucrose synthase exists in a complex with the glucan synthase responsible for cellulose synthesis and allows carbon to be channeled from sucrose into cellulose. This hypothesis has since received considerable experimental support (for example, see Konishi et al. 2004). This ironic story also provides a lesson. The irony is that a project that began with the aim of investigating polysaccharide biosynthesis eventually lead to the discovery that sucrose synthase may be involved in sugar nucleotide biosynthesis (Delmer and Albersheim 1970). This observation eventually resulted in the hypothesis that sucrose synthase may be directly involved in polysaccharide biosynthesis (Amor et al. 1995). The lesson is that biology is full of connections that we don't see, because our knowledge is so fragmented. As our knowledge becomes more complete, it seems likely that we will find many connections that we do not currently understand.

The biochemistry of polysaccharide biosynthesis is still difficult

I want to return briefly to the bigger issue of plant polysaccharide biosynthesis. As noted earlier, progress in understanding polysaccharide biosynthesis has been slow, despite the efforts of many talented investigators. Many good reasons account for the slow progress. Regardless of the explanation, the net result is that during the period when biochemical techniques were the only methods available for studying plant cell wall biosynthesis, none of the many enzymes directly involved in polysaccharide biosynthesis were identified. It was only with the application of molecular biology, molecular genetics and genomic methods that the genes and the proteins responsible for polysaccharide biosynthesis have begun to be identified. Even today, with the availability of massive amounts of DNA sequence information and the tentative identification of hundreds of putative glycosyltransferase and glycan synthase genes from plants, the biochemical functions of very few of their gene products have been established and many important biochemical questions about polysaccharide biosynthesis remain to be answered.

The first cell wall biosynthetic enzyme to be identified was the catalytic subunit of cellulose synthase; this important advance was accomplished by Debby Delmer and her colleagues (Pear et al. 1996). They used sequence information from bacterial cellulose synthase genes to demonstrate that cotton plants had genes (*CesA*) with significant sequence similarity and that these genes were highly expressed during cellulose deposition in cotton fiber cells. More recently, mutants defective in cellulose biosynthesis in both primary and secondary walls have been isolated and characterized, leading to valuable new insights into the process of cellulose biosynthesis (for discussions of recent advances, see Robert et al. 2004; Taylor et al. 2004). Despite recent advances in the ability to study cellulose biosynthesis in vitro (Lai Kee Him et al. 2002) and in the identification of a possible primer for initiating glucan chain biosynthesis (Peng et al. 2002), there are still many important aspects of cellulose biosynthesis that remain to be elucidated. For example, it seems likely that the rosettes involved in cellulose deposition (Doblin et al. 2002) contain more components than the CESA proteins. At a biochemical level, we don't understand how glucan chain length is controlled during cellulose synthesis. The list of important unsolved problems would be very long, so I won't go on.

The biosynthesis of branched polysaccharides is more complicated and requires the action of at least two different types of enzymes (Perrin et al. 2001). The first are processive enzymes that synthesize the sugar backbone; for most plant polysaccharides, they create a homopolymer containing a single type of sugar. While the identity of most of these enzymes remains unclear, one attractive group of candidates is the protein family encoded by the cellulose synthase-like (*Csl*) genes (Richmond and Somerville 2000). These genes have substantial sequence similarity with the cellulose synthase genes, but appear to have different functions (Richmond and Somerville 2000). However, despite considerable effort in the past few years to define the biochemical functions of the CSL proteins, we still have very little information on whether they are glycan synthases. One important advance was the demonstration that a *CslA* gene from guar encodes a mannan synthase Dhugga et al. 2004). More recently, it has been demonstrated that three Arabidopsis genes from the *CslA* family produce mannan synthase activity when expressed in Drosophila cells (Liepman et al. 2005). Interestingly, each CSLA protein produced mannan when provided with GDP-mannose, but produced a glucomannan when provided with a mixture of GDP-mannose and GDP-glucose, thereby indicating that one protein has both activities. The ability to express these wall biosynthetic genes in insect cells that have no background activity for the biosynthesis of plant cell wall polysaccharides opens up the possibility to explore the biochemical details of mannan biosynthesis. It also opens up the possibility to determine whether other *Csl* genes encode other glycan synthases.

The second type of enzyme needed for the biosynthesis of most plant cell wall polysaccarides are glycosyltransferases. These enzymes add the side chain residues to the sugar backbone of complex polysaccharides, such as xyloglucan or arabinoxylan. These enzymes are usually very specific enzymes in that they add only one sugar in a specific linkage to a specific acceptor. Several glycosyltransferases involved in polysaccharide biosynthesis have been identified and characterized in recent years (Scheible and Pauly 2004). However, because it is likely that the biosynthesis of plant cell wall polysaccharides requires more than 100 such enzymes, it is clear that much work still remains to be done. Despite that fact that many predicted plant proteins can be identified as putative glycosyltransferases (for example, see the plant entries in the CAZY database at http://afmb.cnrs-mrs.fr/CAZY/), it is a difficult job to identify the biochemical function of each one. One major difficulty is establishing biochemical assays for such enzymes. One problem derives from their extreme specificity. If a glcosyltransferase is highly specific, it means that a specific carbohydrate structure is needed to serve as an acceptor during in vitro assays. In many cases, the structure of the required acceptor is unknown; but even when this information is available, acquiring adequate quantities of the acceptor can be difficult. When one combines these complexities with the fact that most glycan synthases and glycosyltransferases are integral membrane proteins that are generally present in small quantities, one can begin to understand why progress on understanding the biochemistry of polysaccharide biosynthesis has been so slow.

Despite these difficulties, the future of polysaccharide biosynthesis looks very bright. The genomics revolution has provided many candidate genes to be studied. Powerful new strategies for reverse genetics should allow the biological functions of these genes to be explored. More importantly for understanding the biochemistry of polysaccharide biosynthesis, new procedures for heterologous expression of candidate genes are now available and better ones are being developed all of the time. As novel new strategies for the chemical and enzymatic synthesis of oligosaccharides become available, it should be possible to overcome the problems of developing assays for very specific glycosyltransferases. I predict that the next forty years will be an exiting and productive time for studying the biosynthesis of plant cell wall polysaccharides.

Acknowledgements

Work from the author's laboratory is supported in part by grants from the Energy Biosciences Program at the US Department of Energy and from the Plant Genome Program at the US National Science Foundation.

References

Amor Y, Haigler CH, Johnson S, Wainscott M, Delmer DP (1995) A membrane-associated form of sucrose synthase and its potential role in synthesis of cellulose and callose in plants. Proc Natl Acad Sci USA 92: 9353-9357.

Delmer DP, Albersheim P (1970) The biosynthesis of sucrose and nucleoside diphosphate glucoses in *Phaseolus aureus*. Plant Physiol 45: 782-786.

Biosynthesis

Delmer DP (1972) The purification and properties of sucrose synthetase from etiolated *Phaseolus aureus* seedlings. J Biol Chem 247: 3822-3828.

Delmer DP, Stone BA (1988) Biosynthesis of plant cell walls, In: Stumpf PK, Conn EE (eds) The Biochemistry of Plants, Vol 14, Carbohydrates, Academic Press, New York, pp 373-420.

Delmer DP, Amor Y (1995) Cellulose biosynthesis. Plant Cell 7: 987-1000.

Delmer DP (1999) Cellulose biosynthesis: exciting times for a difficult field of study. Ann Rev Plant Physiol Plant Mol Biol 50: 245-276.

Dhugga KS, Barreiro R, Whitten B, Stecca K, Hazebroek J, Randhawa GS, Dolan M, Kinney AJ, Tomes D, Nichols S, Anderson P (2004) Guar seed β-mannan synthase is a member of the cellulose synthase super gene family. Science 303: 363-366.

Doblin MS, Kurek I, Jacob-Wilk D, Delmer DP (2002) Cellulose biosynthesis in plants: from genes to rosettes. Plant Cell Physiol 43: 1407-1420.

Grimes WJ, Jones BL, Albersheim P (1970) Sucrose synthetase from *Phaseolus aureus* seedlings. J Biol Chem 245: 188-197.

Higashi Y, Strominger JL, Sweeley CC (1967) Structure of a lipid intermediate in cell wall peptidoglycan synthesis: A derivative of a C_{55} isoprenoid alcohol. Proc Natl Acad Sci USA 57: 1878-1864.

Keegstra K, Talmadge KW, Bauer WD, Albersheim P (1973) The structure of plant cell walls III. A model of the walls of suspension-cultured sycamore cells based on the interconnection of the macromolecular components. Plant Physiol 51: 188-196.

Konishi T, Ohmiya Y, Hayashi T (2004) Evidence that sucrose loaded into the phloeim of a poplar leaf is used directly by sucrose synthase associated with various â-glucan synthases in the stem. Plant Physiol 134: 1146-1152.

Lai Kee Him J, Chanzy H, Müller M, Putauz JL, Imai T, Bulone V (2002) In vitro versus in vivo cellulose microfibrils from plant primary wall synthases: structural differences. J Biol Chem 277: 36931-36939.

Liepman AH, Wilkerson CG, Keegstra K (2005) Expression of cellulose synthase-like (Csl) genes in insect cells reveals that *CslA* family members encode mannan synthases. Proc Natl Acad Sci USA, 102: in press (10.1073/pnas.0409179102).

Nordin JH, Kirkwood S (1965) Biochemical aspects of plant polysaccharides. Ann Rev Plant Physiol 16: 393-414.

Pear JR, Kawagoe Y, Schreckengost WE, Delmer DP, Stalker DM (1996) Higher plants contain homologs of the bacterial *celA* genes encoding the catalytic subunit of cellulose synthase. Proc Natl Acad Sci USA 93: 12637-12642.

Peng L, Kawagoe Y, Hogan P, Delmer DP (2002) Sitosterol-â-glucoside as primer for cellulose synthesis in plants. Science 295: 147-150.

Perrin R, Wilkerson C, Keegstra K (2001) Golgi enzymes that synthesize plant cell wall polysaccharides: finding and evaluating candidates in the genomic era. Plant Mol Biol 47: 115-130.

Richmond TA and Somerville CR (2000) The cellulose synthase superfamily. Plant Physiol 124: 495-498.

Robert S, Mouille G, Höfte H (2004) The mechanism and regulation of cellulose synthesis in primary walls: lessons from cellulose-deficient mutants. Cellulose 11:351-364.

Scheible W-R, Pauly M (2004) Glycosyltransferases and cell wall biosynthesis: novel players and insights. Curr Opin Plant Biol 7: 285-295.

Taylor NG, Gardiner JC, Whiteman R, Turner SR (2004) Cellulose synthesis in the Arabidopsis secondary cell wall. Cellulose 11: 329-338.

Villemez CL, JC McNab JC, Albersheim P (1968) Formation of plant cell wall polysaccharides. Nature 218: 878-880.

Wright A, Dankert M, Robbins PW (1965) Evidence for an intermediate stage in the biosynthesis of the Salmonella O-antigen. Proc Natl Acad Sci USA 54: 235-241.

Starting a Life in Science with Debby Delmer: Exploring the Synthesis of Cellulose and the Maize Mixed-linkage (1→3),(1→4)-β-Glucan

Nicholas Carpita

My first and only post-doc experience was with Debby Delmer, then at the MSU-DOE Plant Research Laboratory. Debby was a wonderful mentor in many ways. She had a way of suddenly posing a big question to her group and urging us to take up the challenge of answering with an experiment. Now that I look back, I'm sure she had already figured out the answer but just wanted to see what new feature we could unveil. Regardless of how wild our proposal, she encouraged us to do the experiments she suspected would fail but would eventually lead us to the right answer. The beauty of this philosophy is that it frequently produced an unexpected result that opened a new door to understanding. Such was the case for a small advance towards achieving cellulose synthesis *in vitro*.

Solving the synthesis of cellulose *in vitro* was a major goal of the Delmer lab. The consensus in the late 1970's was that some special feature of the plasma membrane, its site of synthesis, had to be protected. Our model system was an *in vitro* culture of unfertilized cotton ovules, which, in the presence of auxin and gibberellin, produced long fibers with secondary wall cellulose. I suggested that rapid dissipation of turgor pressure causes destruction or disorientation of the synthase complex and that polyethylene glycol (PEG) of sufficient size to not penetrate the cell wall space might mimic turgor *in vitro*. I designed an experiment to excise the cultured cotton fibers in solutions of polyethylene glycol and then feed [^{14}C]glucose, which would be incorporated into cellulose in an *in vitro* synthase system. Debby later confided that she was certain it wouldn't work because I had suggested [^{14}C]glucose as substrate instead of the obvious molecule to use in an *in vitro* assay, UDP-[^{14}C]glucose. The experiment worked. The polyethylene glycol preserved about 50% of the cellulose synthesizing activity compared to *in vivo* activity, and based on autoradiography, the preservation was 100% in about one-half of the cut fibers (Carpita and Delmer 1980). These experiments didn't really constitute an *in vitro* assay but demonstrated a requirement for resealing of the plasma membrane to reconstitute cellulose synthesis in the absence of turgor. We concluded that a membrane potential may be essential for cellulose synthase activity (Carpita and Delmer 1980), and in this system, and in Acetobacter, evidence for this hypothesis was obtained (Bacic and Delmer 1981, Delmer et al. 1982).

Debby also urged us to fiddle with totally new projects that would ultimately launch our independent careers. She let me purchase radiolabeled arabinose and xylose for some experiments on wall synthesis during cell elongation in maize coleoptiles, a classical system for auxin-induced cell elongation. Even though the grass cell wall was known to have more xylose than arabinose, I was amazed to find that about 80% of the arabinose fed to excised sections of the coleoptile partitioned into cell walls, whereas a much smaller fraction of the added xylose was incorporated into the wall. I must admit that I was puzzled by this, whereupon sharing these data with Debby, she gave me another great gem of advice—READ as much scientific literature as you can. Of course, that was great advice, but had I read about the well-known salvage pathways for recovery of arabinose, but not xylose, into the nucleotide sugar pathway, I might never have done the experiments. Reading is sometimes overrated compared to the knowledge gained by doing simple experiments. Debby's willingness and urging of me to do these experiments was most certainly the germ of my career ambition to study the cell walls of grasses.

Setting up my own lab at Purdue, I followed up on the observation that grass species made completely different cell walls than do dicots, and focused on the synthesis of the grass-specific polymer, the mixed-linkage (1→3),(1→4)-β-glucan. This unusual polymer is not a random mixture of linkages but an unbranched polymer of cellotriose, cellotetraosyl, and higher order cellodextrins each connected by single (1→3)-β-linkages (Figure 1). Essentially absent from embryonic and meristematic cells, it accumulates during maximum rates of cell elongation and is mostly hydrolyzed to glucose once growth ceases (Carpita et al. 2001). Our early attempts to assay (1→3),(1→4)-β-glucan synthase activity in enriched maize Golgi membranes from sucrose density centrifugations produced mostly callose, a (1→3)-β-glucan wound polymer. Delmer (1977) had proposed that callose

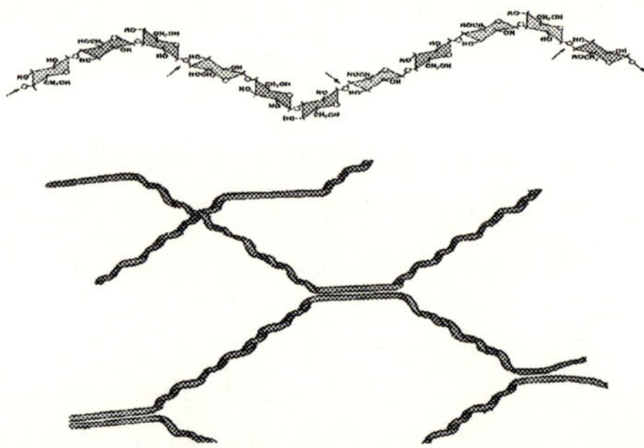

Fig. 1 Structure of the mixed-linkage (1→3),(1→4)-β-glucan from grasses. **Cellotriose and cellotetraose units, in a ratio of about 3:1, are joined by single (1→3)-β-linkages. Longer cellodextrins, up to at least nine residues, are also integrated into the polymer to produce domains that are able to hydrogen bond to other (1→3)-β-glycans (Wood et al. 1994). The arrows show the only sites where the *B. subtilis* endoglucanase is able to cleave the molecule (Anderson and Stone 1975). This action produces primarily a trisaccharide, cellobiosyl-(1→3)-β-glucose, a tetrasaccharide, cellotriosyl-(1→3)-β-glucose, and small amounts of higher order cellodextrin oligomers, which are resolved by high-performance anion-exchange chromatography and provide a diagnostic tool to quantify the amount of true polymer made *in vitro* with Golgi membranes and labeled UDP-glucose (Gibeaut and Carpita 1993; Buckeridge et al. 1999; Urbanowicz et al. 2004).**

was the product of a damaged cellulose synthase, and our finding it in our Golgi membrane preparations indicated that we had significant contamination with plasma membrane. A talented post-doctoral associate, Dr. David Gibeaut, had the idea to use *flotation* centrifugation of homogenates of maize coleoptiles in sucrose density gradients to efficiently enrich Golgi membranes for (1→3),(1→4)-β-glucan synthase reactions *in vitro* (Gibeaut and Carpita 1993). Actually, a couple of tricks were necessary to make study of this synthase possible. First, the homogenization began with the addition of sufficient sucrose in a membrane-stabilization buffer to yield a homogenate of equivalent density of 42% sucrose. Further, activated charcoal powder was added during homogenization to absorb phenolic substances that inhibited the synthase reactions and competed with the synthase for the UDP-glucose substrate. The homogenate was put in the bottom of the centrifuge tube and overlaid with gradient steps of various sucrose concentrations. From this thick black soup, the Golgi membranes would mostly float to the 35%/29% sucrose interface during ultracentrifugation, whereas plasma membrane was trapped in the lower phase. Golgi membranes with active (1→3),(1→4)-β-glucan synthase were collected and the reactions started about an hour after homogenization.

UDP-Glucose is not only the substrate for the (1→3),(1→4)-β-glucan synthase, but it is also the substrate for several other Golgi-derived polysaccharide synthases. A specific assay for (1→3),(1→4)-β-glucan was developed to differentiate it from callose and other β-linked glucans. For this, we used a sequence-specific endo-β-glucanase from *Bacillus subtilis*, which Bruce Stone's group had showed would cleave a (1→4)-β-glucosyl linkage if the preceding linkage were (1→3)-β-glucosyl (Figure 1). The development of high performance anion-exchange liquid chromatography allowed separation of the cellobiosyl-, cellotriosyl-, and higher cellodextrin-(1→3)-glucose oligomers uniquely diagnostic for the polysaccharide. The radioactivity in these oligomers constituted a specific assay for the (1→3),(1→4)-β-glucan synthesized *in vitro*. We were able to confirm that the (1→3),(1→4)-β-glucan synthase resides in the Golgi membranes, and we showed that 250 μM UDP-glucose produced (1→3),(1→4)-β-glucan with the ratio of cellotriose and cellotetraose units similar to that seen in the native polymer (Gibeaut and Carpita 1993). Curiously, callose synthase was still associated with the maize Golgi membranes as well as with the plasma membrane fraction. Whereas the callose synthase of the plasma membrane is activated by calcium ions, that of the Golgi membranes is not (Gibeaut and Carpita 1993). As Delmer (1977) first conjectured for cellulose synthase, we thought callose might result from injured synthases of both cellulose *and* (1→3),(1→4)-β-glucan. If true, then these synthases would not only be catalytically related but also their genes would be evolutionally related.

Debby's group remained focused on the biochemistry of cellulose synthases, but in the 1990s, the attention of the plant cell wall community turned to finding the genes that might encode them. The catalytic domain of a four-gene operon of cellulose synthase and some of the factors involved in the synthesis of cellulose in Acetobacter were established, but progress in plants was slow to follow. The breakthrough came from a collaboration of the Delmer lab with Calgene, a company that had prepared cDNA libraries from cotton fibers during the course of primary to secondary wall formation. Pear et al. (1996) identified the four catalytic motifs and membrane-spanning domains in two cotton fiber cDNAs whose expression was elevated during secondary wall formation. In additional to the catalytic motifs and membrane spans, the plant cellulose synthase also contained zinc-finger domains, proposed to function in protein-protein binding, and two additional plant-specific sequences, including what appeared to be a hypervariable domain. As more cDNAs from EST collections became available, the cellulose synthase (*CesA*) gene family began to take shape (Delmer 1999), and the completion of the Arabidopsis genome sequencing effort revealed the large size of the gene family (Richmond and Somerville 2000).

As *CesA* families of rice and maize were were being assembled, Claudia Vergara, a graduate student in my lab, showed that the so-called 'hypervariable region' was not really hypervariable but representative of different *CesA* sub-classes (Vergara and Carpita 2001). From genetic analyses of *CesA* mutants and biochemical interaction studies, the idea was taking shape that two, and even three, CESA gene products may each be essential to constitute an optimally functional cellulose synthase complex (Taylor et al. 2003). Claudia's work showed that these individual CESA polypeptides came from different sub-classes.

The discovery of the first *CesA* gene by Delmer and her colleagues was also key in the discovery of a large collection of related genes, the cellulose synthase-like genes, which are clustered into six sub-groups in Arabidopsis (Delmer 1999, Richmond and Somerville 2000) and two additional sub-groups specific to rice and the grasses (Hazen et al. 2002). These genes are suspected to be the catalytic domains of many other non-cellulosic cross-linking glycans that possess backbones of (1→4)-β-linked glucose, xylose or mannose residues in which each monosaccharide is inverted roughly 180° with respect to its neighbors (Carpita and Vergara 1998). Debby Delmer announced the identification of the first plant *CesA* cDNA in March, 1996, at a conference devoted to plant cell wall biology (Carpita et al. 1996). At that same meeting, I drew attention once again to an old problem: with the inversion of each sugar unit in a (1→4)-β-linkage the position of hydroxyl group to which the next residue is to be attached is moved several Å in space and presented inverted almost 180°. From the perspective of the non-reducing terminus of the growing chain, a rotation of the last glucose unit 180° would alternate the O-3 and O-4 hydroxyl units with each successive addition of a single glucose unit to the end of the chain (Figure 2A). Thus, to attach each successive (1→4)-β-linked residue, the active site within the catalytic domain must 'toggle' in two alternative conformations, the active site must physically rotate, or the growing chain must rotate. This steric problem is eliminated completely if cellobiosyl units were made coordinately so that the O-4 hydroxyl at the non-reducing terminus would always be presented in the same orientation (Figure 2B). The variation in this mechanism for synthesis of the mixed-linkage glucan, where cellotriose units are the fundamental unit rather than cellobiose units, provides an explanation for why the cellotriose units are connected by (1→3)-β-linkages rather than (1→4)-β-linkages (Carpita et al. 1996).

Dr. Marcos Buckeridge, a visiting scientist from the Botanical Institute of Sao Paulo, Brasil, helped us to obtain experimental evidence for multiple sites of glycosyl transfer within the synthase. Marcos discovered that amounts of cellotriose units synthesized in the polymer increased markedly relative to the cellotetraose units as the concentrations of substrate provided to the Golgi membranes increased (Buckeridge et al. 1999). Another curious finding was that the increases in cellotriose unit compared to cellotetraose units made were paralleled by comparable increases in cellopentaose units. Thus, the synthesis of the odd-numbered units not only drove polymerization, but their synthesis appeared to be independent of the synthesis of the even-numbered units and indicated two types of glycosyl transfer were present in the synthase complex (Buckeridge et al. 1999 and 2001). A precocious undergraduate researcher, Breeanna Urbanowicz, investigated the topology of the synthase at the Golgi membrane. Breeanna found that the mixed-linkage β-glucan synthase not only was susceptible to protease, but that the synthesis of the cellotriose units was also particular sensitive at low doses of protease (Urbanowicz et al. 2004). Subsequent studies with the detergent CHAPS provided the first evidence of a separation and reconstitution of the synthesis of cellotriose unit that indicated a separate, independent glycosyl transferase participated in the synthesis (Urbanowicz et al. 2004).

In our current model of (1→3),(1→4)-β-glucan synthesis we propose that at least two catalytic units are needed: a core synthase that is the topological equivalent of cellulose synthase and makes cellobiose units, and an accessory glycosyl transferase that adds the third glucose to make cellotriose (Figure 3). The more frequently the cellotriose is made, the longer the polymer that is synthesized. The cellotetraose units may arise from failure of the accessory transferase to add the glucose to complete the cellotriose (Buckeridge et al. 2001).

Fascination with cellulose can be traced through more than a century

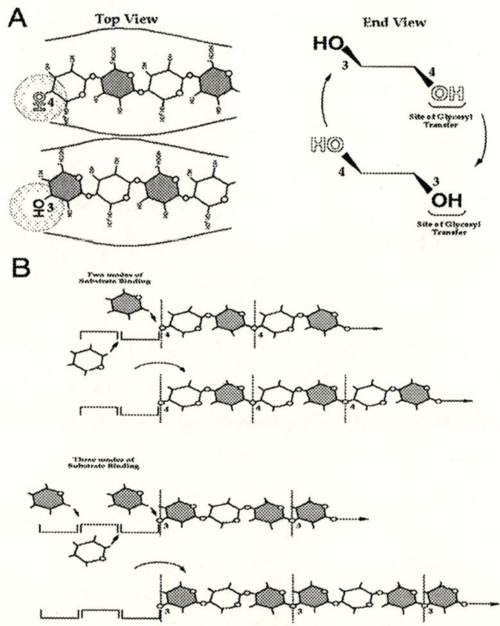

Fig. 2 Models for catalytic mechanisms for the synthesis of cellulose and the (1→3),(1→4)-β-glucan. **A. Iterative addition of a single β-glucosyl residue to the non-reducing terminus of the β-glucan with the O-4 hydroxyl in the active site will invert the next residue nearly 180°, placing the O-3 hydroxyl in the active site. B. If cellobiose units are made by two simultaneous glycosyl transfer events, then the O-4 hydroxyl will always reside at the non-reducing terminus of the growing polymer. Acquisition of a third glycosyl transferase will form cellotriosyl units, the minimum unit size of (1→3),(1→4)-β-glucan, and to connect these iteratively will place the O-3 hydroxyl in the active site instead of the O-4 hydroxyl group. At low substrate concentrations, the associated glycosyl transferase may fail to add a glucose, which, after one iteration, will produce a cellopentaosyl unit, and after two successive iterations will produce a cellotetraosyl unit (Buckeridge et al. 2001).**

of advances, first with knowledge of the fundamental structure of the molecule, credited to Anselme Payen in 1838. Attempts at biochemical synthesis of cellulose *in vitro* date almost one-half century, with the erroneous conclusion that GDP-glucose was the substrate for cellulose synthesis (for this early history, see Delmer 1977). These conclusions were so etched into the biochemical consciousness, that they survive today on commercial metabolic charts. More than twenty years have now elapsed since UDP-glucose has been shown to be the actual substrate for cellulose (Carpita and Delmer 1981). Although her crowning achievement may be judged as her discovery of the gene encoding the catalytic unit of cellulose (Pear et al. 1996), Debby made several other important discoveries concerning the biochemical synthesis (Amor et al. 1995, Kurek et al. 2002, Peng et al. 2001, 2002). However, despite the magnitude of these key contributions to the field, the actual catalytic mechanism of synthesis remains elusive. Does the synthase ratchet along one residue at a time, or is cellobiose the basic unit? Is a single CESA sufficient or do two interacting subunits provide the true catalytic unit for cellobiose units? As Claudia Vergara poses in her paper (Vergara and Carpita 2001), does the cellotriose unit structure of the mixed-linkage glucan synthase provide

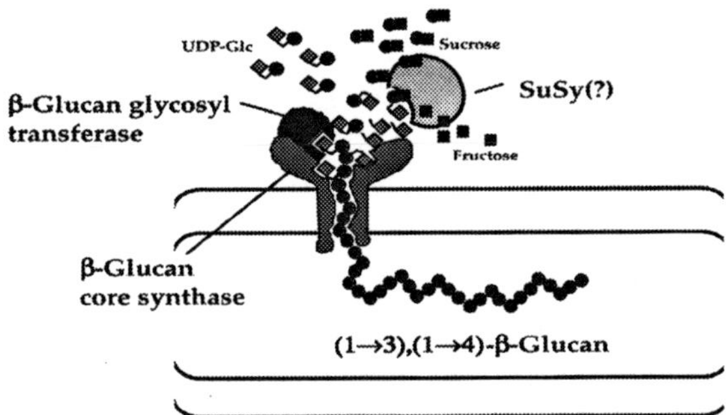

Fig. 3 Model of the proposed topology of the β-glucan synthase at the Golgi membrane. Catalytic dimers of a grass-specific CESA or CSL gene products form an intrinsic membrane protein cellobiosyl-generating core. Associated with this catalytic base on the cytosolic face is an extrinsic glycosyl transferase that associates with the core synthase to generate cellotriosyl units that are processively connected by (1→3)-β-linkages (Buckeridge et al. 2004). The substrate UDP-glucose is added from the cytosolic side directly or, as Amor et al. (1996) suggest for cellulose synthase, by metabolic channeling via SuSy (Buckeridge et al. 1999), and the polymer is extruded into the lumen of the Golgi membrane through a channel formed by the core dimer.

a key lesson to be learned about cellulose synthesis? Thanks to Debby Delmer, a dedicated cadre of young (and not so young) scientists are still seeking the answers to these questions.

References

Amor Y, Haigler CH, Johnson S, Wainscott M, Delmer DP (1995) A membrane-associated form of sucrose synthase and its potential role in synthesis of cellulose and callose in plants. Proc Natl Acad Sci USA 92: 9353–9357.

Bacic A, Delmer DP (1981) Stimulation of membrane-associate polysaccharide synthetases by a membrane potential in developing cotton fibers. Planta 152: 346–351.

Buckeridge MS, Vergara CE, Carpita NC (1999) Mechanism of synthesis of a cereal mixed-linkage (1→3),(1→4)-β-D-glucan: Evidence for multiple sites of glucosyl transfer in the synthase complex. Plant Physiol 120: 1105–1116.

Buckeridge MS, Vergara CE, Carpita NC (2001) Insight into multi-site mechanisms of glycosyl transfer in (1→4)-β-D-glycan synthases provided by the cereal mixed-linkage (1→3),(1→4)-β-D-glucan synthase. Phytochemistry 57: 1045–1053.

Buckeridge MS, Rayon C, Urbanowicz B, Tiné MAS, Carpita NC (2004) Mixed-linkage (1→3),(1→4)-β-glucans of grasses. Cereal Chem 81: 115–127.

Carpita NC, Delmer DP (1981) Concentration and metabolic turnover of UDP-glucose in developing cotton fibers. J Biol Chem 256: 308–315.

Carpita NC, McCann M, Griffing LR (1996) The plant extracellular matrix: News from the cell's frontier. Plant Cell 8: 1451–1463.

Carpita NC, Defernez M, Findlay K, Wells B, Shoue DA, Catchpole G, Wilson RH, McCann MC (2001) Cell wall architecture of the elongating maize coleoptile. Plant Physiol 127: 551–565.

Carpita NC, Vergara CE (1998) A recipe for cellulose. Science 279: 672–673.

Carpita NC (1996) Structure and biogenesis of the cell walls of grasses. Annu Rev Plant Physiol Plant Mol Biol 47: 445–476.

Carpita NC, Delmer DP (1980) Protection of cellulose synthesis in detached cotton fibers by polyethylene glycol. Plant Physiol 66: 911–916.

Delmer DP (1977) Biosynthesis of cellulose and other plant cell wall polysaccharides. Rec Adv Phytochem 11: 105–153.

Delmer DP (1999) Cellulose biosynthesis: exciting times for a difficult field of study. Annu Rev Plant Physiol Plant Mol Biol 50: 245–276.

Delmer DP, Benziman M, Padan E (1982) Requirement for a membrane potential for cellulose synthesis in intact cells of *Acetobacter xylinum*. Proc Natl Acad Sci USA 79: 5282–5286.

Gibeaut DM, Carpita NC (1993) Synthesis of (1→3),(1→4)-β-D-glucan in the Golgi apparatus of maize coleoptiles. Proc Natl Acad Sci USA 90: 3850–3854.

Hazen SP, Scott-Craig JS, Walton JD (2002) Cellulose synthase-like genes of rice. Plant Physiol 128:336-340.

Kurek I, Kawagoe Y, Jacob-Wilk D, Doblin M, Delmer DP (2002) Dimerization of cotton fiber cellulose synthase catalytic subunits occurs via oxidation of the zinc-binding domains. Proc Natl Acad Sci USA 99: 11109–11114.

Pear JR, Kawagoe Y, Schreckengost WE, Delmer DP, Stalker DM (1996) Higher plants contain homologs of the bacterial *celA* genes encoding the catalytic subunit of cellulose synthase. Proc Natl Acad Sci USA 93:12637-12642.

Peng LC, Xiang F, Roberts E, Kawagoe Y, Greve LC, Kreuz K, Delmer DP (2001) The experimental herbicide CGA 325'615 inhibits synthesis of crystalline cellulose and causes accumulation of non-crystalline β-1,4-glucan associated with CesA protein. Plant Physiol 126: 981–992.

Peng LC, Kawagoe Y, Hogan P, Delmer D (2002) Sitosterol-β-glucoside as primer for cellulose synthesis in plants. Science 295: 147–150.

Richmond TA, Somerville CR (2000) The cellulose synthase superfamily. Plant Physiol 124: 495–498.

Taylor NG, Howells RM, Huttly AK, Vickers K, Turner SR (2003) Interactions among three distinct CesA proteins essential for cellulose synthesis. Proc Natl Acad Sci USA 100: 1450-1455.

Urbanowicz BR, Rayon C, Carpita NC (2004) Topology of the maize mixed linkage (1→3),(1→4)-β-D-glucan synthase at the Golgi membrane. Plant Physiol 134: 758-768.

Vergara CE, Carpita NC (2001) β-D-Glucan synthases and the *CesA* gene family: lessons to learned from the mixed-linkage (1→3),(1→4)-β-D-glucan synthase. Plant Mol Biol 47: 145–160.

Unraveling the Secrets of Cellulose Biosynthesis in Plants Using Biochemical Approaches - Prospects for the Future

Vincent Bulone

Introduction

Carbohydrate polymers are quantitatively the most important biomolecules in nature. They have a wide range of functions, from structure and storage to specific signaling. A typical example is cellulose, which accounts for more than 50% of the carbon in the biosphere. This polymer is synthesized by the enzyme cellulose synthase in a large number of organisms including plants, fungi, algae, bacteria and several marine invertebrates known as tunicates (Brown 1996). Plant cell walls represent the largest source of cellulose in nature. The biopolymer has a major structural role and, as such, it is of central importance for the control of the cell shape as well as for the load-bearing function of the cell wall. Cellulose also plays a more dynamic role in normal plant development. In particular, its organized and coordinated deposition in cell walls is one of the most important biochemical processes during cell growth and differentiation.

The basic chemical structure of cellulose is simple since this polysaccharide can be defined as a strictly linear polymer of glucose units linked by (1→4)-β-glycosidic bonds. The (1→4)-β-glucan chains that compose cellulose are however organized as crystalline microfibrils with a complex three-dimensional structure. Cellulose is synthesized naturally in two different forms designated cellulose I and II. Cellulose I (or native cellulose) is synthesized by more than 99% of all cellulose-producing organisms. It consists of microfibrils in which all the chains are organized in a parallel fashion (Koyama et al. 1997). Cellulose I occurs as two different crystalline allomorphs designated I_α and I_β (Atalla and VanderHart 1984). The ratios of these two allomorphs in a given sample are specific of the species producing the polymer. Cellulose II is composed of antiparallel chains and it is synthesized by very few organisms such as the marine alga *Halicystis*, the bacterium *Sarcina* or mutants of the bacterium *Acetobacter xylinum* (Delmer 1999 and references cited therein).

Cellulose II can also be artificially generated from cellulose I by two processes known as regeneration and mercerization. These processes involve the dissolution of cellulose in specific solvents (regeneration) or the swelling of the polymer in aqueous NaOH solutions (mercerization), followed by a recrystallization step (Franz and Blaschek 1990). Mercerization is used to improve the properties of natural yarns and fabrics. Beyond this process, cellulose and its derivatives have many applications (Franz and Blaschek 1990). For instance, cellulose is used as the major constituent of paper whereas several derivatives, e.g. cellulose acetate and mixed organic cellulose ethers, are components of a number of manufactured products (plastics, adhesives, paints, coatings, etc).

Despite the importance of plant cellulose in basic and applied sciences, the mechanisms of biosynthesis of this polymer are not well understood. Thus, it is essential to characterize the enzyme complexes responsible for cellulose synthesis.

The intriguing and fascinating spinning machinery that produces cellulose microfibrils

Cellulose is synthesized at the plasma membrane by enzymatic supramolecular structures that were identified in the alga *Oocystis apiculata* and named terminal complexes (TCs) because of their location at the tips of the elongating microfibrils (reviewed in Brown 1996). Since then, terminal complexes have been observed in many organisms, with essentially two different morphologies (Brown 1996). They can be organized either as linear arrays of synthesizing complexes or as structures called rosettes (Figure 1A). Linear terminal complexes occur in the membranes of most algae and *A. xylinum* where their organization differs by the number of synthesizing units in each row. Rosettes represent the other form of terminal complexes, which are essentially present in higher plants and some green algae. They are composed of six synthesizing subunits organized with a six-fold symmetry. Each subunit possibly contains six catalytic subunits, each producing a single $(1\rightarrow4)$-β-glucan chain (Figure 1B). Thus, it is assumed that each rosette synthesizes a microfibril that contains an average number of glucan chains of 36. This hypothesis contrasts with other results, which indicate that the size of a microfibril is rather consistent with the packing of 18 chains per microfibril or

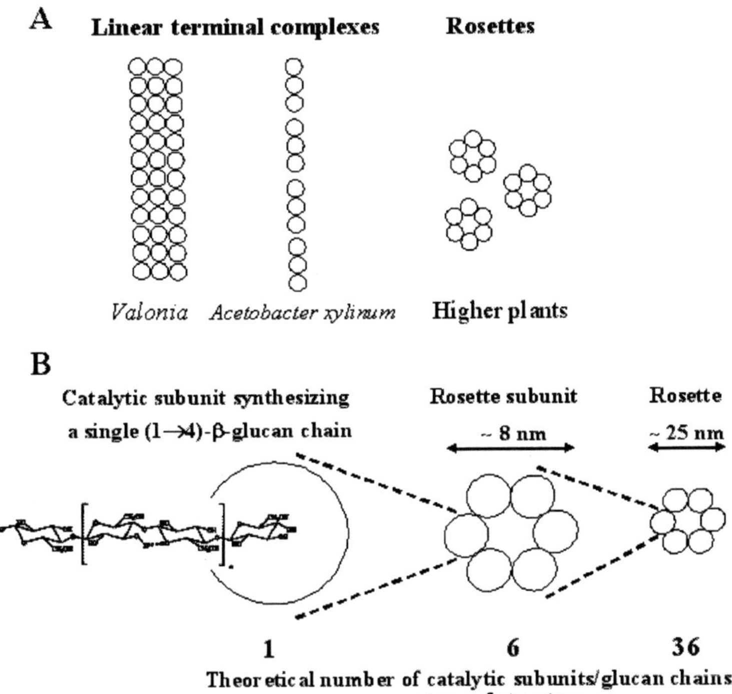

Fig. 1 General organization of terminal complexes (TCs). **(A) The number of rows in linear TCs varies between organisms, as exemplified with the TCs of the alga *Valonia* and the bacterium *Acetobacter xylinum.* In higher plants, TCs are organized as rosettes, which present a six-fold symmetry. (After Brown 1996). (B) Schematic representation of the possible organization of a rosette.**

even less (Ha et al. 1998). In fact, the actual number of catalyzing subunits per rosette has never been experimentally determined. It is only in 1999 that the presence in the rosettes of the catalytic subunits responsible for cellulose polymerization has been firmly demonstrated (Kimura et al. 1999). The use of antibodies suggests that three distinct catalytic subunits coded by different genes interact in the same complex and are required for the correct assembly of the multimeric enzyme (Taylor et al. 2003). It has been proposed that the association of cellulose synthase catalytic subunits occurs under oxidative conditions via a direct interaction between putative N-terminal zinc-finger domains rich in cysteine (Doblin et al. 2002).

Genes coding for catalytic subunits of plant cellulose synthase were first identified in the cotton fiber (reviewed in Doblin et al. 2002). Since then, numerous functional homologues designated *CesA* have been isolated on the basis of sequence similarity and/or through the production and characterization of plant mutants (for an updated list of *CesA* genes see http://cellwall.stanford.edu). It seems that the products of some distinct *CesA* genes interact directly in a given complex to catalyze the coordinated synthesis of cellulose microfibrils, whereas other CESA proteins could be involved in different processes such as, for instance, the synthesis of cellodextrins that would act as primers to initiate the synthesis of the β-glucan chains (Doblin et al. 2002; Peng et al. 2002; Taylor et al. 2003).

If the presence of CESA proteins in rosettes is now established, the precise composition and the stoichiometry of the different subunits in the complexes remain to be determined. In particular, numerous other proteins have been proposed to be associated more or less directly to the cellulose synthesizing machinery. It is for instance the case of a membrane-bound sucrose synthase, Korrigan (a membrane-bound endo-(1→4)-β-glucanase), annexins, actin, tubulin, etc (reviewed in Doblin et al. 2002). Most of these proteins have been identified using molecular genetic approaches. Despite the considerable progress made with this type of strategy, the activity of most candidates, including CesA proteins, has not been firmly demonstrated *in vitro*. Thus, there is a need to further develop biochemical approaches to purify intact cellulose synthase complexes and to confirm or invalidate the role of the different gene products identified so far. Biochemical studies will also be of great importance to identify proteins that are not necessarily attached to the rosettes, but that have an essential function in the regulation and turnover of the complexes.

In addition to the rosette composition, other important molecular aspects need to be elucidated (Figure 2). For instance, it is only recently that a sitosterol-β-glucoside has been identified *in vitro* as a putative primer for cellulose synthesis (Peng et al. 2002). However, the requirement of this molecule or of any other kind of primer for initiation of cellulose polymerization remains to be firmly demonstrated *in vivo*. The direction of chain elongation during cellulose synthesis has been investigated in *A. xylinum*, but the reports available so far are contradictory (Koyama et al. 1997; Han and

Robyt 1998). Our data on the blackberry cellulose synthase are consistent with an elongation of the cellulose chains from the non-reducing end (Lai Kee Him et al. 2002). Another aspect that is not clear concerns the mechanisms by which the polysaccharide is translocated across the plasma membrane during polymerization from the cytoplasmic substrate UDP-glucose. Several models have been proposed to explain this process but none of them has been experimentally demonstrated. In one of these models, the association of eight transmembrane helices born by the CESA protein itself delimits a pore through the membrane to allow the extrusion of the glucan chains (Delmer 1999). In other models, porin-like proteins

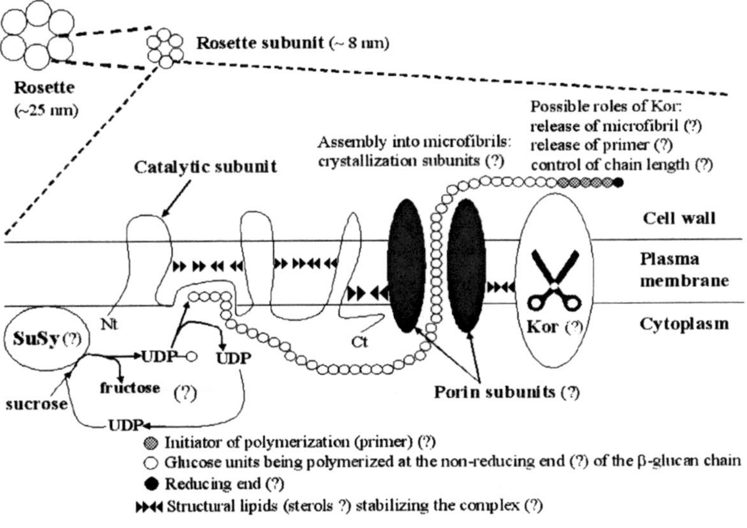

Fig. 2 Hypothetical model for cellulose biosynthesis in plants. **Question marks indicate aspects that remain to be clarified: (a) involvement of a primer to initiate polymerization; (b) orientation of the glucan chain being extruded, with respect to the catalytic site; (c) mechanism of translocation of the glucan chains across the plasma membrane; (d) involvement of crystallization subunits for microfibril formation; (e) involvement of an endo-(1→4)-β-glucanase (Korrigan (Kor)); (f) role of sucrose synthase (SuSy) for channeling the substrate UDP-glucose to the catalytic subunit; (g) role of structural lipids in the stabilization of the enzyme complex. Other proteins such as regulation subunits are not represented. The stoichiometry of the different subunits in the complex is not known. (Adapted from Lai Kee Him et al. 2003).**

have been proposed to be required for the translocation of the chains (Lai Kee Him et al. 2003). Protein subunits involved in the regulation of β-glucan biosynthesis could also be directly associated with cellulose synthase complexes. In *A. xylinum*, proteins have been proposed to be simultaneously responsible for the translocation and crystallization of the (1→4)-β-glucan chains (Saxena et al. 1994). However, there is so far no result showing the existence of proteins with a comparable function in plants. Similarly, the mechanisms responsible for the release of the cellulose microfibrils from the rosettes and for the control of the length of the (1→4)-β-glucan chains are not known.

Overall, considerable progress has been made in the last years in the field of cellulose biosynthesis. However, paradoxically, these advances raise many challenging questions. This is undoubtedly what makes cellulose biosynthesis such a fascinating topic. The next section is focused on the use of *in vitro* approaches to study cellulose biosynthesis. It also presents a number of possible future directions.

The use of *in vitro* synthesis experiments to clarify the mechanisms of cellulose formation

Extraction of cellulose synthase from plasma membranes is a critical step due to the inherent instability of the complexes. When plant membrane-bound proteins are extracted with detergents, the preparations obtained are usually unable to synthesize cellulose from the substrate UDP-glucose. The major *in vitro* product is instead callose, i.e. a strictly linear (1→3)-β-glucan (Okuda et al. 1993; Doblin et al. 2002). It has been proposed that the syntheses of cellulose and callose are co-regulated, and that the same enzyme may actually synthesize either polysaccharides depending on its conformation and/or on the presence of regulation factors (Delmer 1999). However, this hypothesis remains to be demonstrated since, so far, none of the plant callose or cellulose synthases have been purified to homogeneity. In fact, the identification of plant genes that code for putative catalytic subunits of callose synthases indicates that the corresponding polypeptides are different from CESA proteins (see for instance Li et al. 2003). It is nonetheless possible that, apart from the catalytic subunits, cellulose and callose synthase complexes contain some identical polypeptides.

In vitro synthesis of cellulose using plant enzymes has been challenging for many years. The first successful syntheses of the polymer by detergent-extracted enzymes were achieved with preparations from mung bean and cotton (Okuda et al. 1993; Kudlicka et al. 1996). More recently, higher amounts of cellulose could be synthesized *in vitro* using the cellulose synthase from blackberry (Lai Kee Him et al. 2002). This work allowed a complete characterization of the polymer with physical and chemical techniques. However, callose was still the major product in the reaction mixture. These experiments are nonetheless promising for the direct characterization of cellulose synthase using biochemical approaches. Indeed, globular structures, which certainly correspond to the synthesizing machinery, have been observed at the tips of the cellulose microfibrils synthesized *in vitro* (Lai Kee Him et al. 2002). It remains to further optimize the synthesis conditions to isolate higher amounts of these globular structures associated to the *in vitro* products. The development of highly sensitive proteomic techniques adapted to membrane-bound proteins of relatively high molecular weights like CESA represents another important challenge. The optimization of such techniques will greatly facilitate the systematic sequencing of all the proteins present in fractions with high cellulose synthase activity.

As mentioned above, *in vitro* synthesis experiments systematically yield a mixture of callose and cellulose. Thus, it is important to use methods that allow distinction between (1→3) and (1→4) β-linked glucosyl residues in order to quantify and characterize the two *in vitro* products. The methods currently available as well as their advantages and disadvantages have been reviewed (Colombani et al. 2004). However, we have recently developed a new sensitive and straightforward method that has been applied to the identification of *in vitro* products of plant β-glucan synthases (Fairweather et al. 2004). The method relies on the synthesis and use of a substrate containing a glucosyl residue that is uniformly enriched with ^{13}C (UDP-[U-^{13}C]glucose). The *in vitro* products recovered after synthesis with UDP-[U-^{13}C]glucose can be readily analyzed and identified by liquid ^{13}C-NMR spectroscopy. The method is so sensitive that even solid-state NMR spectroscopy, which usually requires tens of mg of polysaccharide, can be used for structural characterization. This NMR technique provides information on the conformation of the glucan chains as well as structural details that cannot be obtained by liquid NMR spectroscopy. It is of particular interest for the analysis of crystalline polymers like cellulose (Atalla and VanderHart 1984).

Our previous work on *in vitro* synthesis of (1→3)-β-glucans has shown that the morphology and degrees of polymerization and crystallinity of the polysaccharides can be dramatically affected depending on the detergent used for enzyme extraction and, for a given detergent, on the species from which the enzymatic fraction was prepared (Lai Kee Him et al. 2001 and 2003; Pelosi et al. 2003; Colombani et al. 2004). It was proposed that the detergents influence the general organization and structure of the enzyme complexes, and indirectly the morphology and structure of the *in vitro* products. Interestingly, when we optimized conditions for *in vitro* synthesis of cellulose from different plant species, we noticed that the choice of the detergent to extract cellulose synthase in an active form is very critical and that it must be determined for each species. From these observations, we can speculate that the callose and cellulose synthases are located within membrane microdomains, which have a lipid composition that varies from one plant species to another. The occurrence of such microdomains in plant plasma membranes has been evidenced (see for instance Peskan et al. 2000). It remains to show experimentally whether microdomains bear glucan synthase complexes. The detergents that preserve cellulose synthase activity probably extract the enzymes as intact complexes, together with structural lipids required for the cohesion and stability of the synthesizing machinery. Some sterols that have been shown to be crucial for cellulose synthesis in plants as well as for elongation and cell wall expansion (Schrick et al. 2004) could be involved in the stability of the rosettes.

Overall, these observations indicate that it is important to take into account the role of lipids in cellulose synthesis. In particular, it will be essential to determine the lipid environment of the cellulose synthase to conceive biomimetic systems in which the enzyme complexes can be reconstituted in an active form. When all the proteins that compose the cellulose synthase machinery are identified, the construction and characterization of active biomimetic systems such as proteoliposomes will help understanding the functional and structural importance of each component in rosettes. If such experiments are completed by the detailed structural characterization of the polysaccharides synthesized by the reconstituted machineries, important information on how β-glucan chains assemble and crystallize will be obtained. These approaches may also help understanding the molecular mechanisms that lead to the formation of cellulose II. It seems that it is only when rosettes are kept intact that cellulose I microfibrils are formed, and that cellulose II is synthesized *in vitro* as a result of the disorganization of the original cellulose synthase complex. In addition, other types of synthesizing systems involving enzymes that are clearly not organized in terminal complexes tend to produce cellulose II (Kobayashi et al. 2000). This hypothesis could be tested *in vitro* if it was possible in the future to prepare artificial membrane structures in which functional cellulose synthases are either incorporated as arrays of organized complexes or as individual polymerizing subunits. On a long term basis, the development and exploitation of such biomimetic systems may open new opportunities to produce tailored polysaccharides as well as new nanocomposites. There is no doubt that we have already stepped in a fascinating period for the study of cellulose biosynthesis, and that many more exciting years of intensive research are still to come.

References

Atalla RH, VanderHart DL (1984) Native cellulose: a composite of two distinct crystalline forms. Science 223: 283-285.

Brown RM Jr (1996) The biosynthesis of cellulose. J Macromol Sci 10: 1345-1373.

Colombani A, Djerbi S, Bessueille L, Blomqvist K, Ohlsson A, Berglund T, Teeri TT, Bulone V (2004) *In vitro* synthesis of (1→3)-β-D-glucan (callose) and cellulose by detergent extracts of membranes from cell suspension cultures of hybrid aspen. Cellulose 11: 313-327.

Delmer DP (1999) Cellulose biosynthesis: exciting times for a difficult field of study. Annu Rev Plant Physiol Plant Mol Biol 50: 245-276.

Doblin MS, Kurek I, Jacob-Wilk D, Delmer DP (2002) Cellulose biosynthesis in plants: from genes to rosettes. Plant Cell Physiol 43: 1407-1420.

Fairweather JK, Lai Kee Him J, Heux L, Driguez H, Bulone V (2004) Structural characterization by ^{13}C-NMR spectroscopy of *in vitro* products synthesized by polysaccharide synthases using ^{13}C-enriched glycosyl donors. Application to a UDP-glucose:(1→3)-β-D-glucan synthase from blackberry (*Rubus fruticosus*). Glycobiology 14: 775-781.

Franz G, Blaschek W (1990) Cellulose, In: Dey PM, Harborne JB (eds) Methods in Plant Biochemistry, Vol. 2, Carbohydrates. Academic Press, New-York, pp 291-322.

Ha MA, Apperley DC, Evans BW, Huxham IM, Jardine WG, Viëtor RJ, Reis D, Vian B, Jarvis MC (1998) Fine structure in cellulose microfibrils: NMR evidence from onion and quince. Plant J 16: 183-190.

Han NS, Robyt JF (1998) The mechanism of *Acetobacter xylinum* cellulose biosynthesis: direction of chain elongation and the role of lipid pyrophosphate intermediates in the cell membrane. Carbohydr Res 313: 125-133.

Kimura S, Laosinchai W, Itoh T, Cui X, Linder CR, Brown RM Jr (1999) Immunogold labeling of rosette terminal cellulose-synthesizing complexes in the vascular plant *Vigna angularis*. Plant Cell 11: 2075-2086.

Kobayashi S, Hobson LJ, Sakamoto J, Kimura S, Sugiyama J, Imai T, Itoh T (2000) Formation and structure of artificial cellulose spherulites via enzymatic polymerization. Biomacromolecules 1: 168-173.

Koyama M, Helbert W, Imai T, Sugiyama J, Henrissat B (1997) Parallel-up structure evidences the molecular directionality during biosynthesis of bacterial cellulose. Proc Natl Acad Sci USA 94: 9091-9095.

Kudlicka K, Lee JH, Brown RM, Jr (1996) A comparative analysis of *in vitro* cellulose synthesis from cell-free extracts of mung bean (*Vigna radiata*, Fabaceae) and cotton (*Gossypium hirsutum*, Malvaceae). Am J Bot 83: 274-284.

Lai Kee Him J, Chanzy H, Müller M, Putaux JL, Imai T, Bulone V (2002) *In vitro versus in vivo* cellulose microfibrils from plant primary wall synthases: structural differences. J Biol Chem 277: 36931-36939.

Lai Kee Him J, Chanzy H, Pelosi L, Putaux JL, Bulone V (2003) Recent developments in the field of *in vitro* biosynthesis of plant β-glucans, In: Gross RA, Cheng HN (eds) Biocatalysis in Polymer Science. ACS Symposium Series N°840, American Chemical Society, Washington D.C., pp 65-77.

Lai Kee Him J, Pelosi L, Chanzy H, Putaux JL, Bulone V (2001) Biosynthesis of (1→3)-β-glucan (callose) by detergent extracts of a microsomal fraction from *Arabidopsis thaliana*. Eur J Biochem 268: 4628-4638.

Li J, Burton RA, Harvey AJ, Hrmova M, Wardak AZ, Stone BA, Fincher GB (2003) Biochemical evidence linking a putative callose synthase gene with (1→3)-β-D-glucan biosynthesis in barley. Plant Mol Biol 53: 213-225.

Okuda K, Li L, Kudlicka K, Kuga S, Brown RM, Jr (1993) β-Glucan synthesis in the cotton fiber. I. Identification of (1→4)-β- and (1→3)-β-glucans synthesized *in vitro*. Plant Physiol 101: 1131-1142.

Pelosi L, Imai T, Chanzy H, Heux L, Buhler E, Bulone V (2003) Structural and morphological diversity of (1→3)-β-D-glucans synthesized *in vitro* by enzymes from *Saprolegnia monoica*. Comparison with a corresponding *in vitro* product from blackberry (*Rubus fruticosus*). Biochemistry 42: 6264-6274.

Peng L, Kawagoe Y, Hogan P, Delmer D (2002) Sitosterol-β-glucoside as primer for cellulose synthesis in plants. Science 295: 147-150.

Peskan T, Westermann M, Oelmüller R (2000) Identification of low-density Triton X-100-insoluble plasma membrane microdomains in higher plants. Eur J Biochem 267: 6989-6995.

Saxena IM, Kudlicka K, Okuda K, Brown RM Jr (1994) Characterization of genes in the cellulose synthesizing operon (*acs* operon) of *Acetobacter xylinum*: implications for cellulose crystallization. J Bacteriol 176: 5735-5752.

Schrick K, Fujioka S, Takatsuto S, Stierhof YD, Stransky H, Yoshida S, Jürgens G (2004) A link between sterol biosynthesis, the cell wall, and cellulose in *Arabidopsis*. Plant J 38: 227-243.

Taylor NG, Howells RM, Huttly AK, Vickers K, Turner SR (2003) Interactions among three distinct CesA proteins essential for cellulose synthesis. Proc Natl Acad Sci USA 100: 1450-1455.

Establishing the Cellular and Biophysical Context of Cellulose Biogenesis

Candace H. Haigler

This article summarizes the fibrillar nature of native cellulose I and the cellular and biophysical mechanisms that control fibril formation. Both the history of emergence of these concepts based on observations in the electron microscope and their implications in an age of nanotechnology are discussed. In particular, it is suggested that the elemental unit of cellulose fibril biogenesis in plants is appropriately called a nanofibril. Major biological concepts discussed include the hierarchical, cell-directed, self-assembly of cellulose, coupled polymerization and crystallization of cellulose, and channeling of carbon from sucrose synthase to cellulose synthase. Research results are included from *Acetobacter xylinum*, *Dictyostelium discoideum*, *Micrasterias denticulata*, *Valonia ventricosa*, *Gossypium hirsutum* cotton fiber, and *Zinnia elegans* tracheary elements.

The fibrillar nature of native cellulose

The properties and functions of native cellulose are ultimately determined by the aggregation of parallel 1,4-β-glucan polymer molecules into partly crystalline fibrils (Figure 1a). These fibrils have remarkable physical properties, for example a high breaking strain energy (5 - 50 x 10^6 J/m^3) in the same range as high tensile steel (20 x 10^6 J/m^3) (Niklas 1992). The ability to form fibrils is associated with the 1,4-β linkage between adjacent glucose residues, which results in a stiff, ribbon-like chain that is ideally suited for close packing. Both cellular and biophysical mechanisms exert control over the final fibrillar form of cellulose. This was inferred in 1948 from observations in the electron microscope of uniform fibrils in plant and bacterial cellulose: "This speaks for a control of cellulose crystallization by the living cytoplasm" (Frey-Wyssling et al. 1948). The alternating lamellae of parallel fibrils in algae such as *Valonia ventricosa* provided additional evidence for biogenesis of cellulose at the plasma membrane under the guidance of the cell (Preston and Kuyper 1951). Although the fibrils were historically called "microfibrils", the primary units of cellulose biogenesis in living organisms are "nanofibrils" with widths between 1 - 20 nm. The smallest 1 - 2 nm diameter nanofibrils are found, for example, as subunits of larger fibrils synthesized by prokaryotic *Acetobacter xylinum*, in quince slime, in wood cambium, and in some other primary walls. Plant cell walls commonly contain 3.5 - 6 nm diameter cellulose nanofibrils, and the walls of certain algae contain nanofibrils up to 25 nm wide (Haigler 1985).

The use of the term 'nanofibril' for native cellulose is novel, but it is appropriate given the emerging science of nanotechnology for materials or objects with nanoscale dimensions (1 - 100 nm). Such materials have unique properties, often arising from a high surface-to-volume ratio, that bridge the quantum mechanics that applies to the atomic scale and the Newtonian physics that applies to larger objects (Roco 2003). Although the width of native cellulose fibrils is in the nanoscale range, their length can be in the micrometer range. However, nanoscale phenomena are undoubtedly important because the interaction of cellulose with other cell wall molecules is largely determined by surface properties. The diameter of cellulose aggregates can be joined with the terminology of nanotechnology to define: (a) nanofibrils, 1 - 100 nm diameter; (b) microfibrils, 0.1 - 100 µm diameter; and (c) macrofibrils, any isolated fibrillar aggregate with diameter > 100 µm. The simple word, fibril, can be used when there is no implication about actual physical size of an isolated cellulose aggregate. To parallel the nomenclature of plant anatomy, the term 'fiber' is best reserved for the composite cell wall of a cotton or wood fiber or the bulk quantities of the same that are used in industry. Use of this terminology in the future would remove ambiguity that often arises between basic researchers and industrial users of cellulose due to different meanings ascribed to words such as "fiber" and "fibril". It will be increasingly important for there to be meaningful communication between these two groups to maximize agricultural productivity and product possibilities within a carbohydrate economy based largely on renewable cellulose.

Hierarchical, cell-directed, self-assembly of cellulose nanofibrils

Because of its fibrillar nature, the production of cellulose by living organisms is appropriately termed biogenesis rather than biosynthesis (Frey-Wyssling 1969). Cellulose biogenesis within the plant cell wall involves biosynthesis of the polymer, co-crystallization of the chains via a mechanism that determines the cellulose I allomorph and fibril size, and the orientation of the fibrils within the cell wall. It is now known through freeze fracture electron microscopy that the cellular control system is manifested by aggregated plasma membrane arrays of cellulose synthase enzymes (Kimura et al. 1999); aggregation of synthases facilitates cellulose I crystallization from clustered glucan chains. In vascular plants and some algae, these complexes were called 'rosettes' because the transmembrane helices of the aggregated CESA proteins were seen in the fractured plasma membrane as a circle of 6, ~ 8 nm diameter, particles (Figure 1b; Mueller and Brown 1980; Giddings et al. 1980). However, the bulk of the CESA proteins including the catalytic sites are thought to reside in the cytoplasm (Doblin et al. 2002). The first researchers who observed rosettes could place them in the context of previous theory about the necessity for synthesizing and assembling cellulose fibrils in association with geometrically arranged cellular components (Preston 1974) and data from algae and *A. xylinum* showing linear particle arrays at the ends of cellulose fibrils (reviewed in Haigler 1985). The correlation between the width of cellulose fibrils and the number of rosettes in closely packed linear rows in the green alga *Micrasterias denticulata* (Giddings et al. 1980) suggested that the biophysics of cellulose self-assembly complemented cellular aggregation of synthases in determining the final properties of native cellulose fibrils.

The details of the relationship between cellulose polymerization and crystallization were revealed by studies of *A. xylinum*, leading to a

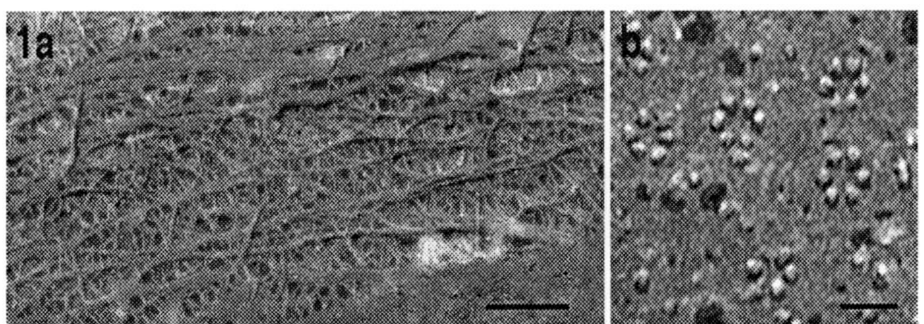

Fig. 1 Electron micrographs of platinum/carbon replicas of differentiating tracheary elements of *Zinnia elegans* processed by freeze fracture. **(a) The first layer of aligned cellulose nanofibrils of the secondary wall is overlaid on the more mesh-like fibrils of the primary wall. Bar = 100 nm. (b) A group of rosettes visualized in the innermost leaflet of the fractured plasma membrane; these indicate the transmembrane helices of cellulose synthase proteins passing through the membrane. Bar = 30 nm.**

model for hierarchical, cell-directed, self-assembly of cellulose. In this prokaryote, cellulose crystallization was altered *in vivo* by addition of cellulose-binding chemicals including fluorescent brightening agents such as Calcofluor White ST (Colour Index # 40622), direct dyes such as Congo Red (Colour Index # 22120), cellulose derivatives such as carboxymethylcellulose, and cell wall polysaccharides such as xylan (Haigler 1991). *A. xylinum* (ATCC 23769) normally synthesizes a 40 - 60 nm extracellular nanofibril of cellulose I (Figure 2a). Depending on the molecular weight and concentration of the cellulose-binding additive, altered bacterial cellulose was induced in the form of: (a) non-crystalline sheets (Brown et al. 1982); (b) 1.5 nm nanofibrils of cellulose IV (with good longitudinal but poor lateral chain order) that arose from one synthetic site (Figure 2b; Haigler et al. 1980); or (c) ~ 6 - 12 nm wide nanofibrils of cellulose I that had assembled from smaller 1.5 nm nanofibrils (Figure 2c; Haigler et al. 1982). The larger cellulose derivatives and xylan could not prevent the assembly of the ~ 6 - 12 nm nanofibrillar subunits of cellulose I, whereas crystallization could be completely prevented by the smaller fluorescent brightening agents and direct dyes. Other researchers elaborated on the potential of cell wall polysaccharides with 1,4-β-linked backbones such as xylan and xyloglucan to modulate the extent of cellulose crystallization in *A. xylinum* and plants (Uhlin et al. 1995; see T. Hayashi, this volume).

Additional experiments with *A. xylinum* showed that crystallization limited the rate of polymerization, and all the data together demonstrated that polymerization and crystallization were consecutive but coupled processes (Benziman et al. 1980). By similar perturbation experiments, consecutive but coupled polymerization and crystallization was also shown to occur during eukaryotic cellulose and chitin biogenesis (reviewed in Haigler 1991). The cross-kingdom role of cell-directed, self-assembly of cellulose was recently emphasized by research on *Dictyostelium discoideum* in which a developmental transition caused a change within the same stalk cells in both intramembrane particle aggregation and cellulose fibril size (Grimson et al. 1996). Since sequencing the genome of *D. discoideum* has revealed only one *CesA*-type gene (Fey et al. 2004), this seems to be a clear case of biophysics rather than genetics as the controlling factor for physical characteristics of cellulose I nanofibrils.

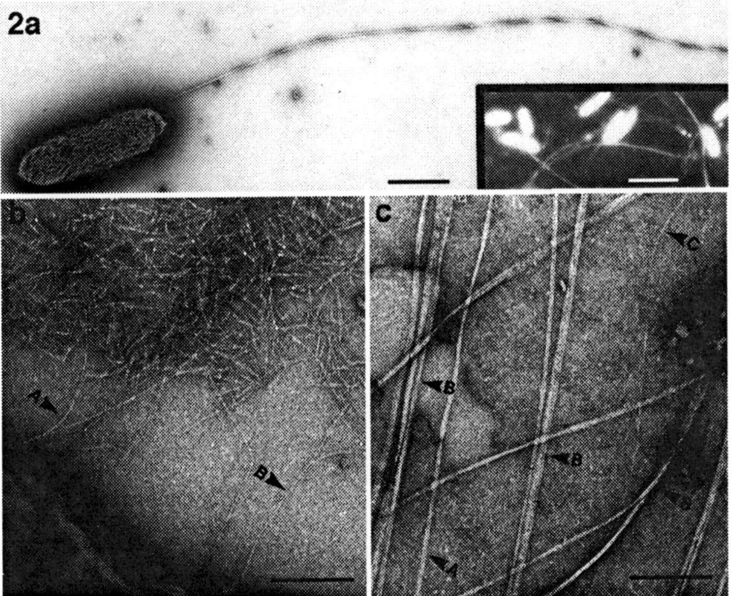

Fig. 2 Normal and altered cellulose synthesized by *Acetobacter xylinum*. (a) A typical composite "ribbon" of cellulose was visualized in the transmission electron microscope after negative staining and, in the inset, with darkfield light microscopy. Bars = 1 μm or 4 μm, in the inset. (b) Addition of a fluorescent brightening agent during cellulose synthesis prevented assembly of the composite ribbon. A and B indicate fibrils that are 3.9 nm and 2.3 nm wide, respectively. Bar = 200 nm. (c) Addition of carboxymethylcellulose during cellulose synthesis also prevented cellulose ribbon assembly, but at a higher level of organization. A and C indicate 11 nm and 2.5 nm wide nanofibrils, respectively. B indicates points where cellulose nanofibrils wider than 20 nm were splitting into smaller subunits. Bar = 200 nm. All micrographs were derived from (Haigler 1982).

Other aspects of cellular control of cellulose biogenesis

It is important to know the complete composition of the cellulose synthesizing complex and how it is integrated functionally with other cellular components (Doblin et al. 2002). For example, it has been known for many decades that microtubules lie close to the plasma membrane at sites of cellulose biogenesis in many cell types and that microtubule antagonists often disrupt cellulose fibril orientation (Baskin 2001). However, establishing a mechanism for this interaction, which seems to be bidirectional between nanofibrils and microtubules in at least some cell types, remains to be accomplished.

Analysis of *Arabidopsis thaliana* mutants with cellulose deficiencies has recently revealed other possible participants in the synthetic complex itself, although relevant mechanisms are still unknown (Doblin et al. 2002; other chapters of this volume). Due to length limitations, only the cellular context for the proposed role of sucrose synthase in channeling UDP-glucose to cellulose synthase of plants will be summarized here.

In a search for a UDP-glucose-binding polypeptide that might be the cellulose synthase, researchers in the Delmer lab instead identified the enzyme sucrose synthase, (SuSy; E.C. 2.4.1.13; sucrose + UDP = UDP-glucose + fructose) within protein extracts of secondary wall stage *Gossypium hirsutum* cotton fibers (Amor et al. 1995). In this research, as in many other cases, the metabolic focus on high rate cellulose biogenesis in secondary wall stage cotton fibers facilitated novel insights into the process. Atypically for this enzyme, the SuSy involved in cellulose biogenesis was strongly associated with membranes and cytoskeletal or other cortical proteins. Semi-permeabilized cotton fibers accepted sucrose as a preferential carbon source for 1,4-β-glucan synthesis compared to exogenous UDP-glucose, and fluorescence immunolocalization showed that SuSy sometimes existed in the same pattern as the helical cellulose fibrils within the cotton fiber secondary wall. Based on these results, a model was developed in which sucrose synthase would preferentially channel UDP-glucose to the cellulose synthases in the plasma membrane.

Although direct channeling between sucrose synthase and cellulose synthase has not been proven biochemically, SuSy is located close to cellulose synthases in the plasma membrane during secondary wall deposition in *Zinnia elegans* tracheary elements and cotton fibers. This was shown by immunolocalization after cryogenic specimen preparation methods that were expected to prohibit protein movement (Salnikov et al. 2001, 2003). In tracheary elements, SuSy was distributed in the pattern of the secondary wall (Figure 3), whereas it was uniformly distributed in cotton fibers with non-patterned secondary wall cellulose biogenesis. In tracheary elements, but not cotton fibers, actin near the cell surface could also be revealed by immunolocalization. Double labeling of tangential sections showed that SuSy was closest to the plasma membrane and adjacent to actin, with microtubules residing slightly deeper in the cortical cytoplasm.

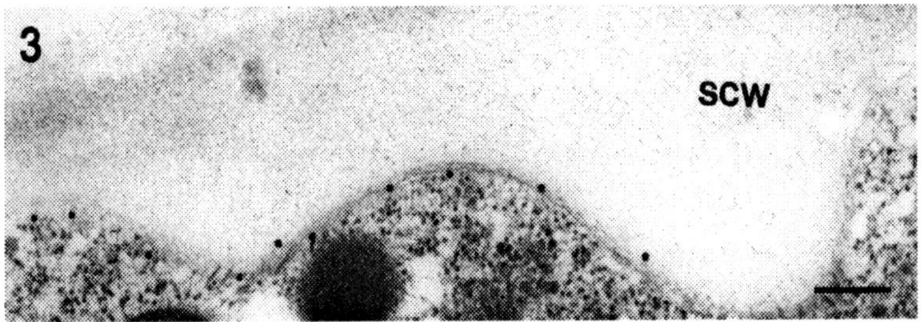

Fig. 3 Thin section of a cryogenically fixed differentiating tracheary element of *Zinnia elegans* with the location of sucrose synthase demonstrated by immunolabeling. **'SCW' indicates the thick secondary wall, and the 20 nm diameter colloidal gold particles indicate that sucrose synthase is close to the plasma membrane where cellulose was being synthesized. Bar = 200 nm.**

These results extend the scope of the complex cellular organization that supports cellulose biogenesis, but they also reinforce our lack of knowledge about mechanistic links between various components of the functional cellulose biogenetic complex.

Future research prospects

In the future, for proteins proposed to be involved in cellulose biogenesis, it will be important to determine their spatial relationships within the cell and to each other. Valid information on this point will be a key to establishing possible mechanisms for roles of multiple proteins in the biogenetic complex. In addition to non-artifactual electron microscopic methods as discussed here, future research can incorporate *in vivo* analyses of functional labeled proteins including analysis of fluorescence resonance energy transfer to establish immediate molecular proximity. For example, it would be of value to know whether or not SuSy is immediately proximal to CESA proteins. It is clear that a full understanding of cellulose biogenesis, which is essentially a structural manufacturing process on the nanoscale, will require continued elucidation of the cellular and biophysical details governing the assembly of cellulose nanofibrils.

In addition to the honoree of this volume, Professor Debby Delmer, this article has mentioned or cited many other great cellulose biogenesis researchers of that generation, for example, Professor Rajai Atalla, Professor Moshe Benziman (deceased), Professor R. Malcolm

Brown Jr., and Dr. Henri Chanzy. It is now particularly important to cultivate another generation of cellulose researchers, especially those who will, like Professor Delmer, maintain a life-long interest in this subject. Repeating her constant refrain, all readers of this volume are invited to join in this exciting area of research. The challenges have always been many, but the rewards will be great as we understand the multifaceted details of cellulose biogenesis and manipulate the process to produce improved renewable products.

Acknowledgments

Dr. Mark Grimson and Dr. Vadim Salnikov are thanked for producing the micrographs in Figures 1 and 3, respectively, while working with the author at Texas Tech University.

References

Amor Y, Haigler CH, Johnson S, Wainscott M, Delmer DP (1995) A membrane-associated form of sucrose synthase and its potential role in synthesis of cellulose and callose in plants. Proc Natl Acad Sci USA 92: 9353-9357.

Baskin TI (2001) On the alignment of cellulose microfibrils by cortical microtubules: a review and a model. Protoplasma 215: 150-171.

Benziman M, Haigler CH, Brown RM, White AR, Cooper KM (1980) Cellulose biogenesis: Polymerization and crystallization are coupled processes in *Acetobacter xylinum.* Proc Natl Acad Sci USA 77: 6678-6682.

Brown RM Jr, Haigler CH, Cooper KM (1982) Experimental induction of altered non-microfibrillar cellulose. Science 218: 1141-1142.

Doblin MS, Kurek K, Jacob-Wilk D, Delmer DP (2002) Cellulose biosynthesis in plants: from genes to rosettes. Plant Cell Physiol 43: 1407-1420.

Fey P, Gaudet P, Just EM, Merchant SN, Pilcher KE, Kibbe WA, ChisholmRL. "dictyBase." http://www.dictybase.org/ (August 2004).

Frey-Wyssling A, Mühlethaler K, Wyckoff RWG (1948) Microfibrillen der pflanzlichen zellwande. Experientia 4: 475-476.

Frey-Wyssling A (1969) The ultrastructure and biogenesis of native cellulose. Fortschr Chem Org Naturst 27: 1-30.

Giddings TH Jr, Brower DL, Staehelin LA (1980) Visualization of particle complexes in the plasma membrane of *Micrasterias denticulata* associated with the formation of cellulose fibrils in primary and secondary walls. J Cell Biol 84: 327-339.

Grimson MJ, Haigler CH, Blanton RL (1996) Cellulose microfibrils, cell motility, and plasma membrane organization change in parallel during culmination in *Dictyostelium discoideum.* J Cell Sci 109: 3079-3087.

Haigler CH, Brown RM Jr, Benziman M (1980) Calcofluor White ST alters the *in vivo* assembly of cellulose microfibrils. Science 210: 903-906.

Haigler CH (1982)Alteration of cellulose assembly in *Acetobacter xylinum* by fluorescent brightening agents, direct dyes, and cellulose derivatives. No: AAI8229383, Diss Abst Int V43-08B: 2432.

Haigler CH, White AR, Brown RM Jr, Cooper KM (1982) Alteration of *in vivo* cellulose ribbon assembly by carboxymethylcellulose and other cellulose derivatives. J Cell Biol 94: 64-69.

Biosynthesis

Haigler CH (1985) The functions and biogenesis of native cellulose, In: Zeronian SH, Nevell TP (eds) Cellulose Chemistry and its Applications. Ellis Horwood, Chichester, England, pp 30-83.

Haigler CH (1991) The relationship between polymerization and crystallization in cellulose biogenesis, In: Haigler CH, Weimer P (eds) Biosynthesis and Biodegradation of Cellulose. Marcel Dekker, New York, pp 99-124.

Kimura S, Laosinchai W, Itoh T, Cui X, Linder R, Brown RM Jr (1999) Immunogold labeling of rosette terminal cellulose-synthesizing complexes in the vascular plant *Vigna angularis*. Plant Cell 11: 2075-2085.

Mueller SC, Brown RM Jr (1980) Evidence for an intermembrane component associated with a cellulose microfibril-synthesizing complex in higher plants. J Cell Biol 84: 315-326.

Niklas KJ (1992) Plant Biomechanics, An Engineering Approach to Plant Form and Function, Univ Chicago Press, Chicago, 607 pp

Preston RD, Kuyper B (1951) Electron microscopic investigation of the walls of green algae. I. A preliminary account of wall lamellation and deposition in *Valonia ventricosa*. J Exp Bot 2: 247-256.

Preston RD (1974) The Physical Biology of Plant Cell Walls, Chapman Hall, London, 491 ppRoco MC (2003) Nanotechnology: Convergence with modern biology and medicine. Curr Op Biotech 14: 337-346.

Salnikov VV, Grimson MJ, Delmer DP, Haigler CH (2001) Sucrose synthase localizes to cellulose synthesis sites in tracheary elements. Phytochem 57: 823-833.

Salnikov VV, Grimson MJ, Seagull RW, Haigler CH (2003) Localization of sucrose synthase and callose in freeze substituted, secondary wall stage, cotton fibers. Protoplasma 221: 175-184.

Uhlin I, Atalla RH, Thompson NS (1995) The influence of hemicelluloses on the aggregation patterns of bacterial cellulose. Cellulose 2: 129-144.

Thermodynamics in Cellulose Biosynthesis: Entropy-driven Synthesis of Cellulose Microfibrils

Yasushi Kawagoe

Why thermodynamics?

Thermodynamics is a potentially useful tool for predicting possible pathways involved in cellulose biosynthesis. Because cellulose microfibrils are insoluble and heterogeneous, describing cellulose biosynthesis in thermodynamic terms has been difficult. However, any synthetic reaction, including cellulose biosynthesis, must conform to the laws of thermodynamics. A major question in cellulose biosynthesis is the nature of force driving the reactions involved in the synthesis of microfibrils. More specifically, do enthalpy changes (ΔH) or entropy changes (ΔS), or both drive the reactions in cellulose synthesis? At present, this question is difficult to answer because we do not know how many steps are involved in microfibril synthesis. However, thermodynamics can allow us to distinguish which reactions will be spontaneous and which are unlikely to occur.

Gibbs free energy (G) is the most useful thermodynamic function in biochemistry. The change in free energy, ΔG, is given by:

$$\Delta G = \Delta H - T\Delta S \quad \textbf{(1)}$$

where ΔG is the change in free energy of a system undergoing a transformation at constant pressure and temperature (T: the absolute temperature, K = °C + 273.15), ΔH is the change in enthalpy of the system, and ΔS is the change in entropy of the system. The microscopic interpretation of entropy is that it is a measure of disorder; the more disorder, the higher the entropy. Reactions can occur spontaneously only if ΔG is negative. Equation (1) indicates that exothermic reactions ($\Delta H < 0$) and reactions that increase entropy (disorder) are favorable (provided also that the energy of activation is overcome usually by intervention of a catalyst such as an enzyme). Conversely, endothermic reactions ($\Delta H > 0$) and those that decrease entropy do not occur spontaneously.

In this paper, I will try to shed light on some unexpected aspects of thermodynamics in cellulose biosynthesis. Alternative pathways in cellulose biosynthesis will also be presented.

Hydrolytic reaction in solution is spontaneous

Most hydrolytic reactions catalyzed by enzymes are highly favorable, e.g., hydrolysis of cellobiose to glucose (equation 2), because such reactions are exothermic ($\Delta H < 0$) and the entropy increases upon hydrolysis ($\Delta S > 0$). Both ΔH and - $T\Delta S$ of the reaction contribute to the negative value of ΔG.

cellobiose + H_2O = 2 glucose (2)

Conversely, we can predict that the reverse reaction, i.e., the synthesis of cellobiose from glucose, does not occur spontaneously because the ΔG for this condensation reaction is positive.

Synthesis of cellobiose from glucose would require a coupled reaction to make an activated glucose, such as UDP-glucose. One known reaction that produces UDP-glucose, shown in equation (3), is catalyzed by sucrose synthase (SuSy). The reaction shown in equation (4) has not been demonstrated to my knowledge and is, thus, a hypothetical reaction. However, if an enzyme that can catalyze this reaction existed, the net reaction, shown in equation (5), would be spontaneous ($\Delta G < 0$) because the net reaction is presumably exothermic ($\Delta H < 0$).

sucrose + UDP = UDP-glucose + fructose (3)
UDP-glucose + glucose = cellobiose + UDP (4)
sucrose + glucose = cellobiose + fructose (5)

Thermodynamics in the formation of insoluble microfibrils

Let us consider a hypothetical reaction in which 36 glucan chains, each with 50 glucose units, assemble into an insoluble microfibril (Figure 1). We know from our experience that this type of reaction is favorable and that the reaction is irreversible under physiological conditions. It is thus safe to assume that the ΔG for the reaction is negative; however, we do not know which of the two terms, ΔH or - $T\Delta S$, or both, contribute lowering the ΔG. If it is ΔH, the reaction should be exothermic. Chemical bond energies that could make this reaction exothermic include (1) both intra- and inter-molecular hydrogen bonding between hydroxyl groups, and (2) hydrophobic or stacking interactions (London-van der Waals interactions) of glucan chains. Cousins and Brown (1997) concluded that van der Waals interactions between glucan chains are a major factor in the crystallization of microfibrils.

The effect of entropy changes in this type of reaction has not been thoroughly considered. Figure 1 shows that a large number of water molecules are displaced from the interacting surfaces between glucan chains as these chains associate. When the glucan chains are separate in water, their hydroxyl groups form hydrogen bonds with water molecules and thus orient them. In addition, the movement of water molecules both above and below the pyranose rings is probably restricted, which also decreases the entropy of water. These considerations suggest that water molecules around each glucan are more ordered compared to the rest of water. Therefore, when glucans form a microfibril, a large number of ordered water molecules around the glucans are released, which may significantly increase the entropy

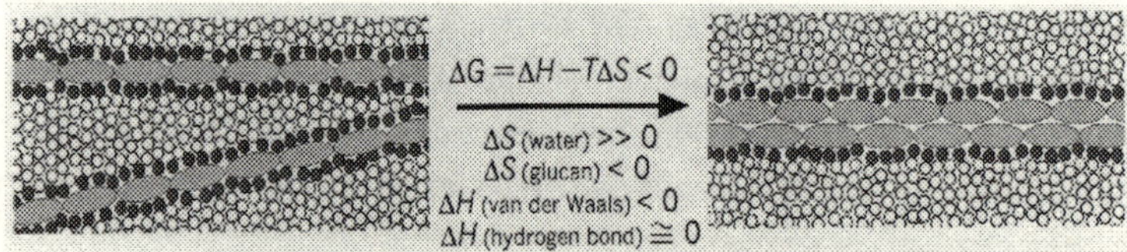

Fig. 1 Diagram depicting the initial stages of glucan chain association that occur in aqueous conditions. **The association of 36 such glucan chains will ultimately lead to the formation of an insoluble cellulose microfibril. The left side of the figure shows a small area during the initial stages of this process, where glucan chains of 50 glucose units are suspended in water. The right side of the figure depicts an intermediate state in which the two glucans are associated. When the two glucans are separate in the initial state (left), the hydroxyl groups (not shown in the figure) form hydrogen bonds with water molecules and thus orient them. In addition, the movement of water molecules on the surface of glucan chains (indicated with solid circles) is probably restricted; therefore, these water molecules are more ordered than the rest of water (open circles). When the two glucans associate, the ordered water molecules around glucans are released from the surfaces, which increase the entropy of water, ΔS (water). In contrast, the entropy of glucan, ΔS (glucan), is expected to decrease but the magnitude is relatively insignificant. The net enthalpy change could be exothermic due to the stacking of glucan chains (London-van der Waals interactions). When glucans associate to form a microfibril, the hydroxyl groups form both intra- and inter-molecular hydrogen bonds. Thus it is likely that the net enthalpy change in hydrogen bonds is small.**

of water, ΔS (water). This could be a typical example of entropy-driven reaction, because it is analogous to the situation in which fatty acids form micelles spontaneously in water. The water molecules around the hydrocarbon chains, when they are separate, form an organized shell and thus are ordered.

Another entropy-driven reaction is seen when the cellulose-binding domain (CBD) of cellulase binds insoluble cellulose microfibrils (this is not the case when CBD binds soluble glucans). Recombinant proteins containing CBD tags, which bind insoluble microfibrils, have been widely used because the CBD domain facilitates protein purification on cellulose columns. Unlike other protein purification tags, including glutathione S-transferase (GST) and maltose-binding protein, CBD binding to cellulose resin is irreversible; once CBD binds the resin, it cannot be eluted with soluble cello-oligosaccharides. The basis for this peculiar property is that the increase in entropy ($-T\Delta S < 0$) is larger than the change in enthalpy, ΔH, and thus lowers the ΔG for the binding reaction (Tomme et al. 1998). Here again we see that the release of water molecules from the interacting surfaces of CBD and cellulose microfibrils is responsible for the increase in entropy of water ($\Delta S > 0$).

Can condensation reaction be spontaneous?

Let us now compare two hypothetical states depicted in Figure 2, which are more relevant to cellulose synthesis. In this hypothetical reversible reaction, soluble cellotetraose is incorporated into an insoluble microfibril, which consists of 9 glucans of 10 glucose units and cellohexaose, and the product of this reaction is an insoluble microfibril consisting of 10 glucan chains of 10 glucose units. It is not immediately apparent whether ΔG for this condensation reaction (forward reaction) is positive or negative. If ΔG is negative, potential factors that could contribute lowering the ΔG include (1) the ΔS of water because ordered water molecules around cellotetraose and microfibril are displaced; (2) ΔH of stacking interactions (London-van der Waals interactions) between glucan chains; and (3) temperature (T) because the higher the temperature, the greater the product of $T\Delta S$. We can predict that the larger the oligosaccharide to be incorporated, the lower the ΔG, because the first two of the three factors mentioned above should contribute more as the size of oligosaccharide increases. The three factors may more than balance the increase in enthalpy (endothermic reaction) due to the formation of a new glycosidic linkage between cellohexaose and cellotetraose. The major diving force in this hypothetical condensation reaction is the large increase in the entropy of water, which is described hereafter as the entropy-driven synthesis of cellulose microfibrils.

Conversely, if the ΔG for the forward reaction is positive, it means that the reverse reaction (hydrolysis) is favorable. It is noted, however, that most, if not all, endoglucanases, when highly purified, cannot hydrolyze cellulose microfibrils to release cello-oligosaccharides into solution. The reason for this is not clear, but both the decrease in the entropy of water upon the solvation of the released cello-oligosaccharide and breaking the stacking interactions between glucan chains may be responsible for preventing these hydrolytic reactions from occurring.

The direction of a chemical reaction is determined by the difference in free energy, ΔG, between the initial and final states. If there is an enzyme for a particular reaction, the enzyme can accelerate the rates of the forward and reverse reactions but cannot change the direction

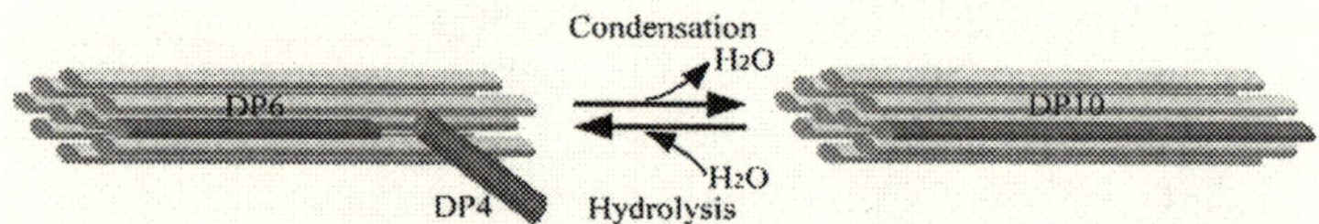

Fig. 2 Either condensation or hydrolysis occurs depending on the difference in free energy (ΔG) between the initial and final states. **In this hypothetical reaction, cellotetraose (DP4, soluble) is incorporated into an insoluble microfibril, which consists of 9 glucans of 10 glucose units (DP10) and cellohexaose (DP6), producing a product of insoluble microfibril consisting of 10 glucan chains of 10 glucose units. If the condensation reaction (forward reaction) is spontaneous, potential contributing factors include the increase in the entropy of water (entropy-driven synthetic reaction, see also Figure. 1) and the stacking of glucan chains (London-van der Waals interactions). The most unfavorable factor for the condensation reaction is the increase in enthalpy (endothermic) upon the formation of a new glycosidic bond between cellohexaose and cellotetraose. On the other hand, if the hydrolytic reaction (reverse reaction) is spontaneous, the major contributing factor would be the decrease in enthalpy (exothermic) upon the hydrolysis of the glycosidic bond, which releases cellotetraose. Although not proven experimentally, these reactions could be catalyzed in both directions by endoglucanases.**

of the reaction. Therefore, any endoglucanase can catalyze condensation reaction if and when the ΔG for the reaction is negative under the existing conditions. As discussed above, the ΔG for a condensation reaction could be negative if the reaction occurs on the surface of insoluble microfibril.

Can rosettes move?

A generally accepted model of cellulose biosynthesis is that rosette terminal complexes are considered as multitasking complexes: (1) multisubunits are synchronized to catalyze polymerization reactions at the catalytic domain, which is presumably positioned in the cytosol, (2) push out glucan chains through an as-yet-unidentified pore to the wall, and (3) the rosette moves directionally in the plasma membrane and leave the product, microfibril, behind the path. In this model, extremely long microfibrils made by the rosettes do not have to be "pushed" into the preexisting network of polysaccharides in the wall because the rosettes, but not the products, are to move. This model is also consistent with the synthesis of cellulose I (parallel) instead of cellulose II (anti-parallel).

Kawagoe and Delmer (1997) reexamined the above-mentioned mobile complex model and raised a number of questions that need to be addressed. The particles aggregated to form rosettes are hypothesized to contain up to 36 CesA catalytic subunits, which together synthesize the approximately 36 glucan chains that form microfibrils. Synchronizing the rates of chemical reactions catalyzed by 36 subunits, or even some of them, would be extremely difficult. It is also important to keep in mind that there has been no direct evidence that the rosettes move in the plasma membrane or that all of the hypothesized CesA catalytic subunits are continuously functional. Furthermore, little is known about the mechanism by which growing chains of glucan pass through the putative pore. At present, it is difficult to envisage a mechanism by which transient interactions between the growing chain and the peptides that constitute the pore keep pace with the rate at which glucose is transferred to the growing chain at the catalytic site in the cytosol.

Interestingly, a functional CESA-GFP fusion protein, IRX3-GFP, was reported to move along the cell in the developing xylem (Gardiner et al. 2003), although some fluorescent signals may have originated from vesicles involved in intracellular trafficking of the secretory pathway. It remains to be seen whether plasma membrane-localized IRX3-GFPs move when they synthesize microfibrils.

Alternative pathways in cellulose biosynthesis

Instead of acting as a mobile synthetic complex, it is possible that the rosette makes cellulose intermediates, e.g., free or lipid-linked cello-oligosaccharides, but does not directly synthesize microfibrils of high molecular weight. In fact, cotton CESA1 synthesizes sitosterol-cellodextrins *in vitro* (Peng et al, 2002). A proposed function of KOR, a membrane bound endoglucanase that is essential for microfibril synthesis in higher plants, is to cleave the sterol to allow further chain elongation.

I propose an alternative pathway of microfibril synthesis, in which endoglucanases, which may include KOR, incorporate cello-oligosaccharides into insoluble microfibrils by the entropy-driven synthesis of cellulose microfibrils as shown in Figure 2. A similar, but distinct, mechanism involving a transglycosylation reaction catalyzed by endoglucanase (CelC) has been proposed in cellulose biosynthesis in *Agrobacterium tumefaciens* (Matthysse et al. 1995). In addition, crystalline cellulose I has been synthesized *in vitro* by a cellulase-catalyzed polymerization of β-cellobiosyl fluoride (Lee et al. 1994), which may suggest that the cellulase also assists the assembly of glucan chains into crystalline cellulose (Kawagoe 1999). A prediction in this line of argument is that if the CBD or the catalytic domain of endoglucanase is engineered to bind preferentially anti-parallel cellulose II, transgenic plants expressing the altered enzyme may produce cellulose II. It should be noted that in this alternative pathway the rosette complexes do not have to be mobile and multitasking.

I think that it is important that we keep up the traditions in the field of cellulose biosynthesis, where scientists are open-minded to different theories and even crazy ideas. In particular, I believe that innovative experimental designs based on thermodynamic considerations are necessary to evaluate the entropy-driven synthesis of cellulose microfibrils.

Acknowledgements

Discussion described in this paper would not be possible if I did not study cellulose synthesis with Debby and the members of her labs at the Hebrew University of Jerusalem and the University of California at Davis. I sincerely thank Debby for years of passionate discussion on cellulose synthesis and her guidance during and after my postdoctoral study. I also thank Eric Roberts for critically reading the manuscript.

References

Cousins SK, Brown RM (1997) X-ray diffraction and ultrastructural analyses of dye-altered celluloses support van der Waals forces as the initial step in cellulose crystallization. Polymer 38: 897-902.

Gardiner JC, Taylor NG, Turner SR (2003) Control of cellulose synthase complex localization in developing xylem. Plant Cell 15: 1740-1748.

Kawagoe Y (1999) Cellulose synthesis in plants: Does rosette really move? Cellulose Communications. 6: 140-144.

Kawagoe Y, Delmer DP (1997) Pathways and genes involved in cellulose biosynthesis. Genet Eng (N Y) 19: 63-87.

Lee JH, Brown RM, Jr, Kuga S, Shoda S, Kobayashi S (1994) Assembly of synthetic cellulose I. Proc Natl Acad Sci USA 91: 7425-7429.

Matthysse AG, Thomas DL, White AR (1995) Mechanism of cellulose synthesis in *Agrobacterium tumefaciens*. J Bacteriol. 177: 1076-1081.

Peng L, Kawagoe Y, Hogan P, Delmer D (2002) Sitosterol-β-glucoside as primer for cellulose synthesis in plants. Science 295: 147-150.

Tomme P, Boraston A, McLean B, Kormos J, Creagh AL, Sturch K, Gilkes NR, Haynes CA, Warren RAJ, Kilburn DG (1998) Characterization and affinity applications of cellulose-binding domains. J Chromatography B 715: 283-296.

Golgi Glucan Synthases

Kanwarpal S. Dhugga

Introduction

Pioneering work from the early researchers streamlined the biochemical assays for a number of the plant cell wall polysaccharide synthases (Ray et al. 1969; Hassid 1972). When compared to some of the other areas of plant biology, such as photosynthesis and lipid metabolism, however, the field of cell wall synthesis has progressed rather slowly, with a molecular lead not coming until relatively recently when Debby Delmer capped her illustrious and pioneering career with a major breakthrough by isolating the first plant cellulose synthase (*CesA*) gene (Pear et al. 1996).

Polymerizing β-glycosyltransferases from a wide variety of prokaryotic and eukaryotic organisms possess several conserved motifs (Saxena et al. 1995). These motifs proved helpful in identifying the cotton *CesA* gene, which was isolated by expressed sequence tag (EST) similarity to the bacterial sequence (Pear et al. 1996; Saxena et al. 1990). A large number of ESTs in the public and private databases were later annotated as being *CesA* or *CesA*-like (*Csl*) because of their homology to the cotton sequence (Cutler and Somerville 1997).

Identification of the Golgi-associated polymerizing glycosyltransferases has been a major challenge. This report describes two attempts in this regard. The first, isolation of a reversibly glycosylated polypeptide (RGP) from pea (Dhugga et al. 1997), was realized during attempts to biochemically purify the glucan synthase I (GS-I) activity, which is believed to make the glucan backbone of xyloglucan (Ray 1979). Isolation of the guar seed mannan synthase (ManS), which makes mannan backbone of galactomannan, a storage polysaccharide in the endosperm walls of some plant species constitutes the second example (Dhugga et al. 2004).

Isolation of RGP

Cellulose and callose are made at the plasma membrane and extruded to the outside of the cell for direct deposition into the cell wall. Matrix polysaccharides, in contrast, are first made in the Golgi, packaged into secretory vesicles, and then exported to the cell wall by exocytosis (Ray et al. 1969; Ray et al. 1976). Two distinct glucan synthase activities, referred to as glucan synthase I and glucan synthase II, were localized to the Golgi and the plasma membrane fractions, respectively (Ray 1979). Before I joined the laboratory of Prof. Peter Ray, he and his group, in their efforts to identify the polypeptides involved in glucan synthase I activity, had discovered a protein of ~40 kD from pea that became labeled with radiolabeled substrate, UDP-[^{14}C]glucose, under the glucan synthase I assay conditions. The labeling could be chased off by an addition of excess unlabeled substrate, suggesting that this protein had enzymatic properties and hence the name RGP. Subsequently, we obtained additional evidence pointing to the involvement of RGP in glucan synthase I activity (Dhugga et al. 1991).

In Coomassie blue-stained SDS-gels derived from a Golgi-enriched membrane fraction, RGP was essentially invisible. Peter Ulvskov, a fellow postdoc in the lab, and I spent nearly a year developing protocols for and running long SDS gels to enhance the resolution of polypeptides around 40 kD, excising the blank-looking zone from the stained gels where we had mapped the radiolabeled polypeptide band, and electroeluting it for antibody production. Peter Ulvskov was able to produce antibodies in mice with a reasonable titer, which allowed him to characterize the protein and make the paper acceptable to the Journal of Biological Chemistry, albeit with some difficulty (Dhugga et al. 1991).

Later on, during attempts to purify glucan synthase I activity from a digitonin-solubilized Golgi-enriched fraction on a UDP-glucose-agarose column, a single polypeptide of ~40 kD was eluted, which turned out to be RGP upon further characterization (Dhugga et al. 1997). The polypeptide that Peter Ulvskov and I had spent a good part of the year obtaining enough of for antibody production in mice now could be purified in large quantities in a short period. This allowed raising high quality antibodies in rabbits and microsequencing of the trypic peptides, both of which proved useful in the isolation and confirmation of a cDNA for RGP (Dhugga et al. 1997).

Although most of RGP was recovered in the Golgi fraction after gradient centrifugation, a small amount was always detected in the plasma membrane, a result that was contrary to its involvement in Golgi polysaccharide formation. Also, a substantial proportion of RGP would come off the particulate fraction upon further resuspension and centrifugation, suggesting that it was a peripheral membrane protein. To address whether RGP was indeed localized to the plasma membrane, Suresh Tiwari, an accomplished cell biologist, obtained clean immunogold-labeled images of pea cells showing specific localization of RGP to the trans Golgi cisternae and none to the plasma membrane (Dhugga et al. 1997). Absence of RGP in the plasma membrane and any of the organelles were in contrast to the earlier reports that had implicated it in starch formation or in plasmodesmatal channel functioning (Singh et al. 1995; Epel et al. 1996). Small amount detected in the cytosol was expected because RGP does not possess a signal peptide and is probably made there (Dhugga et al. 1997). The anti-RGP antibody has since been widely accepted as a Golgi marker for intact cells as well as for isolated membranes (Hinz et al. 1999; Mullen et al. 1999; Nunan and Scheller 2003; Shimada et al. 2002). How RGP finds its way to the trans Golgi cisternae remains unknown.

The glucan synthase I activity was dependent on the activity of RGP but the reverse was not always true. For example, RGP could be labeled with the radiolabeled substrate even when the glucan synthase I activity was completely destroyed with a detergent treatment (Dhugga et al. 1991). To explain this anomaly, we surmised that glucan synthase I consisted of two components: RGP, which acted as a primary sugar acceptor from its UDP-sugar substrate, and a transferase, which then transferred the sugar from RGP to the elongating glucan chain. In addition to UDP-glucose, RGP could also be reversibly labeled with UDP-galactose and UDP-xylose, suggesting that it mediated as a primary sugar acceptor for the three transferase activities that are responsible for xyloglucan formation, i.e., glucan synthase I, xylosyltransferase and galactosyltransferase (Dhugga et al. 1997). Another likely role of RGP could be that it donates sugars to different polymerizing Golgi β-glycan synthases, e.g., glucan, xylan, and galactan synthases.

In native state, RGP occurs as a multimer with a molecular mass exceeding 600 kD (Dhugga et al. 1991 and unpublished data). In the aforementioned models, RGP would constitute the cytosolic portion of the xyloglucan synthase complex, and the transmembrane portion that is made of all three transferases would move the sugars from RGP to the growing xyloglucan polysaccharide (Figure 1). This model is further supported by sequence analysis, which revealed that it contained the conserved motifs for only substrate binding, and not those believed to be essential for polymerizing the substrate into the product (Saxena and Brown 1999). A similar scenario could be envisioned with each of the different synthases if RGP were to participate only in glycan chain formation. Given the size difference between the RGP and the expected functional unit size of a transferase of one or two polypeptides, stoichiometrically, RGP is expected to amplify the substrate capturing capacity of the transferases it interacts with. This would particularly be true for low abundance substrates like UDP-xylose, the *in vivo* concentration of which is 50-fold lower than that of UDP-glucose (Hayashi 1989). The role of RGP in the transport of sugar nucleotides from cytosol to the Golgi lumen has also been proposed (Faik et al. 2000). This model too would require a second, membrane-spanning component that accepts and transports sugars from RGP into the Golgi lumen and transfers them back to UDP, making UDP-sugars (Figure 1). UDP has a short half-life in the Golgi lumen, however, suggesting that the transporter role for RGP is unlikely (Neckelmann and Orellana 1998).

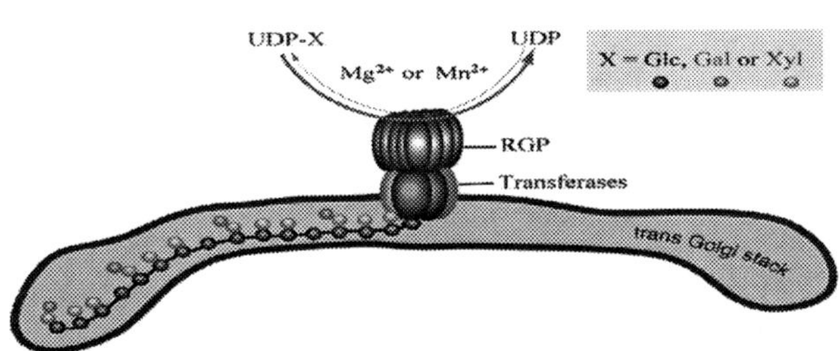

Fig. 1 A model illustrating the possible role of the reversibly glycosylated polypeptide (RGP) in xyloglucan formation.

RGP occurs as a multimer in native state, consisting of 16 or more polypeptides. The sugar residues (shown in different shades) from UDP-sugar substrates are first transferred to RGP. Separate, transmembrane transferases then utilize these residues to make xyloglucan polysaccharide inside the Golgi lumen. Two polypeptides are shown to make a functional unit for glucan backbone formation based on the assumption that two juxtaposed active sites may be needed to form a β-1,4-linkage without having to rotate the chain after each sugar addition. A single polypeptide may be sufficient to add sugars to the backbone.

Mannan synthase from guar (*Cyamopsis tetragonoloba*) seed

After completing purification and gene cloning of RGP, I moved to Pioneer Hi-Bred International, Inc., to pursue production of high-value polysaccharides in crop plants. Pioneer has had a Discovery Research program that allows scientists to carry out research outside of the company goals. I was fortunate to have been awarded a grant to explore the molecular basis of galactomannan formation in guar seeds. Galactomannan is a commercial gum with many industrial applications and guar is one of its major sources (Whistler 1993). For a polysaccharide to be of interest to industry, it is preferable that it commands a large-volume market, has a higher value than the current, large-volume polysaccharides (e.g., starch), and has a simple biosynthetic pathway. Galactomannan fulfilled these criteria.

Galactomannan, which is made in the Golgi by two enzymes, Mannan synthase and α-galactosyltransferase, constitutes nearly all of guar endosperm, offering an elegant system to study Golgi polysaccharide synthesis. Grant Reid and his group had already characterized the Mannan synthase enzyme, which saved us time upfront in developing an assay (Reid et al. 1995).

We first undertook a biochemical approach. We could isolate large quantities of endosperm from guar seeds for the preparation of a membrane fraction that was free of contamination by the particles from the remainder of the seed (embryo and seed coat). As with the other glucan synthases, glucan synthase I and glucan synthase II, Mannan synthase activity could be solubilized in digitonin, a mild, non-ionic detergent (Dhugga and Ray 1994 and unpublished data). Purification of the solubilized enzyme proved difficult, however. The only method that showed some promise was density gradient centrifugation of the digitonin-solubilized particles. Prior experience with the purification of callose synthase, the most stable and highly active of the β-glycan forming plant enzymes, had suggested that the polysaccharide synthases constituted a very minor proportion of the membrane proteins (Dhugga and Ray 1994; Dhugga 2005). Yet, because we had no other option at that time, we proceeded with the identification of the partially purified polypeptides via proteomics. By the time we reached this stage, we had already developed our bias, mainly based on the published works from several laboratories, that ManS was a member of the Csl class of plant sequences. None of the proteins identified by proteomics from our biochemically-purified fractions had the characteristics of a CesA or a Csl, however, so it was about time to explore other approaches.

Genomics approach to identify *ManS*

As we were struggling with biochemistry, Pioneer Hi-Bred merged with DuPont in 1999, which gave us access to the latter's large-scale sequencing facility. We obtained approximately 5,000 ESTs from the guar seed at each of three different stages of development: before, at, and after peak Mannan synthase activity. Out of a total of 15 ESTs that showed similarity to the CesA or Csl sequences, 12 assembled into a single contig. The derived protein from the full-length cDNA contained motifs common to the polymerizing β-glycosyltransferases. The gene was specifically expressed in the endosperm and no other tissue (Dhugga et al. 2004). These results suggested that the *Csl* gene we had isolated encoded the Mannan synthase protein.

To confirm the function of the putative Mannan synthase gene, we transformed it into soybean cells. The common thinking about the polymerizing plant β-glycosyltransferases has been that they consist of large complexes made from unrelated proteins (Dhugga 2005). Just in case the guar Mannan synthase needed other proteins to make a functional enzyme complex, the probability of those proteins being present in the host cell was higher if it was closely related to guar. Soybean was thus selected as a host because, like guar, it is a legume and is a major commercial crop for Pioneer Hi-Bred.

Soybean somatic embryos are routinely used as a surrogate system for studying the functional expression of the genes that would normally be expressed in the seed (see Dhugga et al. 2004 for cited refs.). Mannan synthase activity is nearly absent in the somatic embryos, which provided a clean background for assaying the activity resulting from the introduced gene. The transgenic embryos expressed the guar *Mannan synthase* gene and the membrane particles derived from them contained high levels of the Mannan synthase activity (Dhugga et al. 2004). At first, we found it hard to believe that a single gene could make a functional enzyme in a non-native species. Additional transgenic events containing the guar Mannan synthase gene obtained from multiple, independent transformations consistently expressed

Mannan synthase activity, which convinced us that we had indeed isolated the correct gene.

Upon phylogenetic analysis, the guar Mannan synthase protein matched most closely with the CslA group of sequences from *Arabidopsis* and rice (Dhugga et al. 2004). Plant cell walls contain small amounts of mannose, which is believed to be present in the form of glucomannan (Hassid 1972), although long stretches of 1,4-β-mannan are also known to occur (Handford et al. 2003). Two Csl sequences each from *Arabidopsis* and rice clustered with the guar ManS sequence, suggesting that these might be involved in mannan or glucomannan formation (Dhugga et al. 2004). The remaining Csl sequences formed species-specific clades. It is possible that the CslA sequences that form species-specific clades are involved in the formation of glucan backbone of xyloglucan in dicots and xylan backbone of arabinoxylan in monocots. Like galactomannan, both these polysaccharides are branched and have 1,4-β-linked backbones. Recently, Liepman *et al.* (2005) have determined that the CslA sequences that most closely match the guar Mannan synthase make glucomannan autonomously when expressed in cultured *Drosophila* cells.

For the biotechnological applications, the main objective for undertaking this study, the isolation of the Mannan synthase gene has given us the ability to produce galactomannans in high-yielding crops. Successful production of galactomannans in soybean, for example, might allow the generation of novel gums with diverse chemical properties and thus applications. Locust bean gum is considered to be of much higher quality than guar gum and is thus much more expensive. The only difference between these two is the mannose/galactose ratio, which is ~4 for locust bean and ~2 for guar galactomannan. Transgenic expression of the Mannan synthase and galactosyltransferase genes in soybean might allow the production of gums with varying mannose/galactose ratios. Initial results are encouraging in that the guar Mannan synthase gene is functionally expressed in developing soybean seeds, and mature seeds derived from the transgenic plants contain elevated levels of mannose in comparison to the wildtype seeds (Dhugga et al. 2004).

Conclusions

Although overwhelming evidence indicates a role for RGP in cell wall synthesis, its exact function remains to be determined. Of the 29 *Csl* genes in *Arabidopsis* and 37 in rice, the function of only three has thus far been determined (Liepman et al. 2005). Association of specific biochemical activities with different *Csl* genes holds lots of promise for basic as well as applied research (Dhugga 2005).

References

Cutler S, Somerville C (1997) Cellulose synthesis: Cloning in silico. Current Biol 7: R108-R111.

Dhugga KS (2005) Plant Golgi cell wall synthesis: From genes to enzyme activities. Proc Natl Acad Sci USA 102: 1815-1816.

Dhugga KS, Barreiro R, Whitten B, Stecca K, Hazebroek J, Randhawa GS, Dolan M, Kinney AJ, Tomes D, Nichols S, Anderson P (2004) Guar seed β-mannan synthase is a member of the cellulose synthase super gene family. Science 303: 363-366.

Dhugga KS, Ray PM (1994) Purification of 1,3-β-glucan synthase activity from pea tissue: Two polypeptides of 55 kDa and 70 kDa copurify with enzyme activity. Eur J Biochem 220: 943-953.

Dhugga KS, Tiwari SC, Ray PM (1997) A reversibly glycosylated polypeptide (RGP1) possibly involved in plant cell wall synthesis: Purification, gene cloning, trans Golgi localization. Proc Natl Acad Sci USA 94: 7679-7684.

Dhugga KS, Ulvskov P, Gallagher SR, Ray PM (1991) Plant polypeptides reversibly glycosylated by UDP-glucose: Possible components of Golgi β-glucan synthase in pea cells. J Biol Chem 266: 21977-21984.

Epel BL, vanLent HWM, Cohen L, Kitlizky G, Katz A, Yahalom A (1996) A 41 kDa protein isolated from maize mesocotyl cell walls immunolocalizes to plasmodesmata. Protoplasma 191: 70-78.

Faik A, Desveaux D, Maclachlan G (2000) Sugar-nucleotide-binding and autoglycosylating polypeptide(s) from nasturtium fruit: Biochemical capacities and potential functions. Biochem J 347: 857-864.

Handford MG, Baldwin TC, Goubet F, Prime TA, Miles J, Yu X, Dupree P (2003) Localisation and characterization o cell wall mannan polysaccharides in *Arabidopsis* thaliana. Planta 218: 27-36.

Hassid WZ (1972). Biosynthesis of Polysaccharides from Sugar Nucleotides in Plants. In Biochemistry of the Glycosidic Linkage, R Piras and H G Pontis, eds. (New York: Academic Press), pp. 315-335.

Hayashi T (1989) Xyloglucans in the primary cell wall. Ann Rev Plant Physiol Plant Mol Biol 40: 139-168.

Hazen SP, Scott CJS, Walton JD (2002) Cellulose synthase-like genes of rice. Plant Physiol 128: 336-340.

Liepman AH, Wilkerson C, Keegstra K (2005) Expression of cellulose synthase-like (*Csl*) genes in insect cells reveals that *CslA* family members encode mannan synthases. Proc Natl Acad Sci USA 102: 2221-2226.

Neckelmann G, Orellana A (1998) Metabolism of uridine 5'-diphosphate-glucose in Golgi vesicles from pea stems. Plant Physiol 117: 1007-1014.

Pear JR, Kawagoe Y, Schreckengost WE, Delmer DP, Stalker DM (1996) Higher plants contain homologs of the bacterial celA genes encoding the catalytic subunit of cellulose synthase. Proc Natl Acad Sci USA 93: 12637-12642.

Ray PM (1979) Maize Coleoptile Cellular Membranes Bearing Different Types of Glucan Synthetase Activity. In Plant Organelles, E Reid, ed. (Chichestor, U.K.: Halsted Press/John Wiley & Sons), pp. 135-146.

Ray PM, Eisinger WR, Robinson DG (1976) Organelles involved in cell wall polysaccharide formation and transport in pea cells. Ber Deutsch Bot Ges Bd. 89: 121-146.

Ray PM, Shininger TL, Ray MM (1969) Isolation of β-glucan synthetase particles from plant cells and identification with Golgi membranes. Proc Natl Acad Sci USA 64: 605-612.

Reid JSG, Edwards M, Gidley MJ, Clark AH (1995) Enzyme specificity in galactomannan biosynthesis. Planta 195: 489-495.

Richmond TA, Somerville CR (2000) The cellulose synthase superfamily. Plant Physiol 124: 495-498.

Saxena IM, Brown Jr. RM, (1999) Are the reversibly glycosylated polypeptides implicated in plant cell wall biosynthesis non-processive beta-glycosyltransferases? Trends Plant Sci 4: 6-7.

Saxena IM, Brown Jr. RM, Fevre M, Geremia RA, Henrissat B (1995) Multidomain architecture of beta-glycosyl transferases: Implications for mechanism of action. J Bacteriol 177: 1419-1424.

Saxena IM, Lin FC, Brown Jr. RM, (1990) Cloning and sequencing of the cellulose synthase catalytic subunit gene of Acetobacter xylinum. Plant Mol Biol 15: 673-684.

Singh DG, Lomako J, Lomako WM, Whelan WJ, Meyer HE, Serwe M, Metzger JW (1995) β-Glucosylarginine: a new glucose-protein bond in a self-glucosylating protein from sweet corn. FEBS Lett 376: 61-64.

Whistler RL (1993) Introduction to Industrial Gums. In Industrial Gums: Polysaccharides and Their Derivatives, R L Whistler and J N BeMiller, eds. (San Diego: Academic Press), pp. 1-19.

A Novel Method to Analyze the Biosynthesis of Xyloglucan

David M. Cavalier, Mazz Marry and Alan R. White

Introduction

Over the past five years there has been significant progress in determining the genes that are involved in xyloglucan biosynthesis. Traditional protein purification and sequencing was used to identify the gene that encodes xyloglucan fucosyltransferase (Perrin et al. 1999; Faik et al. 2000). However, the isolation and characterization of xyloglucan glucosyltransferase has been extremely problematic using traditional protein purification techniques because xyloglucan glucosyltransferase is a Golgi membrane localized protein that has a strong propensity to lose activity as purification proceeds (Hayashi et al. 1980; Ray, 1980; Hayashi et al. 1981; Hayashi and Matsuda 1981b; Hayashi and Matsuda 1981a, 1981c; White et al. 1993b; Yang 1998). Alternatively, reverse genetics strategies have been successful in identifying the galactosyltransferase and putative xylosyltransferase genes (Faik et al. 2002; Madson et al. 2003). Indeed, reverse genetics strategies have been successfully employed to identify the genes that encode the enzymes that are responsible for the synthesis of mannan backbones (Dhugga et al. 2004; Liepman et al. 2005) as predicted by the cellulose synthase-like gene hypothesis (Richmond and Somerville 2001). However, the cellulose synthase-like gene hypothesis has yet to be proven in the case of the xyloglucan backbone.

In previous studies, radiolabeled products of xyloglucan biosynthetic reactions were precipitated with ethanol (70% final concentration), and either analyzed directly with scintillation counting or digested with hydrolytic enzymes followed by various types of chromatographic analysis (Hayashi et al. 1980; Ray, 1980; Hayashi et al. 1981; Hayashi and Matsuda 1981b; Hayashi and Matsuda, 1981a 1981c; Walton and Ray 1982; Hayashi et al. 1984; Camirand and Maclachlan 1986; Hayashi et al. 1986; Farkas and Maclachlan 1988; Hayashi et al. 1988; Brummell et al. 1990; Hanna et al. 1991; White et al. 1993b, 1993a; Faik et al. 1997; Yang 1998). In some studies, the precipitation of radio-labeled xyloglucan biosynthetic products has been used to determine enzymatic activities under various substrate, cofactor, and physical conditions by measuring the amount of radioactivity incorporated per unit of time (Hayashi et al. 1981; Hayashi and Matsuda 1981b; Hayashi and Matsuda 1981a, 1981c; Walton and Ray 1982; Hayashi et al. 1984; Camirand et al. 1987; Farkas and Maclachlan 1988; Hanna et al. 1991; White et al. 1993b; Yang 1998). While the isolation of ethanol-insoluble xyloglucan biosynthetic reaction products, by either centrifugation into a pellet or vacuum filtration onto glass-fiber filters, can be done in a short period of time, these procedures may not be the most efficient means of collecting all the biosynthetic products. Consequently, there is a need for methods that are useful in the characterization of xyloglucan biosynthetic products that do not involve a precipitation step.

The hydrolytic enzyme susceptibility assay

While past researchers have used a combination of hydrolytic enzymes and chromatography to characterize enzyme activity (see references above), none of the protocols were designed with the explicit purpose of determining the apparent values of V_{max} and K_m for the enzyme of interest. A methodology called the hydrolytic enzyme susceptibility assay was developed for the purpose of determining the values of V_{max} and K_m of CHAPS-solubilized xyloglucan glucosyltransferase activity. The hydrolytic enzyme susceptibility assay is composed of five distinct steps. Step 1: A large xyloglucan glucosyltransferase incubation was conducted on CHAPS-solubilized Golgi membrane vesicles from *Pisum sativum* and terminated by boiling the sample to decompose any un-reacted UDP-glucose and UDP-xylose. Step 2: The radio-labeled xyloglucan glucosyltransferase products were partitioned over a calibrated Bio-Gel P-2 size-exclusion chromatography column, fractionated, and the amount of radioactivity in each fraction was determined with scintillation counting (Figure 1a). Polysaccharides that are larger than the Bio-Gel P-2 size exclusion limit of approximately 1,800 Da would elute in the void volume, whereas monosaccharides, such as xylose (150.1 Da) and glucose, would elute in the included volume. Oligosaccharides, such as XXXG (dp 7, 1063.1 Da), XXLG (dp 8, 1225.3 Da), and XLLG (dp 9, 1387.5 Da), that have molecular weights smaller than the Bio-Gel P-2 exclusion limit would elute in the partially included volume. Step 3: Fractions containing radioactivity that corresponded to the void,

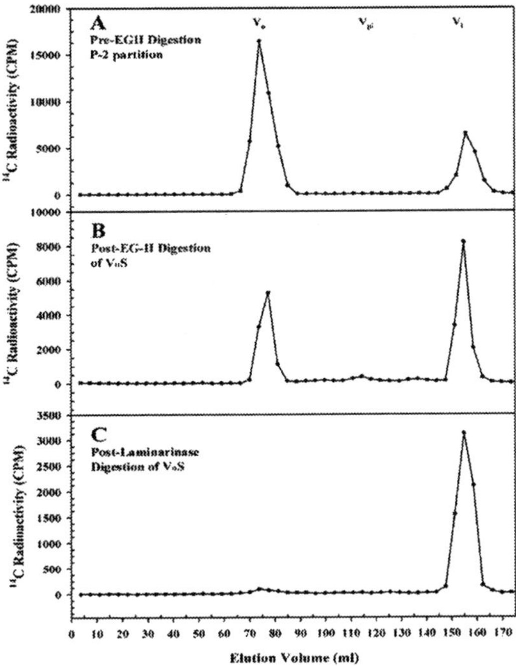

Fig. 1 The hydrolytic enzyme susceptibility assay analysis of the 1.5 mM UDP-glucose sample from the 0.5 - 3.0 mM UDP-glucose range. The hydrolytic enzyme susceptibility assay analysis of the 1.5 mM UDP-glucose sample is representative of all the samples in the 0.5-3.0 mM UDP-glucose range. (A) The pre-endoglucanse II digestion P-2 partition radioactivity profile of the 1.5 mM UDP-glucose sample indicated a large peak of radioactivity that corresponded to the void volume (V_0) and a smaller peak of radioactivity that corresponded to the included volume (V_i). Unlike the samples in the 0.010 - 0.200 mM UDP-glucose range, the partially included volume (V_{pi}) of the pre-endoglucanse II digestion P-2 partition of each sample in the 0.5-3.0 mM UDP-glucose range contained an insignificant amount of radioactivity with respect to the void and partially included volumes. (B) The post-endoglucanse II digestion P-2 radioactivity profile of the void volume sample showed that 62.73% of the radioactivity were susceptible to endoglucanse II activity and eluted in the partially included (10.79%) and the included (51.94%) volumes, whereas 37.27% of the radioactivity were resistant to endoglucanse II digestion and eluted in the void volume. Fractions corresponding to the void of the post-endoglucanse II digestion P-2 partition of void volume sample were pooled, dried down, and digested with laminarinase. (C) The post-laminarinase digestion P-2 radioactivity profile of the radioactivity that was resistant to endoglucanse II digestion showed that 96.67% of the radioactivity was susceptible to laminarinase digestion and eluted in the partially included (4.28%) and the included (92.39%) volumes, which would indicate the presence of 1,3-β-glucan.

partially included, and included volumes were pooled and dried down, respectively. Each sample was digested with 1,4-β-endoglucanase, which hydrolyzed the 1,4-β-glucan linkages between the anomeric carbon of an unsubstituted glucosyl residue and the 4-carbon of the adjacent glucosyl residue. Step 4: Endoglucanse II digested void and partially included volume samples were partitioned with a second P-2, and the elution profile of the endoglucanse II-digested products was determined with scintillation counting (Figure 1b and 1c). Step 5: Analysis of the void, partially included, and included volume samples with an HPLC equipped with an in-line scintillation counter was conducted to determine if any radio-labeled xyloglucan glucosyltransferase products had been substituted with xylosyl residues by xyloglucan xylosyltransferase activity. The amount of xyloglucan glucosyltransferase products was calculated from the amount of radioactivity that had shifted from one size-exclusion chromatography elution volume to another and from amount of radioactivity that was shown to be incorporated into xyloglucan oligosaccharides as determined by HPLC.

Practical application of hydrolytic enzyme susceptibility assay: The determination of an apparent K_m for xyloglucan glucosyltransferase activity, and the differentiation of xyloglucan glucosyltransferase, xyloglucan synthase, and callose synthase activities

The utility of the hydrolytic enzyme susceptibility assay was investigated with an attempt to calculate the apparent values of V_{max} and K_m of CHAPS-solubilized xyloglucan glucosyltransferase for UDP-glucose. The apparent values of V_{max} and K_m were calculated from non-linear regression analysis of Henri-Michaelis-Menten plots of hydrolytic enzyme susceptibility assay-derived catalytic activities for xyloglucan glucosyltransferase over a total substrate concentration range of 0.010 - 3.0 mM UDP-glucose. An initial survey of hydrolytic enzyme susceptibility assay derived values for xyloglucan glucosyltransferase catalytic activity versus UDP-glucose concentration over a 0.010 - 0.200 mM UDP-glucose range indicated first-order kinetics that prevented the accurate calculation of the apparent values for V_{max} and K_m of xyloglucan glucosyltransferase. However, non-linear regression analysis of Henri-Michaelis-Menten plots of the catalytic activity for xyloglucan glucosyltransferase activity over a 0.5 - 3.0 mM UDP-glucose range indicated curvilinear kinetics that permitted the calculation of an apparent V_{max} value of 26.06 nkatals and an apparent K_m value of 5.71 mM UDP-glucose (Figure 2). The apparent values of V_{max} and K_m for xyloglucan glucosyltransferase were calculated from the amount of radioactivity that had shifted from either the void or partially included volumes of the pre-endoglucanse II digestion P-2 to the partially included and included volumes of the post-endoglucanse II digestion P-2. However, there was a significant presence of unknown radio-labeled entities that remained in either the void or partially included volumes (For example see Figure 1b).

When Golgi microsomes are solubilized with CHAPS and partitioned with isopycnic centrifugation on a 20-50% sucrose gradient and fractionated, xyloglucan glucosyltransferase and xylosyltransferase

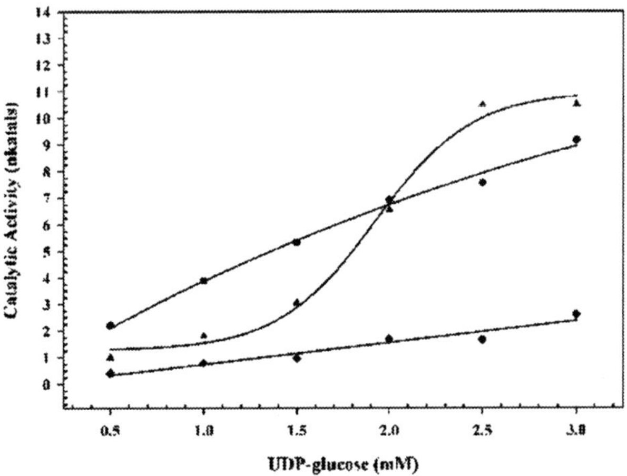

Fig. 2 A composite graph of the hydrolytic enzyme susceptibility assay-derived catalytic activities for xyloglucan glucosyltransferase (circles), xyloglucan synthase (diamonds), and callose synthase (triangles) as a function of UDP-glucose concentration. **The callose synthase displayed sigmoidal kinetics in the 0.5-3.0 UDP-glucose range that were indicative of sub-optimal assay conditions that contained EDTA and lacked Ca^{2+}, as previously published by Hayashi et al. (1987). Under the xyloglucan glucosyltransferase reaction conditions, the callose synthase became saturated at around 2.5 mM UDP-glucose, whereas the xyloglucan glucosyltransferase and xyloglucan synthase activities still displayed curvilinear and first-order kinetics, respectively.**

activities are found in the same fractions. Indeed, HPLC analysis of endoglucanse II resistant fractions and endoglucanse II resistant fractions digested with Driselase indicated that the presence of radiolabeled nascent xyloglucan oligosaccharides. While xyloglucan glucosyltransferase and xylosyltransferase activities have been solubilized from pea microsomes (Yang 1998; Faik et al. 2002; Marry et al. manuscript under revision), further steps to purify xyloglucan glucosyltransferase have languished because there is a significant loss of activity at subsequent purification steps. While there could be many reasons why there is a loss of xyloglucan glucosyltransferase activity as purification proceeds, we can speculate on some intriguing possibilities. First, xyloglucan glucosyltransferase may become inactive because it is partitioned away from either a required primer or the enzymes that synthesize the primer. Therefore, xyloglucan glucosyltransferase would either lack the primer or have a finite pool of primer substrates that would not be replenished by the primer biosynthetic enzymes. Second, it has been proposed that the xyloglucan biosynthesis is accomplished by a xyloglucan synthase complex that consists of xyloglucan glucosyltransferase and xylosyltransferase enzymes working in concert to synthesize the XXXG repeating unit of dicot xyloglucan (Hayashi and Matsuda 1981b). Perhaps the concomitant activities of xyloglucan glucosyltransferase and xylosyltransferase are required to produce a xyloglucan product that remains soluble *in vivo*. While unsubstituted 1,4-β-glucans show a significant decrease in solubility starting with cellohexaose, large xyloglucan polymers remain soluble. Conceivably, the removal of xylosyltransferase activity during xyloglucan glucosyltransferase purification may cause a significant decrease in xyloglucan glucosyltransferase processivity because the 1,4-β-glucan products would become essentially insoluble at a relatively low degree of polymerization. Therefore, the nascent radio-labeled xyloglucan products were probably produced by xyloglucan synthase complexes. The catalytic activities of xyloglucan synthase over the 0.010 - 0.200 and 0.5 - 3.0 mM UDP-glucose ranges were calculated from radio-labeled products that were either partially resistant or resistant to endoglucanse II digestion during hydrolytic enzyme susceptibility assay analysis. The Henri-Michaelis-Menten plot of xyloglucan synthase catalytic activity versus UDP-glucose concentration indicated first-order kinetics over both UDP-glucose concentration ranges (data not shown). However, it is premature to draw any conclusions regarding the apparent values for V_{max} and K_m for xyloglucan synthase activity because the experiments were designed to measure xyloglucan glucosyltransferase activity and not xyloglucan synthase activity. Indeed, the substrate requirements for the xyloglucan synthase complex also include UDP-xylose, which was kept at a constant concentration (15 µM) over the UDP-glucose concentration ranges used to measure xyloglucan glucosyltransferase activity with the hydrolytic enzyme susceptibility assay. Further studies that use hydrolytic enzyme susceptibility assay to investigate the interactions of UDP-glucose and UDP-xylose on xyloglucan synthase activity are warranted. Regardless, results indicated that the hydrolytic enzyme susceptibility assay is sensitive enough to differentiate between xyloglucan glucosyltransferase and xylosyltransferase activities.

As the concentration of UDP-glucose increased in the 0.5 to 3.0 mM concentration range, the amount of radioactivity that was resistant to endoglucanse II digestion increased. The endoglucanse II resistant material from the void and partially included volumes was pooled, dried down, and digested with laminarinase, which digests 1,3-β-glucans. The results from the post-laminarinase digestion P-2 size-exclusion chromatography partition indicated that the majority of the endoglucanse II-resistant radioactivity was susceptible to laminarinase digestion. hydrolytic enzyme susceptibility assay analysis of the samples in the 0.5 - 3.0 mM UDP-glucose range using a combination of endoglucanse II and laminarinase indicated that over 95% of the ^{14}C-glucose was incorporated into polysaccharides that were either 1,4-β-glucans and nascent xyloglucan or callose; furthermore, as the UDP-glucose concentration increased there was a decrease in the amount of 1,4-β-glucan and nascent xyloglucan, and an increase in the amount of callose. Non-linear regression analysis of a Henri-Michaelis-Menten plot of the catalytic activity for callose synthase over the 0.5 - 3.0 mM UDP-glucose range indicated sigmoidal kinetics (Figure 2), which was indicative of suboptimal callose synthase assay conditions that contained EDTA and lacked Ca^{2+} (Hayashi et al. 1987). Indeed, the callose was synthesized under assay conditions for xyloglucan glucosyltransferase that contained 1 mM EDTA (w/v) and lacked Ca^{2+}.

The hydrolytic enzyme susceptibility assay has advantages over other methodologies that have been traditionally used to study xyloglucan biosynthesis. An ethanol precipitation step is not needed to separate the unreacted UDP-[^{14}C]glucose from the reaction products of interest, so the potential problems of non-specific precipitation of products not related to xyloglucan biosynthesis and variable precipitation efficiency are avoided. The specificity of endoglucanse II for 1,4-β-glucans and the ability of chromatography to partition endoglucanase II susceptible moieties allows the quantification of xyloglucan glucosyltransferase products even if the enzyme preparation contains other synthases that have a UDP-glucose substrate requirement but produce different products. Finally, we propose that the hydrolytic enzyme susceptibility assay can be adapted to study synthase systems provided that there is a corresponding hydrolytic enzyme that is specific for the synthase products and that the hydrolytic enzyme susceptible products can be separated from hydrolytic enzyme resistant products with chromatography. Although the hydrolytic enzyme susceptibility assay was initially developed to measure xyloglucan glucosyltransferase enzymatic activity associated with CHAPS-solubilized Golgi microsomes, it was shown to be sensitive enough to differentiate between xyloglucan glucosyltransferase, xyloglucan synthase (a combination of xyloglucan glucosyltransferase and xyloglucan xylosyltransferase activities), and callose synthase activities.

The authors would like to congratulate Debbie Delmer on her long and productive career studying polysaccharide synthesis. She has always encouraged researchers to push the limits of their methods and to try new things. It is in that spirit that we embarked on this project to see how far we could go in determining apparent V_{max} and K_m for a complex mixture of enzyme activities that together produce oligosaccharides with mixed glycosyl linkages.

References

Brummell DA, Camirand A, MacLachlan GA (1990) Differential distribution of xyloglucan glycosyl transferases in pea Golgi dictyosomes and secretory vesicles. J Cell Biology 96: 705-710.

Camirand A, Brummell DA, MacLachlan GA (1987) Xyloglucan fucosyltransferase activity is localized in Golgi dictyosomes. Penn State Symposium in Plant Physiology

Camirand A, Maclachlan G (1986) Biosynthesis of the fucose-containing xyloglucan nonasaccharide by pea microsomal membranes. Plant Physiol 82: 379-383.

Dhugga KS, Barreiro R, Whitten B, Hazebroek J, Randhawa G, Dolan M, Kinney A, Tomes D, Nichols S, Anderson P (2004) Guar seed beta-mannan synthase is a member of the cellulose synthase super gene family. Science 16: 363-366.

Biosynthesis

Faik A, Bar-Peled M, DeRocher AE, Zeng W, Perrin RM, Wilkerson C, Raikhel NV, Keegstra K (2000) Biochemical characterization and molecular cloning of an alpha-1,2-fucosyltransferase that catalyzes the last step of cell wall xyloglucan biosynthesis in pea. J Biol Chem 275: 15082-15089.

Faik A, Chileshe C, Sterling J, Maclachlan G (1997) Xyloglucan galactosyl- and fucosyltransferase activities from pea epicotyl microsomes. Plant Physiol 114: 245-254.

Faik A, Price NJ, Raikhel NV, Keegstra K (2002) An Arabidopsis gene encoding an alpha-xylosyltransferase involved in xyloglucan biosynthesis. Proc Nati Acad Sci USA 99: 7797-7802.

Farkas V, Maclachlan G (1988) Fucosylation of exogenous xyloglucans by pea microsomal membranes. Archi Biochem Biophy 264: 48-53.

Hanna R, Brummell DA, Camirand A, Hensel A, Russell EF, Maclachlan GA (1991) Solubilization and properties of GDP-fucose: xyloglucan 1,2-alpha-L-fucosyltransferase from pea epicotyl membranes. Archi Biochem Biophys 290: 7-13.

Hayashi T, Kato Y, Matsuda K (1980) Xyloglucan from suspension-cultured soybean cells. Plant & Cell Physiol 21: 1405-1408.

Hayashi T, Kato Y, Matsuda K (1981) Biosynthesis of xyloglucan in suspension-cultured soybean cells. An assay method for xyloglucan xylosyltransferase and attempted synthesis of xyloglucan from UDP-D-xylose. J Biochem 89: 325-328.

Hayashi T, Koyama T, Matsuda K (1988) Formation of UDP-xylose and xyloglucan in soybean Golgi membranes. Plant Physiol 87: 341-345.

Hayashi T, Matsuda K (1981a) Biosynthesis of xyloglucan in suspension-cultured soybean cells. Evidence that the enzyme system of xyloglucan synthesis does not contain ß-1,4-glucan 4-ß-D-glucosyltransferase. Plant Cell Physiol 22: 1571-1584.

Hayashi T, Matsuda K (1981b) Biosynthesis of xyloglucan in suspension-cultured soybean cells. Occurrence and some properties of xyloglucan 4-beta-D-glucosyltransferase and 6-alpha-D-xylosyltransferase.J Biol Chem 256: 11117-11122.

Hayashi T, Matsuda K (1981c) Biosynthesis of xyloglucan in suspension-cultured soybean cells. Synthesis of xyloglucan from UDP-glucose and UDP-xylose in the cell -free system. Plant Cell Physiol 22: 517-523.

Hayashi T, Nakajima T, Matsuda K (1984) Biosynthesis of xyloglucan in suspension-cultured soybean cells. Processing of the oligosaccharide building blocks. Agric Biol Chem 48: 1023-1027.

Hayashi T, Polonenko DR, Camirand A, Maclachlan G (1986) Pea xyloglucan and cellulose. IV. Assembly of ß-glucans by pea protoplasts. Plant Physiol 82: 301-306.

Hayashi T, Read SM, Bussel M, Thelen M, Lin FC, Brown RMJ, Delmer DP (1987) UDP-glucose: (1'!3)-ß-glucan synthases from Mung bean and cotton. Plant Physiol 83: 1054-1062.

Liepman A, Wilkerson C, Keegstra K (2005) Expression of cellulose synthase-like (Csl) genes in insect cells reveals that CslA family members encode mannan synthases. Proc Natl Acad Sci USA 102: 2221-2226.

Madson M, Dunand C, Li XL, Vanzin GF, Caplan J, Shoue DA, Carpita N, Reiter WD (2003) The *mur3* gene of Arabidopsis encodes a xyloglucan galactosyltransferase that is evolutionarily related to animal exostosisn. Plant Cell 15: 1662-1670.

Perrin RM, DeRocher AE, Bar-Peled M, Zeng W, Norambuena L, Orellana A, Raikhel NV, Keegstra K (1999) Xyloglucan fucosyltransferase, an enzyme involved in plant cell wall biosynthesis. Science 284: 1976-1979.

Ray PM (1980) Cooperative action of beta-glucan synthetase and UDP-xylose xylosyl transferase of Golgi membranes in the synthesis of xyloglucan-like polysaccharide. Biochim et Biophys Acta 629: 431-444.

Richmond T, Somerville C (2001) Integrative approaches to determining *Csl* function. Plant Mol Biol 47: 131-143.

Walton JD, Ray PM (1982) Inhibition by light of growth and Golgi-localized glucan synthetase in maize mesocotyl. Planta 156: 302-308.

White AR, Xin Y, Pezeshk V (1993a) Separation of membranes from semiprotoplasts of suspension-cultured sycamore maple (*Acer pseudoplantanus*) cells. Physiologia Plantarum 87: 31-38.

White AR, Xin Y, Pezeshk V (1993b) Xyloglucan glucosyltransferase in Golgi membranes from Pisum sativum (pea).Biochem 294: 231-238.

Yang Z (1998) Partial purification and characterization of xyloglucan glucosyltransferase from *Pisum sativum* (pea) Golgi membranes. M.S. thesis. North Dakota State University, Fargo, ND.

Explorations on the Cell Walls of Grasses and Cereals

Bruce Stone

The grass family: a brief background

The grasses and cereals (Poaceae, Polaes) belong to the most highly evolved major group of monocotyledons, the commelinids. One feature that distinguishes the commelinids from non-commelinid monocotyledons, and from dicotyledons, is their primary wall polysaccharide composition. Glucuronoarabinoxylans, esterified with ferulic acid and to a lesser extent *p*-coumaric acid, are their abundant and characteristic polysaccharide components. Pectic polysaccharides, xyloglucans and glucomannans may also be present but usually as minor components. Uniquely among commelinids, walls of the graminoid, grasses and cereals, contain (1→3),(1→4)-β-glucans in low proportions in mature vegetative primary walls, but more abundantly in walls of endosperm cells, notably in those of barley and oats.

Of course we did not know all that when I first became interested in polysaccharides and polysaccharide hydrolases during my first research on cell walls.

Non-starchy endosperm polysaccharides and their occurrence in cell walls

Polysaccharide chemists of the 1950's used a 'lucky-dip' approach in isolating their target polymers from aqueous and alkaline extracts of homogenized grain. They assumed that the (1→3),(1→4)-β-glucans and arabinoxylans were located in the walls of the endosperm cells, but could they be sure? To understand the disposition of non-starch polysaccharides of the endosperm we isolated walls of barley and wheat in as near-native-state as possible, without loss of water-soluble components, using carefully-prepared endosperm flours as the starting material and 70% ethanol as the isolation medium. Wall preparations were fractionated by successive extraction, first with water, and then with a series of alkaline reagents. Depending on the cereal source, varying amounts of arabinoxylans were found in both water-extractable and water-unextractable, but 1M NaOH-extractable, fractions. Cellulose and glucomannans were present in the dilute alkali residues (Table 1). Significantly, there was little difference in the basic chemistry or size distribution of water-extractable and water-unextractable (1M NaOH-soluble) arabinoxylans from walls of wheat endosperm cells.

The underlying reasons for the differences in extractability of the non-cellulosic polysaccharides lie in the form of non-covalent and/or covalent associations between wall polymers.

Composition, organization and development of endosperm walls
Transmission electron microscopy of successively extracted walls revealed the organization of polysaccharides (Mares and Stone 1973a). Water-soluble and NaOH-soluble polymers, chiefly arabinoxylans, in walls from wheat endosperm, and arabinoxylans and (1→3),(1→4)-β-glucans in ryegrass (*Lolium multiflorum*) walls, overlaid the cellulose microfibrils, which were themselves associated with 4M KOH-soluble glucomannans.

The location of specific polysaccharides in endosperm walls was explored using specific probes for wall components. An FITC (RITC)-

Table 1 Composition of walls of cells of the starchy endosperm and aleurone of wheat and barley.

Cereal	Source of Walls	Major Polysaccharides
Barley	Aleurone	71% arabinoxylan 26% (1→3),(1→4)-β-glucan 2% cellulose 2% glucomannan
	Starchy Endosperm	75% (1→3),(1→4)-β-glucan 20% arabinoxylan 2% cellulose 2% glucomannan
Wheat	Aleurone	65% arabinoxylan 29% (1→3),(1→4)-β-glucan 2% glucomannan 2% cellulose
	Starchy Endosperm	70% arabinoxylan 20% (1→3),(1→4)-β-glucan 7% glucomannan 4% cellulose

labelled *Bacillus subtilis* (1→3),(1→4)-β-glucan hydrolase was used to follow the deposition of wheat (1→3),(1→4)-β-glucan in aleurone walls during their development and their dissolution during germination (Joyner 1985). Monoclonal antibodies to callose, (1→3),(1→4)-β-glucan, arabinoxylan and galactomannan proved potent in combination with gold-labelling in electron microscopy for locating polysaccharides in walls of pollen tubes and coconut endosperm as well as for their quantitation by ELISA assay.

Unsuccessful attempts to obtain explants of developing wheat led us to become interested in the timing and process of endosperm wall development. We followed the truly remarkable series of events leading to the cellularization of the multinucleate primary (3n) endosperm cell in the five or six days following anthesis (Mares et al. 1977). The periclinal walls of first-formed cells are deposited between nuclei lining the periphery of the endosperm mother cell without nuclear division or phragmoplast formation (Mares et al. 1977). This mode of wall initiation is quite unlike that in vegetative cells where the new wall is deposited centripetally at the cell plate situated in a phragmoplast between the daughter nuclei (Otegui et al. 2001). The endosperm cellularization process in cereals has now been described in detail by Brown et al. (1996). In the syncytial endosperm of rice, callose was shown to be a major wall component in the developing periclinal walls using the (1→3)-β-glucan specific monoclonal antibody (Brown et al. 1997). Extension of these observations to developing barley endosperm using three specific monoclonal antibodies has shown that the initially deposited callose in periclinal walls is replaced by (1→3),(1→4)-β-glucan, and later arabinoxylan and glucomannan are added to the wall (Doblin et al. 2004). Although the cytology of periclinal wall development in the nuclear endosperm is different from normal wall formation, a transient deposition of callose is common to both types (Otegui et al. 2001).

Cereal aleurone and its cell walls

The bilayered walls of aleurone cells are a barrier to the release of hydrolytic enzymes into the starchy endosperm during germination, and in the early stages the outer layer is preferentially degraded. The aleurone walls are composed chiefly of (1→3),(1→4)-β-glucan and arabinoxylan (Table 1). Due to the high content (1.8 %) of monomeric hydroxycinnamic acids esterified to the arabinoxylans, the aleurone walls are strongly autofluorescent (Rhodes and Stone 2002a). In contrast to the aleurone walls, the esterified hydroxycinnamic acids in the pericarp-seed coat fraction are partly dimerised and cross-link the arabinoxylans. Proteins comprise 1% of the aleurone wall and are of three types: glycine-rich (37-86 %), proline-rich (11-39 %) and serine-rich (up to 23 %) (Rhodes and Stone 2002b). Removal of arabinoxylan and (1→3),(1→4)-β-glucan using specific hydrolases leaves a residue enriched in protein (4.5 %) and highly-branched arabinoxylan that retains its UV-induced autofluorescence even after alkali treatment. On the basis of these observations we proposed that the ferulic acid esterified to the arabinoxylan is cross-linked to wall protein through ferulate-tyrosine bridges as suggested by Geissmann and Neukom (1973). Such covalent associations involving ester linkages had been implicated earlier by the observation that dilute alkalis, and more specifically, neutral hydroxamic acid treatment, liberates a substantial fraction of the arabinoxylan from water-extracted, wheat endosperm walls (Mares and Stone 1973b). A comprehensive analysis of these interactions still remains to be provided.

Arabinogalactan-peptides and -proteins in endosperm of grasses

Glycosidically-linked galactose is present in small amounts in walls of wheat endosperm walls and originates from the arabinogalactan portion of a proteoglycan belonging to the arabinogalactan-protein family (Fincher et al. 1983). The wheat proteoglycan co-extracts with, and could be separated from, the larger water-soluble arabinoxylan. The arabino-3,6-galactan chains are attached by alkali-stable galactosyl hydroxyproline residues to a short peptide. More recently the 18-residue peptide backbone from the wheat arabinogalactan-peptide was isolated and sequenced and shown to contain three hydroxyproline residues (Van den Bulck et al. 2004). Arabinogalactan-peptides are found in other cereal endosperms and in each case the peptide sequence is found at the N-terminal region of a protein precursor associated with grain softness. The peptides are thought to be processing products of grain softness protein synthesis (Van den Bulck et al. 2004). An arabinogalactan-protein was also located histochemically using the Yariv reagent in ryegrass endosperm, in suspension-cultured ryegrass endosperm cells and at the aleurone protoplast-cell wall interface and around the starch granules of mature barley endosperm (Anderson et al. 1977). The arabinogalactan-protein found on the barley aleurone protoplast surface has been proposed to provide a signal for α-amylase production in germination (Suzuki et al. 2002).

Biosynthesis of (1→3),(1→4)-β-glucans

The (1→3),(1→4)-β-glucans of cereals and grasses have structural features that are unique among wall polysaccharides from higher plants. The two linkage types are non-randomly, but yet not regularly, distributed in the linear chain; the (1→3)-linkages are always solitary, and the predominating cellotriose and cellotetraose sequences are independently distributed, but runs of (1→4)-linked glucose residues up to 7 or longer are present (Woodward et al. 1983). These unusual structural features raised the intriguing question of the mechanism of their biosynthesis.

We addressed their biosynthesis using microsomal membrane preparations which showed robust incorporation of [^{14}C]glucose from UDP-[^{14}C]glucose into ethanol-precipitable β-glucans. The reaction products were shown by specific enzymatic digestion to comprise not only (1→3),(1→4)-β-glucan, but also (1→4)-β-glucans and (1→3)-β-glucans (Smith and Stone 1973a). The yield of (1→3),(1→4)-β-glucan increased with increasing UDP-glucose concentration but was always lower than (1→3)-β-glucan (Smith and Stone 1973a; Henry and Stone 1982; Meikle et al. 1991) Significantly the ratio of cellotriose to cellotetraose in the (1→3),(1→4)-β-glucan product increased when the concentration of UDP-glucose in the reaction mixture increased from 10 μM to 1 mM (Henry and Stone, 1982). (1→3),(1→4)-β-Glucan and callose synthase activities showed differences in pH dependence and detergent extractability. The mechanism of formation of (1→3),(1→4)-β-glucan still awaits a definitive description although various proposals have been made (Stone 1984; Urbanowicz et al. 2004).

Although callose is not a usual cell wall component in ryegrass endosperm cells it was by far the most abundant reaction product. The original observation was made by Feingold et al. (1958) for *Phaseolus aureus* and subsequently for many other plant membrane systems (Clarke and Stone 1993). Product-entrapped preparations made microfibrillar callose and significantly, from a mechanistic viewpoint, chain extension appears to follow a repetitive pattern from the non-reducing end (Henry and Stone 1985). Carboxylic acid amino acids are essential for synthase activity as shown by the complete inhibition of callose synthase by carbodi-imide reagents (Bulone et al. 1998). The specificity of the activation of the callose synthase by β-glucosides was explored using a range of synthetic glycosides which showed that those with hydrophobic aglycons were the best activators (Ng et al. 1996). Attempts to covalently attach activating β-glucosides to the callose synthase using photactivatable analogues failed. However active callose synthase preparations were labeled with a novel ^{128}I-labeled UDP-glucose photoaffinity derivative (Meikle et al. 1991) and led to a identification of several UDP-glucose-labeled proteins in the range 30-31 and 54-58 kDa and also, in literal hindsight, at ~200 kDa. The finding by Kudlicka and Brown (1997) that the active callose synthase in mung beans had a short amino acid sequence also found in the 210 kDa (1→3)-β-glucan synthase (FSK) proteins of yeasts and fungi, led to a re-examination of the polypeptides in active ryegrass membrane fractions from sucrose gradients. A 200 kDa protein is present in the most active ryegrass (Wardak et al. 1999) and barley (Li et al. 2003) fractions and sequences of tryptic peptides from the barley protein matched the sequences in the protein encoded by the *fsk* gene (Li et al. 2003). A ryegrass cDNA encoding a protein with sequence homology to the *fks*-encoded proteins and other putative plant callose synthases was cloned by Andrew Jacobs (Wardak et al. 1999).

From an evolutionary point of view it was quite surprising that the (1→3)-β-glucan (curdlan) synthase from a Gram negative *Agrobacterium sp.* which we investigated in parallel (Karnezis et al. 2003) turned out to be a close relative of members of the cellulose synthase glucosyl transferase Family 2 from other bacteria and from embryophytes and quite unlike that found in the plant, yeast and fungal (1→3)-β-glucan synthases, classified in glucosyl transferase Family 45 (http://afmb.cnrs-mrs.fr/CAZY/).

Cell walls of grasses in ruminant and monogastric nutrition

It has long been apparent that the organization of the component polymers in walls of cells in leaves and stems of grasses, the predominant dietary components of herbivorous ruminants and non-ruminants, limits the access of polysaccharide degrading enzymes to their substrates in these walls. Unlignified walls of parenchymatous cells are readily digested, but the walls of fibres and tracheary elements whose wall polysaccharides are overlain by lignin are not. By following compositional changes of walls in developing internodes of wheat and the parallel changes in their *in vitro* digestibility it was apparent that lignin accumulation and the decrease in digestibility did not follow one another as expected. There is, however, a clear relationship between accumulation of wall-bound hydroxycinnamic acids and the decrease in digestibility (Lam et al. 1990). Detailed examination of the forms of wall-bound hydroxycinnamic acids, and in particular, ferulic acid, led to the discovery that some of the ferulic acid (and dehydrodiferulic acid) esterified to wall polysaccharides (arabinoxylan) was also etherified (to lignin) forming ester-ether bridges between the two wall polymers (Iiyama et al. 1990). A correlation between ester-ether bridge content, but not lignin content, and digestibility was shown for internodes of the pasture grass (*Phalaris aquatica*) (Lam et al. 2003).

Acknowledgements
It is a pleasure to acknowledge the central roles played by my mentor, Eric Crook, at University College London, postgraduate students, postdoctoral fellows and colleagues at the University of Melbourne and La Trobe University who took part in the work described. The account would not be complete without recognizing the stimulus given to many aspects of the research by my interactions with international cell wall colleagues, not the least by *my very good friend Debby Delmer* to whom this volume is dedicated.

References

Anderson RL, Stone BA (1978) Studies on the *Lolium multiflorum* endosperm in tissue culture III. Structure of the cell walls. Aust J Biol Sci 31: 573-578.

Brown RC, Lemmon BE, Olsen OA (1996) Development of the endosperm in rice (*Oryza sativa* L.): cellularization. J Plant Res 109: 301-313.

Brown RC, Lemmon BE, Stone BA, Olsen OA (1997) β-Glucans during early grain development in rice (*Oryza sativa*). Planta 202: 414-426.

Bulone V, Fincher GB, Stone BA (1995) β-Glucan by a ryegrass (*Lolium multiflorum*) endosperm (1→3)-β-glucan synthase enriched by product entrapment. Plant J 8: 213-225.

Bulone V, Lam BT, Stone BA (1998) β-Glucan synthase from Italian ryegrass (*Lolium multiflorum*) endosperm. Phytochem 50: 9-15.

Doblin M, Wilson S, Burton R, Shirley N, Stone B, Fincher G, Newbegin E, Bacic T (2004) Immunocytochemical analysis of early endosperm development in *Hordeum vulgare* (barley). In Xth Cell Wall Meeting, Sorrento, Italy, pp 178.

Feingold DS, Neufeld EF, Hassid WZ (1958) β-1,3-Linked glucan by extracts of *Phaseolus aureus* seedlings. J Biol Chem 233: 783-788.

Fincher GB, Stone BA, Clarke AE (1983) Arabinogalactan-proteins: structure, biosynthesis, and function. Ann Rev Plant Physiol 34: 47-70.

Geissmann T, Neukom H (1973) Ferulic acid as a constituent of water-insoluble pentosans of wheat flour. Cereal Chem 50: 414-416.

Henry RJ, Stone BA (1982) β-Glucan synthesis by particulate enzymes from suspension-cultured *Lolium multiflorum* endosperm cells. Plant Physiol 69: 632-636.

Henry RJ, Stone BA (1985) β-glucan chain elongation by ryegrass (*Lolium multiflorum*) enzymes. Carbohydr Polym 5: 1-12.

Iiyama K, Lam, TB-T, Stone BA (1990) Phenolic acid bridges between polysaccharides and lignin in wheat internodes. Phytochem 29: 733-737.

Joyner SJ (1985) The histochemistry and ultrastructure of wheat aleurone cell wall. Masters Thesis. La Trobe University, Melbourne

Karnezis T, Epa VC, Stone BA, Stanisich VA (2003) Topological characterization of an inner membrane (1→3)-β-D-glucan (curdlan) synthase from *Agrobacterium sp.* strain ATCC31749. Glycobiology 13: 693-706.

Kudlicka K, Brown RM (1997) Cellulose and callose biosynthesis in higher plants 1. Solubilization and separation of (1→3)-and (1→4)-β-glucan synthase activities from mung bean. Plant Physiol 115: 643-656.

Lam TB-T, Iiyama K, Stone BA (1990) Distribution of free and combined phenolic acids in wheat internodes. Phytochem 29: 429-433.

Lam TBT, Iiyama K, Stone BA (2003) *Phalaris aquatica* and *Lolium perenne* and their relation to *in vitro* wall digestibility. Phytochemistry 64: 603-607.

Li J, Burton RA, Harvey AJ, Hrmova M, Wardak AZ, Stone BA, Fincher GB (2003) Biochemical evidence linking a putative callose synthase gene with (1→3)-β-D-glucan biosynthesis in barley. Plant Mol Biol 53: 213-225.

Mares DJ, Stone BA (1973a) Wheat endosperm. I. Chemical composition and ultrastructure of the cell walls. Aust J Biol Sci 26: 793-812.

Mares DJ, Stone BA (1973b) Wheat endosperm. II. Properties of the wall components and studies on their organization in the wall. Aust J Biol Sci 26: 813-830.

Mares DJ, Stone BA, Jeffrey C, Norstog K (1977) Early stages in the development of wheat endosperm. II. Ultrastructural observations on cell wall formation. Aust J Bot 25: 599-613.

Meikle PJ, Ng KF, Johnson E, Hoogenraad NJ, Stone BA (1991) β-glucan synthase from *Lolium multiflorum*. Detergent solubilization, purification using monoclonal antibodies, and photoaffinity labeling with a novel photoreactive pyrimidine analog of uridine 5'-diphosphoglucose. J Biol Chem 266: 22569-22581.

Ng K, Johnson E, Stone BA (1996) β-Glucoside activators of ryegrass (1→3)-β-glucan synthase and the synthesis of some potential photoaffinity activators. Plant Physiol 111: 1227-1231.

Otegui MS, Mastronarde DN, Kang BH, Bednarek SY, Staehelin LA (2001) Three-dimensional analysis of syncytial-type cell plates during endosperm cellularization visualized by high resolution electron tomography. Plant Cell 13: 2033-2051.

Rhodes DI, Sadek M, Stone BA (2002) Hydroxycinnamic acids in walls of wheat aleurone cells. J Cereal Sci 36: 67-81.

Rhodes DI, Stone BA (2002) Proteins in walls of wheat aleurone cells. J Cereal Sci 36: 83-101.

Smith MM, Stone BA (1973a) β-Glucan synthesis by cell-free extracts from *Lolium multiflorum* endosperm. Biochim Biophys Acta 313: 72-94.

Smith MM, Stone BA (1973b) Chemical composition of the cell walls of *Lolium multiflorum* endosperm. Phytochem 12: 1361-1367.

Stone BA (1984) β-glucans in cell walls. Struct., Funct., Biosynth. Plant Cell Walls, Proc Annu Symp Bot, 7th: 52-74.

Suzuki Y, Kitagawa M, Knox JP, Ymaguchi I (2002) α-Amylase production in barley aleurone cells. Plant J 29: 733-741.

Urbanowicz BR, Rayon C, Carpita NC (2004) Topology of the maize mixed linkage β-D-glucan synthase at the Golgi membrane. Plant Physiol 134: 758-768.

Van den Bulck K, Swennen K, Loosveld A-MA, Courtin CM, Brijs K, Proost P, Van Damme J, Van Campenhout S, Mort A, Delcour JA (2004) Isolation of cereal arabinogalactan-peptides and structural comparison of their carbohydrate and peptide moeities. J Cereal Sci 41: 59-67.

Wardak AZ, Jacobs AK, Anderson MA, Fincher GB, Stone BA (1999) The glucan synthase from suspension-cultured *Lolium multiflorum* (rye grass) endosperm. *In* Proceedings ASBMB Combio99, Gold Coast, Queensland.

Woodward JR, Fincher GB, Stone BA (1983) β-D-Glucans from barley (*Hordeum vulgare*) endosperm. II. Fine structure. Carbohydr Polym 3: 207-225.

Secondary Wall Formation in an *In Vitro* System

Maureen C. McCann

My first experience of research was as an undergraduate in the laboratory of Professor Don Northcote at Cambridge University, UK. I didn't appreciate at the time what a huge impact this would have on my scientific career! Don, a kindly mentor, was interested in the biosynthesis of plant cell walls. I became fascinated both with how these complex macromolecular structures allow cells to grow while protecting the cell within a rigid box, and also with how cell walls become specialized for their functions within the plant. Plant cells are surrounded by primary cell walls that are able to expand, as new material is deposited, to accommodate cell growth (Carpita and McCann 2000). Vascular cells, including phloem fibers, xylem and transfer cells, elaborate within their primary walls, a secondary wall, as part of the process of cell differentiation, providing mechanical support and transport pathways for solutes and water. The construction of these specialized secondary walls is not well understood, either in terms of their composition or the patterning of wall deposition. Don's lab was using a truly remarkable model system, the zinnia mesophyll cell system, to study the biosynthesis of secondary walls during formation of xylem cells (Fukuda 1996).

In higher plants, some cells become specialized as cells that are capable of conducting a column of water from the roots to hundreds of feet in the air for some trees (Fukuda 1992, McCann 1997a). This is a critical feature of land plants that delivers water to every living cell. The cells in a file must divide and elongate, before patterned deposition of cell wall material is made to reinforce the cell against the compressive forces of the surrounding tissues created by the suction forces of transpiration. This material is stiffened and waterproofed by deposition of phenolic compounds. Finally, the end walls of the cell are broken down and the cell contents are destroyed. The resulting hollow water-conducting tubes are called xylem vessels or tracheids, and the individual cells that form them are called vessel or tracheary elements. In intact plants, only a few cells at any time are actively engaged in the process of forming xylem cells. However, this entire developmental pathway can be reprised in the zinnia mesophyll cell system (Fukuda and Komamine 1980).

In the mid-1970s, a paper from a German group had noted an astonishing phenomenon: photosynthetic cells that were mechanically isolated from the leaves of zinnia, an ornamental garden plant, could be induced to change their cell fate to that of tracheary elements (Kohlenbach and Schmidt 1975). They had taken pairs of young leaves and gently ground them in a mortar and pestle to release isolated single mesophyll and palisade cells from inside the leaves. The cells of the epidermis and the vasculature remained attached to themselves, and could easily be removed by filtering through muslin. Then, Kohlenbach and Schmidt (1975) put the isolated mesophyll and palisade cells in liquid medium together with two plant growth factors, auxin and cytokinin. Over a week, some of the cells changed their fate and acquired the cytological features of tracheary elements. It was as if a liver cell had been induced to become a heart cell or a kidney cell! However, it was Fukuda and Komamine (1980) that really refined the culture conditions to improve both the final numbers of tracheary elements obtained and also the synchrony of differentiation, such that the zinnia mesophyll cell system became a suitable model system for biochemical and molecular approaches. In 4 days, half of the cell population that were already specialized as photosynthetic cells in the leaf could be induced to form tracheary elements (Fukuda and Komamine 1980). The zinnia system is unique among plant systems for two reasons. First, the entry into a new developmental pathway is induced by adding plant growth factors that act like a molecular switch to turn on the process of *trans*-differentiation. Second, the cells synchronously undergo *trans*-differentiation, making it possible to precisely stage the events involved in building a tracheary element (Milioni et al. 2002). In addition, the zinnia system is amenable to studies involving the addition of inhibitors or other substances to the culture medium in order to perturb the system.

Don recommended that I apply for a PhD studentship with his former PhD student, Professor Keith Roberts, at the John Innes Centre, Norwich UK, a government-funded institute in plant and microbial sciences. My studentship work was to look directly at the architecture of cell walls using electron microscopy, but I still had fond memories of the magical zinnia system! After a couple of years of post-doctoral work, also at John Innes Centre, I suggested to Keith that we re-establish the zinnia system in the lab as an interesting model system to look at events involved in building secondary walls.

The time courses for differentiation are somewhat variable in the various laboratories using the zinnia system, perhaps as a consequence of the different cultivars and cell isolation procedures adopted. Therefore, I felt it was important to define the sequential cellular events with respect to markers related to developmental state rather than simply the time in culture (Stacey et al. 1995). Also, although differentiation is semi-synchronous in the zinnia system, other processes, such as wound response, cell division and cell elongation are occurring, and the timing of tracheary element formation may vary considerably among different cell preparations. Given that building a secondary wall is a key event in the differentiation of a tracheary element, and because the expectation is for a large number of cell-wall-related genes to be involved, the synchrony of the system is critical. We began to investigate the precise times in the differentiation time-course at which the auxin and cytokinin were required in the culture medium. If the auxin and cytokinin were added to the culture 48 h after cell isolation, we observed an enhanced level of differentiation (60 to 80% of living cells), with the culture apparently more synchronous than if the cells are cultured continuously in inductive medium (about 50% of living cells) (Stacey et al. 1995). Thus, we had a time-course where the majority of cells would have trans-differentiated within 48 h after addition of the inductive signals (Figure 1).

By differential screening methods, other labs had identified cDNA clones of transcripts that are up-regulated in inductive medium at different stages of the developmental pathway to tracheary element fate (Demura and Fukuda 1993, Ye and Varner 1993). This approach had resulted in the identification of about 30 differentiation-related gene sequences that were extremely useful molecular markers (Fukuda 1996, McCann 1997a). However, only a few of these were seemingly related to secondary wall formation. We decided to use a similar approach to identify transcripts involved in secondary wall formation using our modified time-course (McCann 1997a). The very first gene that we cloned encoded a pectate lyase, a potent cell wall-degrading enzyme (Domingo et al. 1998). Expression of the pectate lyase gene in *E. coli* confirmed that the protein had the correct enzyme activity, and *in situ* hybridization showed that the gene was transcribed in cells associated with vascularization in the zinnia plant. At first, we were surprised – pectate lyases were known to be highly destructive enzymes secreted by plant pathogens, the major cause of potato soft rot when the crop is attacked by *Erwinia* (Bartling et al. 1995). However, we found other reports of pectate lyases in plants, a pollen allergen from Japanese cedar (Taniguchi et al. 1995) and a pectate lyase gene expressed during banana fruit ripening (Dominguez-Puigjaner et al. 1997). The zinnia recombinant enzyme had a very alkaline pH optimum – perhaps it had very low levels of activity when secreted into the acid environment of the plant cell wall. Its function in xylem differentiation remains mysterious. However, the experience had whetted my appetite for using the zinnia system to uncover cell wall-related genes.

The improved synchrony of the time-course of xylogenesis in the zinnia system raised the possibility of identifying very early events in the process, the signals that initiate the process of secondary wall

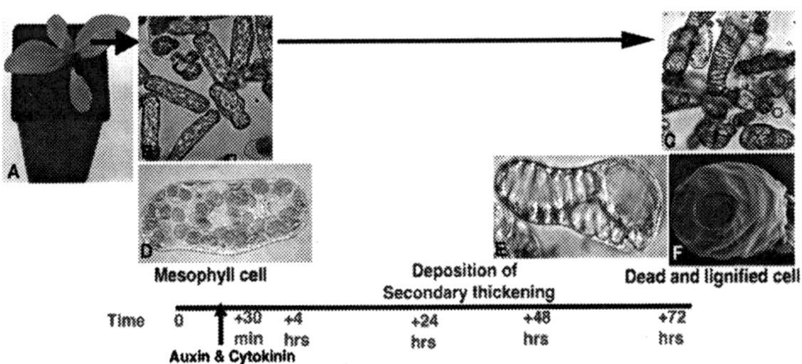

Fig. 1 (A) Leaves of a zinnia plant are gently ground in a mortar and pestle to release single isolated mesophyll and palisade cells (B). **Forty-eight hours after auxin and cytokinin are added to the culture medium, 60 to 80% of cells have trans-differentiated to form tracheary elements (C). (D) A single palisade cell 48 h after cell isolation prior to addition of auxin and cytokinin. (E) Between 24 and 48 h after addition of auxin and cytokinin, characteristic secondary wall patterns are deposited. (F) Scanning electron micrograph of a fully differentiated tracheary element, at 72 h after addition of auxin and cytokinin, showing the characteristic hole pierced in one end of the cell.**

deposition as well as all of the biosynthetic and hydrolytic enzymes involved. In short, we required a broad-based screen. The approach that we finally decided upon was an RNA fingerprinting technique, called complementary DNA-amplified fragment length polymorphism (cDNA-AFLP). Keith Roberts and I pooled our resources to recruit an extremely talented post-doctoral molecular biologist, Dr Dimitra Milioni, to head this effort. We selected 5 different time-points during formation of tracheary elements in the zinnia system (Milioni et al. 2001, 2002) reflecting the transcript populations made when auxin and cytokinin were first added to the culture medium (0 h), early changes in gene expression (30 min) and immediate downstream events (4 h), the time at which secondary wall synthesis begins (24 h), and then the time at which Tracheary elements begin to deposit lignin and autolyse their cellular contents (48 h) after the addition of the inductive signals, auxin and cytokinin. cDNAs are synthesized from mRNA populations isolated from the zinnia cultures at each of the five time-points, digested with a pair of restriction enzymes, adaptor-ligated and amplified by PCR to produce the primary templates. A subset of this population of fragments was selectively amplified using degenerate primers with two selective nucleotides, and then analyzed on polyacrylamide gels. We selected over 600 genes whose transcription showed overt changes in abundance over time. We then obtained partial sequences that were compared with public databases that allowed us to assign an identity to about one-half of the predicted gene products, including about 10% that encode cell wall biosynthetic enzymes, hydrolytic enzymes or structural proteins. However, there were some notable absences – in particular, although we had one sequence encoding a secondary wall cellulose synthase (Milioni et al. 2001), sequences similar to the other two cellulose synthase polypeptides thought to be components of a heteromeric complex (Taylor et al. 2003) were absent – confirming that we do not yet have a comprehensive list of secondary wall-related genes. The approach was validated by the detection of zinnia sequences with good identity to genes previously implicated in vascular development (Milioni et al. 2002). Also, RNA *in situ* localization confirmed that some of the genes of unknown function were expressed in vascular cells in zinnia stems (Milioni et al. 2002). By using one pair of restriction enzymes (*ApoI* and *MseI*) and all 512 possible primer combinations, we estimated that we had sampled two-thirds of the original mRNA populations - we inspected the sequence of 47 zinnia mRNAs and found that 66% contained suitable *ApoI-MseI* fragments between 50 and 450 Kb. From this, we estimate that about 1000 genes could be involved in building a tracheary element.

Completion of the genome sequence of Arabidopsis revealed that many cell wall-related genes are members of large multi-gene families (Arabidopsis Genome Initiative 2000). The autonomy that plant cells display during specified developmental programs generating specific patterns of temporal and spatial expression may account for some of the multiplicity, but in some instances several genes of the same family are co-expressed (Im et al. 2000). Some may serve a specific channeling function to direct substrates to distinct metabolic pathways (Seifert et al. 2003), whereas others may comprise heteromeric complexes that together form a functional unit (Turner et al. 2001), and still others may have completely different enzymic functions or substrate specificities despite their gene sequence similarities. In particular, the identification of secondary wall-specific cellulose synthase genes among a 10-member gene family indicates that the construction of these highly specialized cell walls requires the coordinate expression of specific gene family members. In our cDNA-AFLP screen, we identified 63 partial sequences with similarity to 50 cell wall-related genes, some sequence fragments having derived from the same transcript. These now form a set of candidate genes involved specifically in secondary wall formation.

However, most interesting is the possibility of identifying novel cell wall-related genes among the subset of sequences currently annotated as similar to sequences of unknown function or hypothetical proteins. After all, if 10% of the annotated genes are similar to cell-wall-related genes then perhaps 10% of the unknowns, about 30 sequences, are cell wall-related. But how are they to be identified?

A major difficulty has been to establish methods of stable transformation for zinnia with which to examine gene function. Two transient expression assays can be used, one using transformation of isolated mesophyll cells by *Agrobacterium* before inducing Tracheary elements formation (Ito and Fukuda 2002) and the other by transforming callus cells and then inducing root formation (Igarashi et al. 1998). Another option is to use the power of Arabidopsis molecular genetics to address function. Sequence similarities with Arabidopsis sequences (Arabidopsis Genome Initiative 2000) for almost 70% of the sequences from the cDNA-AFLP screen encourage the use of transgenic approaches and insertion mutants from collections such as SAIL and SALK T-DNA insertion collection libraries (available from the Arabidopsis Biological Resource Center) for potential homologues. Seventeen of the partial cDNA sequences from the cDNA-AFLP screen had high sequence similarities to Arabidopsis hypothetical proteins or proteins of unknown function that have signal peptides – putting their gene products somewhere in the secretory pathway out to the wall. The Arabidopsis genes also belonged to large gene families, with predicted signal peptide sequences. If we have indeed identified novel cell wall-related genes through the zinnia system, it is likely that other gene family members will carry out their functions in primary walls. As we assemble mutants in each of the members of these gene families, we are beginning to look for cell wall phenotypes specifically in xylem cells for the putative homologues of each of these 17 genes, and then extending this analysis to primary walls of mutants in other gene family members. Another article in this book describes our use of infrared spectroscopy as a tool to probe for altered cell wall composition and architectures very rapidly in Arabidopsis mutants (Carpita and McCann, this volume) and, with the addition of an infrared microscope, in specific cell types (McCann et al. 1997b).

Debby Delmer made her keynote discovery of the sequence of cellulose synthase in the differentiating cotton fiber system, realizing the immense value in choosing an appropriate system in which to tackle the biological question of interest. While Arabidopsis is a superb genetic system, my hope is that the zinnia system will also prove to be an engine of gene discovery for secondary wall formation. To end on a personal note, Debby is one of the smartest and nicest people in science – I can't imagine a better role model to follow.

Acknowledgments

Many thanks to all of the good colleagues that have worked with me on the zinnia project – post-doctorals Dimitra Milioni, Preeti Dahiya, Concha Domingo and Daniel Fulton, PhD students Maria Mourelatou, Pierre-Etienne Sado, Guy Wheeler, technician Nicola Stacey – and collaborators John Doonan, Nick Carpita and Clive Lloyd. Most important of all, many thanks to my favorite colleague and mentor, Keith Roberts for his unfailing generosity, support and good advice.

References

Arabidopsis Genome Initiative (2000) Analysis of the genome sequence of the flowering plant *Arabidopsis thaliana*. Nature 408: 796-815.

Bartling S, Wegener C, Olsen O (1995) Synergism between *Erwinia* pectate lyase isoenzymes that depolymerize both pectate and pectin. Microbiol 141: 873-881.

Carpita NC, McCann MC (2000) Chapter 2: The cell wall.In: Biochemistry and Molecular Biology of Plants (Eds. BB Buchanan, W Gruissem, R Jones), American Society Plant Physiologists, Rockville, MD, pp 52-109.

Demura T, Fukuda H (1994) Novel vascular cell-specific genes whose expression is regulated temporally and spatially during vascular system development. Plant Cell 6: 967-981.

Domingo C, Roberts K, Stacey NJ, Connerton I, Ruíz-Teran F, McCann MC (1998) A pectate lyase from *Zinnia elegans* is auxin inducible. Plant J 13: 17-28.

Dominguez-Puigjaner E, Llop I, Vendrell M, Prat S (1997) A cDNA clone highly expressed in ripe banana fruit shows homology to pectate lyases. Plant Physiol 114: 1071-1076.

Fukuda H, Komamine A (1980) Establishment of an experimental system for the tracheary element differentiation from single cells isolated from the mesophyll of *Zinnia elegans*. Plant Physiol 52: 57-60.

Fukuda H (1992) Tracheary element formation as a model system of cell differentiation. Int Rev Cytol 136: 289-332.

Fukuda H (1996) Xylogenesis: initation, progression and cell death. Ann Rev Plant Physiol Plant Mol Biol 47: 299-325.

Igarashi M, Demura T, Fukuda H (1998) Expression of the *Zinnia TED3* promoter in developing tracheary elements of transgenic *Arabidopsis*. Plant Mol Biol 36: 917-927.

Im KH, Cosgrove DT, Jones AM (2000) Subcellular localization of expansin mRNA in xylem cells. Plant Physiol 123: 463-470.

Ito J, Fukuda H (2002) ZEN1 is a key enzyme in the degradation of nuclear DNA during programmed cell death of tracheary elements. Plant Cell 14: 3201-3211.

Kohlenbach HW, Schmidt B (1975) Cytodifferenzierung in Form einer direkten Umwandung isolierter Mesophyll-Zellen zu Tracheiden. Z Pflanzenphysiol 75: 369-374.

McCann MC (1997a) Tracheary element formation: building up to a dead end. Trends Plant Sci 2: 333-338.

McCann MC, Chen L, Roberts K, Kemsley EK, Sene C, Carpita NC, Stacey NJ, Wilson RH (1997b) Infrared microspectroscopy: sampling heterogeneity in plant cell wall composition and architecture. Physiol Plant 100: 729-738.

Milioni D, Sado PE, Stacey NJ, Domingo C, Roberts K, McCann MC (2001) Differential expression of cell-wall-related genes during the formation of tracheary elements in the Zinnia mesophyll cell system. Plant Mol Biol 47: 221-238.

Milioni D, Sado P, Stacey NJ, Roberts K, McCann MC (2002) Early gene expression associated with the commitment and differentiation of a plant tracheary element is revealed by cDNA-Amplified Fragment Length Polymorphism analysis. Plant Cell 14: 2813-2824.

Seifert GJ, Barber C, Wells B, Dolan L, Roberts K (2002) Galactose biosynthesis in Arabidopsis: Genetic evidence for substrate channeling from UDP-D-galactose into cell wall polymers. Curr Biol 12: 1840-1845.

Stacey NJ, Roberts K, Carpita NC, Wells B, McCann MC (1995) Dynamic changes in cell surface molecules are very early events in the differentiation of mesophyll cells from *Zinnia elegans* into tracheary elements. Plant J 8: 891-906.

Taniguchi Y, Ono A., Sawatani M, Nanba M, Kohno K, Usui M, Kurimoto M, Matuhasi T (1995) Cry j I, a major allergen of Japanese cedar pollen, has pectate lyase enzyme activity. Allergy 50: 90-93.

A Brief History of the XTH Gene Family

Kazuhiko Nishitani

Prehistory

It is possible to trace the history of studies on cell walls back to the time of Nehemiah Grew, who dissected plant organs and tissues, under both the naked eye and a microscope, and exhibited his observations to the Royal Society of London from May 15, 1672 to April 2, 1674 (Grew 1682). He described plant organs as being composed of cells in the same way that a tower is constructed with various types of bricks. Furthermore, he disclosed that individual cells are characterized by their respective structures, which are now termed cell walls.

Three hundred years passed before Peter Albersheim and his colleagues determined the molecular structure of the walls of suspension-cultured sycamore cells by enzymatic fragmentation, followed by a methylation analysis, and advanced the structural model for the plant cell wall (Keegstra et al. 1973). This model envisaged xyloglucan molecules as strongly attached to cellulose microfibrils by hydrogen bonds, some of which are attached through certain polymers to other cellulose microfibrils. Consequently, cellulose microfibrils are effectively cross-linked through matrix polysaccharides to form a super molecular framework. The biological implication of this model is that cross-links between cellulose microfibrils serve as load-bearing linkages of the cell wall, and that breakage and reconnection of the cross-links are required for the cell wall to loosen and expand during cell growth and differentiation. Whereas new findings challenged the details of the original structural model, its essence has withstood the test of time. The current view of the cell wall structure largely stems from the original model (Hayashi 1989; Carpita 1993; Cosgrove 2000).

The debut of xyloglucan

The first biochemical evidence suggesting a role for xyloglucan in cell wall expansion was obtained by Peter Ray and his colleagues. They showed that the metabolic turnover of xyloglucan, as studied by pulse-chase experiments using ^{14}C-labeled glucose, was enhanced during auxin- and acid-induced cell wall expansion in pea epicotyl sections (Labavitch and Ray 1974). Their study was followed by the finding that molecular weight changes in xyloglucan decreased when auxin and acidic pH caused cell extension growth in epicotyl sections of the azuki bean (*Vigna angularis*) (Nishitani and Masuda 1981, 1982 and 1983). Similar molecular-weight changes in xyloglucans were observed in various plant species, including monocotyledonous plants and gymnosperms (Reviewed in Nishitani 1997). These observations provide strong evidence for the hypothesis that splitting the load-bearing xyloglucan molecules by hydrolase is the key step controlling stress relaxation of the cell wall.

However, cleavage of load-bearing linkages between cellulose microfibrils cannot account for the process of cell wall changes through which newly synthesized cellulose microfibrils are integrated into a preexisting cellulose xyloglucan framework (Nishitani 1997). Without the deposition of the cell wall, the wall would not maintain thickness and would lose the mechanical strength to resist turgor pressure. To explain this paradox, Albersheim (1976) postulated the involvement of "an endotransglycosylase that transfers a portion of a polysaccharide to itself" in the restructuring process of plant cell walls. The existence of this hypothetical enzyme activity was later demonstrated by three research groups, those of Reid, Fry and our own work, with each group aimed at a different goal (Reviewed in Nishitani 1997). Finally, the protein responsible for the endotransglycosylation between xyloglucan molecules was isolated from the cell wall space of the azuki bean (Nishitani and Tominaga 1992). This protein is currently termed xyloglucan endotrans-glucosylase/hydrolase (XTH). In this chapter, I deal with our studies on XTH and the perspective of the multigene family that encodes XTH proteins.

Discovery of the XTH family

We noticed the first sign of the existence of endotransglycosylation activity in the cell wall during the research we undertook to explore hydrolase, but not transferase, acting on xyloglucans. In an attempt to detect enzyme activity to hydrolyze xyloglucans in the cell wall, we prepared protein fractions from an apoplastic solution of azuki bean epicotyls and assayed its activity to degrade xyloglucan molecules with defined molecular weight distributions. Contrary to our prediction, the reaction generated xyloglucan components with molecular weights both higher and lower than that of the initial substrate (Nishitani and Tominaga 1991). This surprising result was immediately explained as being the action of an endo-type transglycosylase that mediated the transfer of a large segment of a xyloglucan molecule to another xyloglucan polymer. To verify our hypothetical scheme, we synthesized a fluorescent-labeled xyloglucan oligosaccharide, which we used as an acceptor substrate to demonstrate the molecular grafting reaction of the apoplastic enzyme (Nishitani 1992). With the aid of the enzyme assay system, we finally succeeded in purifying the protein from the cell wall and characterized its mode of enzyme action. We designated it endoxyloglucan transferase (EXT or EXGT) based on its unique mode of enzyme action (Nishitani and Tominaga 1992). Identification of the enzyme protein was followed by molecular cloning of the cDNA that encoded it. This was undertaken using several plant species, which showed that the EXGT protein is ubiquitous, at least among angiosperms (Okazawa et al. 1993).

However, Fry's group independently detected transglycosylation activity in plant extracts using a radioactive xyloglucan oligomer as an acceptor substrate, and referred to it as xyloglucan endotransglycosylase (XET) activity (Fry et al. 1992). The transglycosylation activity was also detected by other research groups (Farkas et al. 1992, Fanutti et al. 1993). The use of two different names for the same class of enzymes created confusion for the nomenclature of the gene family. A decade later, this confusion was resolved with the introduction of a unified nomenclature (Rose et al. 2002).

Molecular cloning of genes encoding the EXGT proteins from five plant species revealed that they are homologous to *Tropaeolum majus* endo-1,4-β-glucanase (de Silva et al. 1993) and other proteins with unknown protein functions, such as *Arabidopsis* Meri-5 (Medford et al. 1991), the *Glycine max* BRU1 protein (Zurek and Clouse 1994) and the *Zea mays* wusl 1005 gene product (Peschke and Sachs 1994). These findings indicate that EXGT is a member of a gene family that encodes a class of enzymes capable of mediating the hydrolysis of a xyloglucan molecule and/or molecular grafting between xyloglucan molecules (Nishitani 1997). The most notable point is that most of the reactions required for the formation and rearrangement of the cellulose xyloglucan framework are easily explained by the actions of this class of enzymes alone. These reactions include: (1) simple cleavage of load-bearing xyloglucans by hydrolytic activity, (2) the cleavage of xyloglucan by transferring the split end of the donor-xyloglucan molecule to a soluble acceptor-xyloglucan oligomer or polymer present in the wall space, (3) integration of new cellulose/xyloglucan components into the preexisting cell wall framework by transferase activity, and (4) interchange of xyloglucan cross-links and the various manipulations of the xyloglucan chains by molecular grafting activity. Because of its versatile functions, this class of enzymes is thought to play key roles in cell-wall construction and modification.

A unified nomenclature for the XTH family

The fact that members of the EXGT/XET family exhibit endotransglucosylase activity and/or endohydrolase activity added further confusion and contradiction to its nomenclature. To resolve such discrepancies in the names of individual family genes, a new unified systematic nomenclature for this family was proposed and accepted at the Ninth International Cell Wall Meeting held at Toulouse, France in 2001 (Yokoyama and Nishitani 2001). According to the unified nomenclature, any member of this class of genes/proteins would be referred to as xyloglucan endotransglucosylase/hydrolase (XTH). The two different enzymatic activities of XTH proteins would be referred to as xyloglucan endotransglucosylase (XET) activity and xyloglucan endohydrolase (XEH) activity (Rose et al. 2002). Since the term XET and XEH is defined as functions of XTH gene products, which is currently definitely defined based on phylogenetic analysis of this family of proteins, XEH is distinguished from any other enzymes exhibiting hydrolytic activities.

Biological implication of the XTH gene families

With reference to the genome sequence database of *Arabidopsis thaliana* released in December 2000 by Arabidopsis Genome Initiative, we have identified 33 XTH genes that are widely dispersed across the five chromosomes (Yokoyama and Nishitani 2001). Some are found as solitary genes and others as clusters of a few genes occurring in tandem. The presence of so many multiple copies of the XTH genes raised the question of whether individual members are redundant in terms of sharing the same function. To address this question, we examined mRNA expression profiles of all members of this gene family using quantitative real-time RT-PCR. What we found is that most members exhibit distinct expression profiles in terms of spatial expression profiles and responses to hormonal signals (Yokoyama and Nishitani 2001). For example, the *AtXTH-1* gene is expressed in the silique, *AtXTH-9* in the flower, *AtXTH-17* in the root, *AtXTH-32* in the stem, etc. However, several genes such as *AtXTH-2*, *AtXTH-4* and *AtXTH-27* are expressed in the five organs. In addition, individual genes respond differently to plant hormones such as IAA, gibberellin, brassinolide and ABA. Therefore, most genes exhibit distinct expression profiles in terms of both the site of expression and the response to hormones. These results show that members of the XTH gene family are involved in a specific process in a specific tissue at a specific stage of development. They are also regulated by different sets of plant hormones. This means that members of the XTH family are not redundant and have their own specific roles in *Arabidopsis* (Nishitani 2002).

The XTH family in monocots

Commelinoid monocotyledons, which include rice (*Oryza sativa*), have a so-called type II cell wall, which is distinct from the type I cell wall found in dicotyledonous plants, as represented by Arabidopsis. Most studies report that type II walls in rice have relatively little xyloglucan, and the predominant glycan that cross-links the cellulose microfibrils is, instead, glucuronoarabinoxylan (GAX) and (1→3;1→4)-β-glucan (mixed-linkage glucan) (Carpita and Gibeaut 1993). Moreover, the relatively small amounts of xyloglucans in the rice wall are structurally quite different from those found in type I walls in terms of their size and branching patterns, and are typically described as not being involved in cross-linking cellulose microfibrils. Given the structural and functional differences of xyloglucans in the two distinct types of cell wall, it would be logical to predict that the XTH gene families would also have evolved quite differently in the two plant species, and that XTHs would be less abundant in rice.

Based on the genomic sequence of *Oryza sativa* subsp. Japonica recently released by the International Rice Genome Sequencing Project (IRGSP), we revealed a large rice XTH (*OsXTH*) gene family with 29 members. Contrary to expectations, a number of these are similar to the *AtXTH* gene family. To examine the functions of the rice XTHs further, we characterized the expression patterns of the whole complement of this gene family using DNA array expression profiling and quantitative real-time RT-PCR. The results indicated that most of the OsXTH ORFs are actually transcribed and exhibit organ- and growth-stage-dependent expression patterns. It is surprising that not only is the number of OsXTH family members comparable to that of the AtXTH family, but that most members of the OsXTH family are expressed and in a pattern similar to that seen in *Arabidopsis*. More interestingly, a certain member (*OsXTH 19*) was specifically expressed in the dividing/elongating zone of the shoot and internode, which suggested that it played an important role in cell expansion in the shoot organs of rice plants (Yokoyama et al. 2004). These new findings imply that the biochemical characterization of individual proteins, as well as comparative genomics of the XTH family between the two extreme plant species with distinct types of cell walls, will be promising for elucidating the mechanism of cell wall dynamics. Such an approach, in combination with analysis of the wall polysaccharides, may offer an opportunity to explore new aspects of XTH function in plants, the elucidation of which might break fresh ground in the field of the plant cell wall biology.

References

Albersheim P (1976) The primary cell wall In: Plant Biochemistry, Bonner J, Varner JF (eds) Academic Press, New York, pp. 225-274.

Carpita NC, Gibeaut DM (1993) Structural models of primary cell walls in flowering plants: Consistency of molecular structure with the physical properties of the walls during growth. Plant J 3: 1-30.

Cosgrove DJ (2000) Expansive growth of plant cell walls. Plant Physiol Biochem 38: 109-124.

de Silva J, Jarman CD, Arrowsmith DA, Stronach MS, Chengappa S, Sidebottom C, and Reid JSG. (1993) Molecular characterization of a xyloglucan-specific endo-(1→4)-β-D-glucanase (xyloglucan endo-transglycosylase) from nasturtium seeds. Plant J 3: 701-711.

Fanutti C, Gidley MJ, Reid J (1993) Action of a pure xyloglucan endo-transglycosylase (formerly called xyloglucan-specific endo-(1→4)-beta-D-glucanase) from the cotyledons of germinated nasturtium seeds. Plant J 3: 691-700.

Farkas V, Sulova Z, Stratilova E, Hanna R, Maclachlan G (1992) Cleavage of xyloglucan by nasturtium seed xyloglucanase and transglycosylation to xyloglucan subunit oligosaccharides. Arch Biochem Biophys 298: 365-370.

Fry SC, Smith RC, Renwick KF, Martin DJ, Hodge SK, Matthews KJ (1992) Xyloglucan endotransglycosylase, a new wall-loosing enzyme activity from plants. Biochem J 282: 821-828.

Grew N (1682) The Anatomy of plants. Printed by W. Rawlins for the author, London.

Hayashi T (1989) Xyloglucans in the primary cell wall. Ann Rev Plant Physiol Mol Biol 40: 139-168.

Keegstra K, TalmadgeK T, BauerWD, Albersheim P. (1973) The structure of plant cell walls. III. A model of the walls of suspension-cultured sycamore cells based on the interconnections of the macromolecular components. Plant Physiol 51: 188-196.

Labavitch JM, Ray PM (1974) Relationship between promotion of xyloglucan metabolism and induction of elongation by indoleacetic acid. Plant Physiol 54: 499-502.

Medford JI, Elmer JS, Klee HJ (1991) Molecular cloning and characterization of genes expressed in shoot apical meristems. Plant Cell 3: 359-370.

Nishitani K (1992) A novel method for detection of endo-xyloglucan transferase. Plant Cell Physiol. 33: 1159-1164.

Nishitani K, Masuda Y (1981) Auxin-induced changes in the cell wall structure: Changes in the sugar compositions, intrinsic viscosity and molecular weight distributions of matrix polysaccharides of the epicotyls cell wall of *Vigna angularis*. Physiol Plant 52: 482-494.

Nishitani K, Masuda Y (1982) Acid pH-induced structural changes in cell wall xyloglucans in *Vigna angularis* epicotyl segments. Plant Sci Lett 28: 87-94.

Nishitani K, Masuda Y (1983) Auxin-induced changes in the cell wall xyloglucans: effects of auxin on the two different subfractions of xyloglucans in the epicotyl cell wall of *Vigna angularis*. Plant Cell Physiol 24: 345-355.

Nishitani K, Tominaga R (1991) *In vitro* molecular weight increase in xyloglucans by an apoplastic enzyme preparation from epicotyls of *Vigna angularis*. Physiol Plant 82: 490-497.

Nishitani K, Tominaga R (1992) Endo-xyloglucan transferase, a novel class of glycosyltransferase that catalyzes transfer of a segment of xyloglucan molecule to another xyloglucan molecule. J Biol Chem 267: 21058-21064.

Okazawa K, Sato Y, Nakagawa T, Asada K, Kato I, Tomita E, Nishitani K (1993) Molecular cloning and cDNA sequencing of endoxyloglucan transferase, a novel class of glycosyltransferase that mediates molecular grafting between matrix polysaccharides in plant cell walls. J Biol Chem 268: 25364-25368.

Peschke VM, Sachs MM (1994) Characterization and expression of transcripts induced by oxygen deprivation in maize (*Zea mays* L) Plant Physiol 104: 387-394.

Rose JKC, Braam J, Fry SC, Nishitani K (2002) The XTH family of enzymes involved in xyloglucan endotransglucosylation and endohydrolysis: Current perspectives and a new unifying nomenclature. Plant Cell Physiol 43: 1421-1435.

Yokoyama R, Nishitani K (2001) A comprehensive expression analysis of all members of a gene family encoding cell-wall enzymes allowed us to predict cis-regulatory regions involved in cell-wall construction in specific organs of Arabidopsis. Plant Cell Physiol 42: 1025-1033.

Yokoyama R, Rose JKC, Nishitani K. (2004) A surprising diversity and abundance of XTHs (xyloglucan endotransglucosylase/hydrolases) in rice: classification and expression analysis. Plant Physiol 134: 1088-1099.

Zurek DM, Clouse SD (1994) Molecular cloning and characterization of a brassinosteroid-regulated gene from elongating soybean (*Glycine max* L.) epicotyls. Plant Physiol 104: 161-170.

Using Cotton Fiber Development to Discover How Plant Cells Grow

Barbara A. Triplett and Hee Jin Kim

Molecular aspects of cotton fiber development

The fundamental work conducted by our group started in 1985 when a new research group was founded at the USDA-ARS, Southern Regional Research Center in New Orleans, Louisiana with a mission to focus on a molecular genetic approach to fiber quality improvement. The Center has had a 64 year history in cellulose and fiber structure, textile chemistry, and cotton utilization research. It was evident in the mid-1980s that advances in plant molecular biology would have a significant impact on crop production in the United States. Since cotton is a key commodity produced in the U.S., worth more than $120 billion annually to the economy (National Cotton Council of America), harnessing this new technology for cotton improvement was a worthwhile objective for our agency.

In the late 1990s we started a new project to identify differentially expressed cotton fiber genes using a technique called differential display (Liang and Pardee 1992). Two screens, one developmental and one genetic, were used to identify genes important for cotton fiber development. From the beginning it was clear that a large number of candidate genes could be identified by this approach. Since our team was small, we restricted our attention to a few genes and began the laborious process of testing and validating gene function. Our differential display screen was conducted just one year after the landmark announcement of cloning the first plant cellulose synthase subunits, now known as *GhCesA1* and *GhCesA2* from cotton fiber by the Delmer and Calgene groups (Pear et al. 1996). We were intrigued by finding a similar, but not identical, sequence from our differential display screen that we named *GhCesA4*. The relationship between *GhCesA1* and *GhCesA4* was unclear. The deduced proteins of the two genes were different by 26 amino acids. Were the sequence differences between the two genes due solely to the different genotypes used by our respective groups or was there another explanation?

The genetic diversity of commercial cotton varieties in the U.S. is quite low; therefore we hypothesized that the differences between *GhCesA1* and *GhCesA4* were probably not due to varietal differences. In our effort to distinguish these genes, we prepared a series of primers for quantitative, reverse transcription PCR (Q-RT-PCR), also called real-time PCR (Figure 1A). One set of primers was designed in a region where the two *CesA* sequences do not diverge. Amplification from this region was expected to yield the sum of transcripts encoded by *GhCesA1* and *GhCesA4*. Gene-specific primer pairs were designed in a nearby region where there were two nucleotide differences. This degree of sequence divergence is sufficient to prevent priming by a non-identical primer. The specificity of each primer pair was verified by dissociation curve analysis at the end of amplification. Results from this analysis show that each gene-specific primer pair was capable of amplifying cDNA prepared from fiber from 14-20 days post anthesis (DPA) (Figure 1B). Transcript abundance in reactions using the non-specific primer pair was additive, suggesting that there are two genes and that they are expressed simultaneously in the same tissue.

Additional clues about the potential relationship between these two genes emerged after additional sequence information of *CesA* genes from other *Gossypium* species became available (Cronn et al. 2002). Cotton is an allotetraploid resulting from hybridization of two diploid progenitors, one designated as the A genome progenitor and the other designated as the D genome progenitor. Alignment of *GhCesA4* with *GhCesA1* and similar genes from modern descendents of the A and D genome plants reveals that *GhCesA4* is more similar to the *CesA* genes from an A genome species and *GhCesA1* is more similar to sequences in the D genome (Figure 2). Therefore, we proposed that *GhCesA4* is homologous to *GhCesA1*.During polyploid formation, numerous epigenetic phenomena can take place, sometimes leaving only one of the genomes transcriptionally active.For the *CesA* genes in cotton fiber, such genome silencing mechanisms appear to be absent or selective since the *GhCesA2* gene and its homolog are also reported to be transcribed at nearly equivalent levels (Adams et al. 2003). It is intriguing to speculate that all *CesA* sequences are expressed due to the high level of cellulose synthesis required to form mature cotton fibers for textile applications.

Several laboratories have now shown that, in *Arabidopsis,* distinct

CesA genes are expressed when primary cell walls are forming (Arioli et al. 1998; Fagard et al. 2000; Scheible et al. 2001) and another distinct set of three *CesA* genes are expressed in tissues producing secondary walls (Taylor et al. 1999; 2000; 2003). We expect that

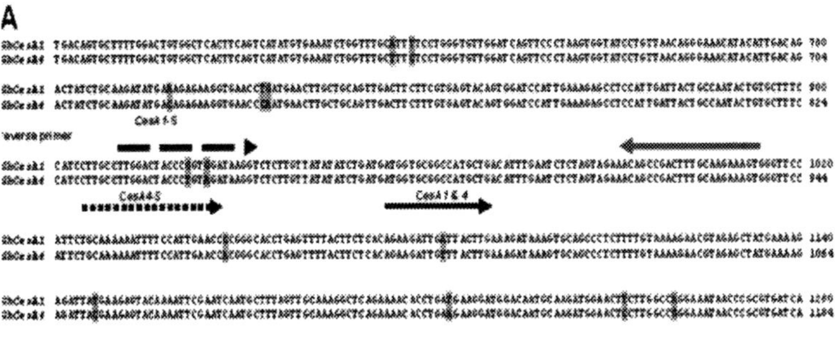

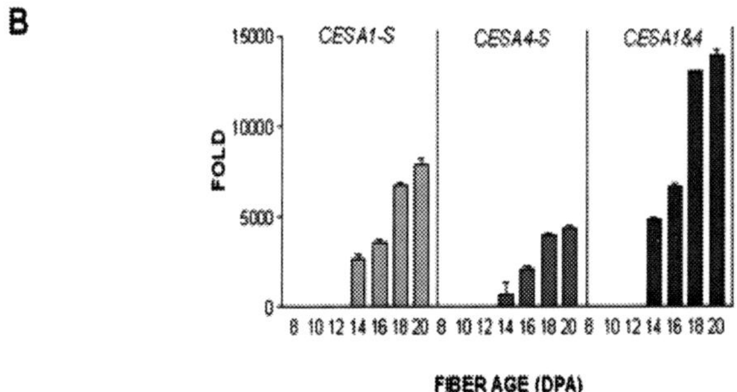

Fig. 1 Relative transcript abundance of *GhCesA1* and *GhCesA4* determined by quantitative, reverse transcription PCR using gene-specific primers.

A. Gene-specific primers were designed from regions that differentiated *GhCesA1* and *GhCesA4*. The long-dashed arrow indicates the location of the forward *GhCesA1*-specific primer (*CesA1*-S) and the short-dashed arrow indicate the location of the forward *GhCesA4*-specific primer (*CesA4*-S). A third forward primer designed from a region where the DNA sequences of *GhCesA1* and *GhCesA4* are identical is denoted by the black arrow. The same reverse primer was used for all three Q-RT-PCRs (gray arrow). B. Relative transcript abundance for genes amplified by the gene-specific primer pairs (CesA1S and CesA4) or both sequences (CesA1 & 4). DNA-free total RNA was isolated from field-grown fiber (8-20 DPA) and analyzed by quantitative, reverse transcription PCR using SYBR® Green Master Mix in the ABI Prism 7900HT Sequence Detection System. All results are representative of duplicate experiments beginning with RNA isolation. Transcript levels were normalized with respect to the level of α-tubulin 4, *GhTua4* (AF106570).

```
GhCesA4          MMESGVPVCHTCGEHVGLNVNGEPFVACHECNFPICKSCFEYDLKEGRKACLRCGSPYDE 60
AAD33796 (A)     ---------HTCGEHVGLNVNGEPFVACHECNFPICKSCFEYDLKEGRKACLRCGSPYDE 51
AAD33795 (A)     ---------HTCGEHVGLNVNGEPFVACHECNFPICKSCFEYDLKEGRKACLRCGSPYDE 51
GhCesA1          MMESGVPVCHTCGEHVGLNVNGEPFVACHECNFPICKSCFEYDLKEGRKACLRCGSPYDE 60
AAD33798 (D)     ---------HTCGEHVGLNVSGEPFVACHECNFPICESCFEYDLKEGRKACLRCGSPYDE 51
AAD33797 (D)     ---------HTCGEHVGLNVNGEPFVACHECNFPICKSCFEYDLKEGRKACLRCGSPYDE 51

                                     *                      *     *
GhCesA4          NLLDDVEKTTGDQSTMAAHLSKSQDVGIHARHISSVSTLDSEMTGDNGNPIWKNRVESWK 120
AAD33796 (A)     NLLDDVEKATGDQSTMAAHLSKSQDVGIHARHISSVSTLDSEMTEDNGNPIWKNRVESWK 111
AAD33795 (A)     NLLDDVEKATGDQSTMAAHLSKSQDVGIHARHISSVSTLDSEMTEDNGNPIWKNRVESWK 111
GhCesA1          NLLDDVEKATGDQSTMAAHLNKSQDVGIHARHISSVSTLDSEMAEDNGNSIWKNRVESWK 120
AAD33798 (D)     NLLDDVEKATGDQSTMAAHLNKSQDVGIHARHISSVSTLDSEMAEDNGNSIWKNRVESWK 111
AAD33797 (D)     NLLDDVEKATGDQSTMAAHLNKSQDVGIHARHISSVSTLDSEMAEDNGNSIWKNRVESWK 111

GhCesA4          EKKNKKKKPATTKVEREAEIPPEQQMEDKPAPDASQPLSTIIPIPKSRLAPYRTVIIMRL 180
AAD33796 (A)     EKKNKKKKPATTKVEREAEIPPEQQMEDKPAPDASQPLSTIIPIPKSRLAPYRTVIIMRL 171
AAD33795 (A)     EKKNKKKKPATTKVEREAEIPPEQQMEDKPAPDASQPLSTIIPIPKSRLAPYRTVIIMRL 171
GhCesA1          EKKNKKKKPATTKVEREAEIPPEQQMEDKPAPDASQPLSTIIPIPKSRLAPYRTVIIMRL 180
AAD33798 (D)     EKKNKKKKPATTKVEREAEIPPEQQMEDKPAPDASQPLSTIIPIPKSRLAPYRTVIIMRL 171
AAD33797 (D)     EKKNKKKKPATTKVEREAEIPPEQQMEDKPAPDASQPLSTIIPIPKSRLAPYRTVIIMRL 171

GhCesA4          IILGLFFHYRVTHPVDSAFGLWLTSVICEIWFAFSWVLDQFPKWYPVNRETYIDRLSARY 240
AAD33796 (A)     IILGLFFHYRVTNPVDSAFGLWLTSVICEIWFAF-------------------------- 205
AAD33795 (A)     IILGLFFHYRVTNPVDSAFGLWLTSVICEIWFAF-------------------------- 205
GhCesA1          IILGLFFHYRVTNPVDSAFGLWLTSVICEIWFAFSWVLDQFPKWYPVNRETYIDRLSARY 240
AAD33798 (D)     IILGLFFHYRVTNPVDSAFGL--------------------------------------- 192
AAD33797 (D)     IILGLFFHYRVTNPVDSAFGLWLTSVICEIWFAF-------------------------- 205
```

Fig. 2 Alignment of the deduced protein sequences from *Gossypium hirsutum CesA1 and CesA4* genes with *CesA* genes from Gossypium species. ***CesA4* (*G. hirsutum*, AF13210), AAD33796 (*G. hirsutum*), AAD33795 (*G. herbaceum*), *CesA1* (*G. hirsutum*, T10797), AAD33798 (*G. hirsutum*), AAD33797 (*G. ramondii*). (A) denotes A-subgenome and (D) denotes D-subgenome. Asterisks mark amino acids that differentiate the two subgenomes.**

cotton fiber will have at least six *CesA* genes, three from the D genome and three from the A genome, that are expressed during secondary wall production, but are not expressed while the cell has only a primary cell wall (Kim and Triplett, unpublished). Transcript profiles of *CesA* genes expressed during the cell elongation phase of fiber development are incomplete and await additional sequence information to permit the design of gene-specific primers. The notion of unique sets of cellulose synthases for primary and secondary walls was foreshadowed by our earlier work on cellulose molecular weight distributions (Timpa and Triplett, 1993). Gel permeation chromatography, using a unique, nondegrading solvent system composed of lithium chloride and dimethylacetamide, verified that the cellulose molecular weight in cotton fiber primary walls was smaller than that of secondary walls. This example shows the power of combining chemical and molecular genetic approaches for analyzing plant cell walls and development.

What triggers plant secondary wall biosynthesis?

The Delmer group found that in cotton fibers two small G-proteins,

Rac9 and *Rac13*, were abundantly expressed at the transition from primary to secondary wall formation (Delmer et al., 1995). In mammals, Rac GTPases stabilize the assembly of NADPH oxidase, an enzyme complex that generates superoxide in the presence of NADPH and flavin adenine dinucleotide (Dalton et al. 1999; Moldovan et al. 1999). Superoxide is rapidly converted to H_2O_2 by superoxide dismutases. When cotton *Rac13* was transiently expressed in plant cell cultures, production of reactive oxygen species nearly doubled (Potikha et al.1999). Polarized light microscopy showed that birefringence of fiber secondary walls increased after exogenous 50 μM H_2O_2 treatment and decreased with diphenyleneidonium (DPI), an inhibitor of NADPH oxidase. From these data and other observations, a model was advanced proposing that the cellular H_2O_2 content mediated by Rac GTPase through NADPH oxidase might induce secondary wall differentiation (Potikha et al. 1999; Delmer et al. 1999). Several aspects of the proposed H_2O_2-model seemed uncertain. First, H_2O_2 levels in detached 26 days post anthesis fibers were very low, while levels of H_2O_2 in whole 26 days post anthesis locules (seeds with attached fibers) were high. This difference suggests that a transient elevation in H_2O_2 level does not occur within the cotton fruit (boll), *i.e.* there was no oxidative burst. Secondly, the H_2O_2-model suggested that NADPH oxidase was the enzyme generating superoxide, there was no direct evidence to rule out other generators of potential reactive oxygen species such as amine oxidases, apoplastic peroxidases, or oxalate oxidases. Finally, diphenyleneidonium was shown to not be a specific inhibitor of plant NADPH oxidase (Delledonne et al.1998), so the reported effects could have been due to inhibition of other flavohemoproteins.

Plant secondary wall formation can also be monitored *in vitro* by the *Zinnia* tracheary element differentiation system (McCann et al. 2001). Since both cotton fibers and *Zinnia* tracheary elements are producing secondary walls *in vitro*, the factors inducing secondary wall formation might be similar in the two systems. Investigators failed to find an oxidative burst during tracheary element differentiation, although low levels of H_2O_2 accumulated (Groover et al. 1997). Addition of H_2O_2 to *Zinnia* cultures failed to induce differentiation, but led instead to necrosis. Millimolar levels of H_2O_2 were used in these experiments, and the argument could be made that micromolar concentrations might be more appropriate. Nevertheless, the three factors known to be required for differentiation of *Zinnia* tracheary elements from mesophyll cells are wounding, auxin, and cytokinin (McCann et al. 2001).

With the availability of reverse transcription PCR (Q-RT-PCR) primers for specific *CesA* genes and cotton ovule cultures, we decided to test whether the addition of H_2O_2 (50-1000 µM) to cultures could selectively stimulate the expression of the *CesA* genes involved in fiber secondary wall synthesis and not induce the expression of genes known to be constitutively expressed or expressed predominantly during cell elongation (Figure. 3). In addition to testing H_2O_2, we also tested 1 µM diphenyleneidonium (an inhibitor of NADPH oxidase and other flavohemoproteins), 500 µM TEMPO (a H_2O_2 scavenger), and 1 mM SHAM (a H_2O_2 producer). Ovule cultures were initiated with day of anthesis ovules in basal medium containing 5 µM indole acetic acid and 1.0 µM GA_3 using standard protocols in our lab (Kim and Triplett, 2001). Twelve days after culture initiation, H_2O_2 or other test compounds were added to replicate cultures (Figure 3A). Cultures at 12 days post anthesis have not yet started secondary wall formation with its associated high levels of cellulose biosynthesis. After 24 hr exposure to the chemical treatments, fibers were harvested for RNA isolation and analysis by reverse transcription PCR (Q-RT-PCR). The expression of six different "indicator" genes was monitored in each experiment.

Among the treatments, only 50 µM H_2O_2 slightly increased (2X) the levels of *GhCesA1* and *GhCesA2* transcripts (Figure 3B), whereas higher levels of H_2O_2 inhibited expression of *GhCesA1* and *GhCesA2* possibly due to phytotoxicity (Kim and Triplett, unpublished). If the H_2O_2-model is correct, expression of *GhCesA1* and *GhCesA2* should be down-regulated in the presence of diphenyleneidonium; however, diphenyleneidonium had no effect on *GhCesA1* or *GhCesA2* transcript abundance. Instead, diphenyleneidonium slightly increased the transcript levels of *GhExp1* (expansin), *GhTua5* (α-tubulin 5), and *GhAct1* (actin), genes that are abundantly expressed during fiber elongation. Treatment with 500 µM TEMPO, a H_2O_2 scavenger, or 1 mM SHAM, a H_2O_2 producer, had little influence on the levels of *GhCesA1* or *GhCesA2* transcripts. Although low concentrations of H_2O_2 slightly increased transcript abundance for *GhCesA1* and

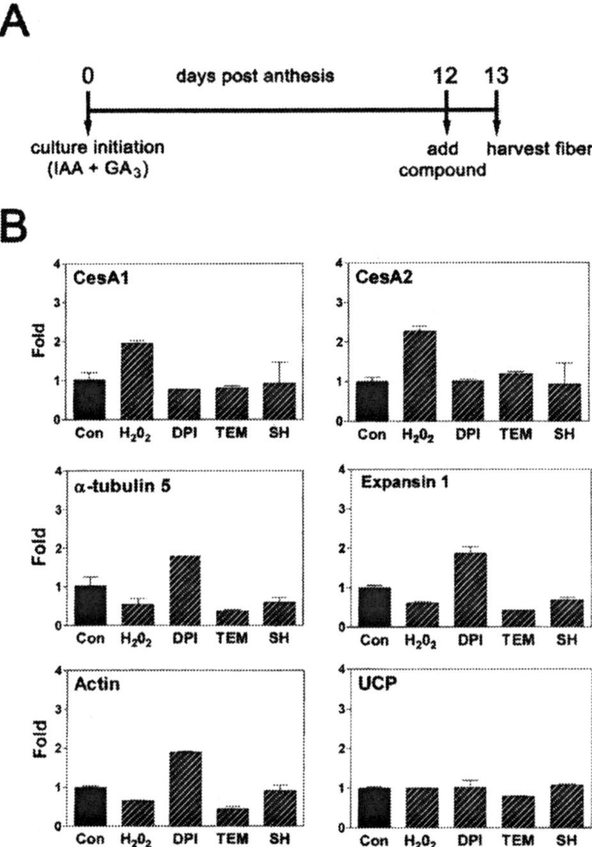

Fig. 3 Relative transcript abundance of six genes expressed in ovule culture-produced fibers.
A. Ovule culture conditions: 12 DPA ovules (*Gossypium hirsutum* L., DPL90, unfertilized ovules cultured in basal media for 12 days with 5 µM IAA and 1 µM GA_3) were either untreated (Con) or treated for 24 hr with 50 µM H_2O_2; 1 µM DPI (diphenyleneidonium) an NADPH oxidase inhibitor (DPI); 500 µM TEMPO (2,2,6,6-tetramethyl-1-piperidinyloxy, free radical) a hydrogen peroxide scavenger (TEM); or 1 mM SHAM (salicylhydroxamic acid) a hydrogen peroxide producer (SH). B. DNA-free total RNA was isolated from 13 DPA fibers and analyzed by quantitative, reverse transcription PCR using SYBR® Green Master Mix in the ABI Prism 7900HT Sequence Detection System as previously described (Kim and Triplett, 2004). Gene-specific primers were designed to cellulose synthase 1, *GhCesA1* (U58283); cellulose synthase 2, *GhCesA2* (U58284); α-tubulin 5, *GhTua5* (AF106571); expansin 1, *GhExp1* (AF512539); actin, *GhAct1* (D88414); and ubiquitin conjugating protein, *GhUCP1* (AI730710). All results are representative of duplicate experiments beginning with RNA isolation. Transcript levels were normalized with respect to the level of 18S RNA (U42827).

GhCesA2, we believe that the magnitude of the response is insufficient to account for the massive rates of cellulose synthesis that must occur during secondary cell wall formation in cotton fiber. These results suggested to us that NADPH oxidase activity may not be the key regulator of genes involved in secondary wall synthesis.

Taking our clues from the *Zinnia* tracheary element culture system where phytohormone addition can induce the cells to differentiate and produce secondary walls, we began testing a wide range of phytohormones and plant growth regulators for their effects on cotton fiber *GhCesA1* and *GhCesA2* expression. Our results are encouraging that some similarities exist between the cotton fiber and *Zinnia* tracheary element systems in that certain phytohormones are very effective in stimulating *GhCesA* expression (Triplett and Kim 2004). Reactive oxygen species are important second messengers in phytohormone signaling (Kwak et al. 2003), therefore H_2O_2 may be involved indirectly in a phytohormone-mediated stimulation of secondary wall cellulose production.

Acknowledgements

This project was supported by the USDA-Agricultural Research Service, the National Aeronautic and Space Administration, and the Louisiana State Support Program of Cotton Incorporated.

References

Adams KL, Cronn R, Percifield R, Wendel JF (2003) Genes duplicated by polyploidy show unequal contributions to the transcriptome and organ-specific reciprocal silencing. Proc Natl Acad Sci USA 100: 4649-4654.

Arioli T, Peng L, Betzner AS, Burn J, Wittke W, Herth W, Camilleri C, Hofte H, Plazinski J, Birch R, Cork A, Glover J, Redmond J, Williamson RE (1998) Molecular analysis of cellulose biosynthesis in *Arabidopsis.* Science 279: 717-720.

Cronn RC, Small RL, Haselkorn T, Wendel JF (2002) Rapid diversification of the cotton genus (*Gossypium*: Malvaceae) revealed by analysis of sixteen nuclear and chloroplast genes. Am J Bot 89: 707-725.

Dalton TP, Shertzer HG, Puga A (1999) Regulation of gene expression by reactive oxygen. Annu Rev Pharmacol Toxicol 39: 67-101.

Delledonne M, Xia Y, Dixon RA, Lamb C (1998) Nitric oxide functions as a signal in plant disease resistance. Nature 394: 585-588.

Delmer DP (1999) Cellulose biosynthesis in developing cotton fibers. In "Cotton Fibers" Basra AS, ed. Food Product Press, New York p. 86-112.

Delmer DP, Pear JR, Andrawis A, Stalker DM (1995) Genes encoding small GTP-binding proteins analogous to mammalian rac are preferentially expressed in developing cotton fibers. Mol Gen Genet 248: 43-51.

Fargard M, Desnos T, Desprez T, Goubet F, Refregier G, Mouille G, McCann MC, Rayon C, Vernhettes S, Hofte H (2000) PROCUSTE1 encodes a cellulose synthase required for normal cell elongation specifically in roots and dark-grown hypocotyls of Arabidopsis. Plant Cell 12: 2409-2423.

Groover A, Dewitt N, Heidel A, Jones A (1997) Programmed cell death of plant tracheary elements differentiating *in vitro*. Protoplasma 196: 197-211.

Kim HJ, Triplett BA (2001) Cotton fiber growth *in planta* and *in vitro*. Models for plant cell elongation and cell wall biogenesis. Plant Physiol 127: 1361-1366.

Kim HJ, Triplett BA (2004) Characterization of *GhRac1* GTPase expressed in developing cotton (*Gossypium hirsutum* L.) fibers. Biochim Biophys Acta 1679: 214-221.

Kwak JM, Mori IC, Pei ZM, Leonhardt N, Torres MA, Dangl JL, Bloon RE, Bodde S, Jones JD, Schroeder JI (2003) NADPH oxidase *AtrbohD* and *AtrbohF* genes function in ROS-dependent ABA signaling in *Arabidopsis*. EMBO J 22: 2623-2633.

Liang P, Pardee A (1992) Differential display of eukaryotic messenger RNA by means of the polymerase chain reaction. Science 257: 967-971.

McCann MC, Stacey NJ, Dahiya P, Milioni D, Sado P, Roberts K (2001) Zinnia. Everybody needs good neighbors. Plant Physiol 127: 1380-1382.

Moldovan L, Irani K, Moldovan NI, Finkel T, Goldschmidt-Clermont PJ (1999) The actin cytoskeleton reorganization induced by Rac1 requires the production of superoxide. Antiox Redox Signal 1:29-43.

Pear JR, Kawagoe Y, Schreckengost WE, Delmer DP, Stalker DM (1996) Higher plants contain homologs of the bacterial celA genes encoding the catalytic subunit of cellulose synthase. Proc Natl Acad Sci USA 93: 12637-12642.

Potikha TS, Collins CC, Johnson DI, Delmer DP, Levine A (1999) The involvement of hydrogen peroxide in the differentiation of secondary walls in cotton fibers. Plant Physiol 119: 849-858.

Scheible WR, Eshed R, Richmond T, Delmer D, Somerville C (2001) Modifications of cellulose synthase confer resistance to isoxalen and thiazolidinone herbicides in *Arabidopsis ixr1* mutants. Proc Natl Acad Sci USA 98: 10079-10084.

Taylor NG, Howells RM, Huttly AK, Vickers K, and Turner SR (2003) Interactions among three distinct CesA proteins essential for cellulose synthesis. Proc Natl Acad Sci USA 100: 1450–1455.

Taylor NG, Laurie S, Turner SR (2000) Multiple cellulose synthase catalytic subunits are required for cellulose synthesis in *Arabidopsis*. Plant Cell 12: 2529-2539.

Taylor NG, Scheible WR, Cutler S, Somerville CR, Turner SR (1999) The irregular xylem3 locus of *Arabidopsis* encodes a cellulose synthase required for secondary cell wall synthesis. Plant Cell 11: 769-779.

Timpa JD, Triplett BA (1993) Analysis of cell-wall polymers during cotton fiber development. Planta 189: 101-108.

Triplett BA, Kim HJ (2004) Regulation of gene expression in the transition from cell elongation to secondary wall formation in cotton fiber. X Cell Wall Meeting, Sorrento, Italy, p. 93.

Proline-Rich Cell Wall Proteins – Building Blocks for an Expanding Cell Wall?

Christine Bernhardt and Mary L. Tierney

The extracellular matrix of eukaryotic cells is formed through a process referred to as self-assembly. In many animal systems, matrix proteins are secreted in a manner that results in the assembly of matrix structure(s) necessary for maintaining cell shape and function. In plants, the matrix is composed of non-cellulosic polysaccharides and structural proteins secreted into the growing wall where they assemble into several independent networks. While hemicelluloses and pectins are secreted into the extracellular matrix of most plant cells, various families of structural proteins appear to show cell- and organ-type specific expression patterns. Thus, these proteins are candidates for molecules that modify matrix structure within the cell wall in unique ways and facilitate the generation and maintenance of cell shape and cell function during growth.

Proline-rich proteins, or PRPs, represent one family of wall proteins that are likely to play important roles in defining matrix structure within the plant cell wall. Based on primary sequence motifs and their pattern of proline hydroxylation, PRPs are considered members of a supergroup of hydroxyproline-rich cell wall proteins (HRGPs) that also include many extensins and arabinogalactan proteins (Kieliszewski and Lamport 1994). Extensins and PRPs are first secreted into the cell wall where they assemble within the matrix and subsequently become cross-linked. The role of these proteins in cell wall assembly and overall matrix structure may thus be two fold. Initially, PRPs and extensins may play a role in modeling matrix structure during wall assembly while later their insolubilization within the matrix may help create a stable cell wall structure necessary to maintain cellular shape and function.

A feature that may serve to functionally distinguish PRPs from other members of the HRGP superfamily is their degree of glycosylation. Many extensin and arabinogalactan proteins are heavily glycosylated with sugars representing up to 70% of their molecular mass. In contrast, PRPs identified to date show little to modest levels of glycosylation (Averyhart-Fullard et al. 1988; Kleis-SanFrancisco and Tierney 1990). The glycosylation of a number of HRGPs has been shown to stabilize their secondary structure and is likely to be an important factor in defining their structural contributions to the plant cell wall. This difference in the extent of glycosylation between PRPs, extensins and AGPs suggests that the physical interactions between various members of the HRGP family and other matrix components may differ in part based on the extent to which the surface properties of these proteins are dictated by sugar linkages vs. primary amino acid sequence.

Genes encoding PRPs have been isolated from a number of plant species and analysis of their primary sequence has shown that they can be divided into three classes. The first of these consist of single domain proteins that are composed predominantly of tandem copies of a conserved pentapeptide (ProHypVal(Tyr/Glu)Lys). PRPs containing this simple repetitive motif are common in legumes and were first defined by p33 and DcPRP1 in carrot (Chen and Varner 1985; Hong et al. 1987, 1990; Ebener 1993). The other two structural classes of PRPs consist of two domains – a proline-rich domain containing copies of the conserved, signature pentapeptide found in the single domain proteins and a novel domain (Sheng et al. 1991; Fowler et al. 1999; Menke 2000; Smart et al. 2000). The tyrosine and/or glutamate residues within the POVYK sequence are likely to be involved in the cross-linking of these proteins to other components within the wall through an oxygen radical-based mechanism. However, whether this crosslink serves to form a protein-based or protein-carbohydrate network within the cell wall remains to be determined.

PRPs are expressed in a variety of developmental contexts, including hypocotyl, root and root hair growth (Hong et al. 1989; Suzuki and Tierney 1993), early stages of legume nodule development (Franssen et al. 1987), early stages of vascular differentiation (Wyatt et al. 1992; Ye et al. 1991, guard cell development (Smart et al. 2000) and seed coat development (Lindstrom and Vodkin 1991) and this expression is often linked to stages of cell expansion. After their secretion into the wall of growing cells, crosslinking of PRPs occurs coincident with cell maturation during plant development but is also induced in response to physical damage, pathogen infection, elicitor treatment (Kleis-SanFrancisco and Tierney 1990; Bradley et al. 1992;

Brisson et al. 1994). These studies suggest that the insolubilization of PRPs may also play a generalized defense role by strengthening the wall in response to adverse environmental conditions. Questions stemming from these studies include what cell types express PRPs?, how is PRP expression regulated? what is the nature of the PRP cross-link?, is this conserved in all cell types in which PRPs are expressed?, is the insolubilization of PRPs in the wall essential for generating cell-type specific matrix structure necessary for cell function?, what are these structures?

Arabidopsis provides an excellent system in which to answer many of these questions. Four PRPs have been identified in arabidopsis (Fowler et al. 1999). These genes encode two classes of two-domain PRPs. *AtPRP1* and *AtPRP3* encode two-domain proteins with a N-terminal PRP-like domain and a highly-charged C-terminal domain. The expression of these two genes is associated with root tissue (Fowler et al. 1999). Furthermore, *AtPRP3* expression is controlled by developmental pathways leading to root hair formation (Bernhardt and Tierney 2000). In contrast, *AtPRP2* and *AtPRP4* predict proteins having a non-repetitive N-terminal domain followed by a PRP-like C-terminal domain (Fowler et al. 1999). During the vegetative stage of growth, both genes are expressed in the hypocotyl, cotyledons, and true leaves. After transition to reproductive growth, expression of *AtPRP2* and *AtPRP4* can be detected in stems, cauline leaves, developing flowers and siliques. In addition, *AtPRP4* is expressed in lateral roots. In each case, expression is strongest during periods of active growth and diminishes in mature organs (Fowler et al. 1999).

Gene expression studies have demonstrated that more than one class of the HRGPs are often expressed within a given cell type during plant development, indicating that the matrix structure of mature cells is likely to require the combined action of a number of structural cell wall proteins. For example, *AtPRP3* and *LRX1*, a leucine-rich extensin, are both expressed in root hairs in arabidopsis (Bernhardt and Tierney 2000; Baumberger et al. 2001). In a similar fashion, expression of *AtPRP4* and a tobacco extensin (*HRGPnt3*) have been linked to lateral root development (Keller and Lamb 1989; Fowler et al. 1999). In order to determine the stage(s) in which *AtPRP4* may contribute to lateral root cell wall structure and to contrast this with the lateral root expression of *HRGPnt3*, we analyzed AtPRP4::GUS expression at various stages of lateral root development in 12-day-old seedlings grown in liquid culture.

The earliest expression of *AtPRP4* during lateral root growth was detected some distance away from the primary root tip in a row of pericycle cells (Figure 1a). These cells are characterized by unusually closely spaced cell walls compared to the pericycle cells on the opposite side of the stele. This increased frequency of anticlinal cell divisions at one side of the pericycle cylinder represents the first visible step of lateral root initiation (Malamy and Benfey 1997; Dubrovsky et al. 2000). High AtPRP4::GUS activity occurred during subsequent stages of lateral root primordia formation (Figure 1b). As the lateral roots grew through the root epidermis, AtPRP4::GUS expression was concentrated at the base of the emerging root. No AtPRP4::GUS activity was detected in the tip of the newly formed lateral after emergence from the parent root (Figure 1c and d), distinguishing *ATPRP4* expression from that of HRGPnt3. This same pattern of AtPRP4::GUS expression was reiterated when new lateral roots developed on already existing laterals. These studies illustrate that while both *AtPRP4* and *HPRGnt3* are expressed in pericycle prior to lateral root development, their patterns of gene expression diverge as the lateral root primordia is established and an independent meristem is formed suggesting that these proteins may play non-overlapping roles in modeling the extracellular matrix within lateral root primordia.

Lateral root development has been shown to require auxin, which is normally transported from the shoot to the root (Celenza et al. 1995). To further characterize the relationship between AtPRP4 expression and lateral root development, we investigated the influence of 1-NAA (α-naphthaleneacetic acid), an active auxin, and NPA (N-1-naphthylphthalamic acid), an inhibitor of polar auxin transport, on AtPRP4::GUS expression in roots. Seedlings were grown for 4 d in liquid culture, effectors were added, and seedlings were grown for an additional 3 d. The seedlings were then scored for lateral root formation and analyzed for GUS activity using histochemistry and a quantitative fluorometric assay.

In the presence of 50 nM 1-α-naphthaleneacetic acid, both the number of lateral roots and the amount of AtPRP4::GUS activity increased, while treatment with 500 nM NPA resulted in a decrease in

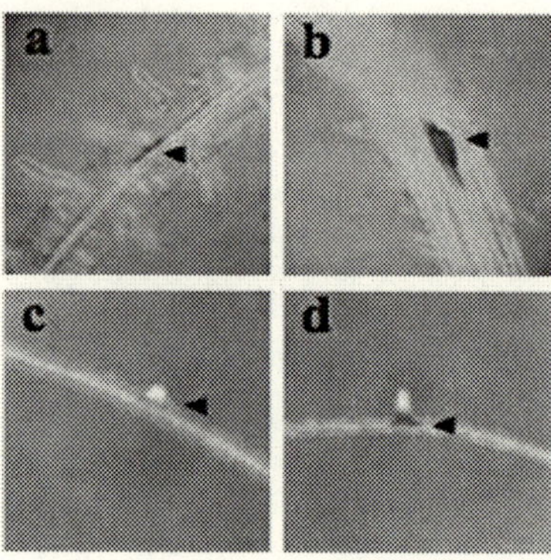

Fig. 1 *AtPRP4* is expressed during early stages of lateral root formation.

AtPRP4::GUS transgenic seedlings were grown in liquid culture under continuous white light for a period of 7d. AtPRP4::GUS activity was then analyzed histochemically at different stages of lateral root development. (a) anticlinal cell divisions in the pericycle representing the first visible step of lateral root formation; (b) lateral root primordium; (c, d) young lateral root containing a lateral root meristem.

AtPRP4::GUS expression and lateral root number (Table 1). This suppression of *AtPRP4* expression and lateral root initiation by N-1-naphthylphthalamic acid could be reversed when N-1-naphthylphthalamic acid and 1-α-naphthaleneacetic acid were added simultaneously to the medium. The specificity of these reactions was confirmed by demonstrating that treatment of seedlings with 2-α-naphthaleneacetic acid, a physiologically inactive form of auxin, did not result in an increase in *AtPRP4* expression or lateral root initiation (Table 1). Finally, co-addition of 2-α-naphthaleneacetic acid did not rescue *AtPRP4* expression or lateral root formation in NPA-treated seedlings. In each of these treatments, there was a close correlation between AtPRP4::GUS activity and the number of laterals formed. These results confirm that *AtPRP4* expression is linked to pathways controlling lateral root formation in arabidopsis and implicate the direct or indirect involvement of auxin in the regulation of *AtPRP4* expression.

In summary, analysis of patterns of PRP gene expression such as those described above and those of other members of the HRGP superfamily have provided us with a partial profile of the cell types

Table I. The level of AtPRP4 expression in roots is modulated by auxin-regulated pathways.
AtPRP4::GUS transgenic seedlings were grown in liquid culture for 4 d, effectors were added, and the seedlings were grown for an additional 3 d. Root tissue was harvested and scored for lateral root formation and GUS activity. NAA: naphthaleneacetic acid; NPA: 1-naphthylphthalamic acid

	GUS activity	Lateral roots
	(% of control)	
Control	100	100
1-NAA	344	316
2-NAA	107	98
NPA	20	19
NPA + 1-NAA	186	263
NPA + 2-NAA	26	23

in which these proteins play a role in modeling extracellular matrix structure. From an evolutionary point of view, it will be important to determine whether members of a gene family that are expressed within the same cells but have different structural features have been recruited for similar or distinct functions in determining plant cell wall structure. Some of the challenges for the future involve determining the molecular basis of the interaction(s) between these proteins and other structural components within the cell wall and to identify the molecular structures within the matrix of different cell types that are essential for maintaining the diversity of cell form and function necessary for plant growth and development.

References

Averyhart-Fullard V, Datta K, Marcus A (1988) A hydroxyproline-rich protein in the soybean cell wall. Proc Natl Acad Sci U S A. 85:1082-5.

Baumberger N, Ringli C, Keller B (2001) The chimeric leucine-rich repeat/extensin cell wall protein LRX1 is required for root hair morphogenesis in *Arabidopsis thaliana*. Genes Dev 15:1128-39.

Bernhardt C, Tierney ML (2000) Expression of AtPRP3, a proline-rich structural cell wall protein from Arabidopsis, is regulated by cell-type-specific developmental pathways involved in root hair formation. Plant Physiol 122: 705-14.

Bradley DJ, Kjellbom P, Lamb CJ (1992) Elicitor- and wound-induced oxidative cross-linking of a proline-rich plant cell wall protein: a novel, rapid defense response. Cell 70:21-30.

Brisson LF, Tenhaken R, Lamb C. (1994) Function of Oxidative Cross-Linking of Cell Wall Structural Proteins in Plant Disease Resistance. Plant Cell 6:1703-1712.

Celenza JL, Jr., Grisafi PL, Fink GR (1995) A pathway for lateral root formation in *Arabidopsis thaliana*. Genes Dev 9: 2131-42.

Chen J, Varner JE (1985) Isolation and characterization of cDNA clones for carrot extensin and proline-rich 33-kDa protein. Proc Natl Acad Sci USA 82: 4399-4403.

Dubrovsky JG, Doerner PW, Colon-Carmona A, Rost TL (2000) Pericycle cell proliferation and lateral root initiation in Arabidopsis. Plant Physiol 124: 1648-57.

Ebener W, Fowler TJ, Suzuki H, Shaver J, Tierney ML (1993) Expression of DcPRP1 is linked to carrot storage root formation and is induced by wounding and auxin treatment. Plant Physiol 101: 259-65.

Fowler TJ, Bernhardt C, Tierney ML (1999) Characterization and expression of four proline-rich cell wall protein genes in Arabidopsis encoding two distinct subsets of multiple domain proteins. Plant Physiol 121: 1081-92.

Franssen HJ, Nap J-P, Gloudemans T, Stiekema W, van Dam H, Govers F, Louwerse J, vanKammen A, Bisseling T (1987) Characterization of cDNA for nodulin-75 of so;ybean: a gene product involved in early stages of root nodule development. Proc Natl Acad Sci USA 84: 4495-4499.

Hong JC, Nagao RT, Key JL (1987) Characterization and sequence analysis of a developmentally regulated putative cell wall protein gene isolated from soybean. J Biol Chem 262: 8367-76.

Hong JC, Nagao RT, Key JL. (1989) Developmentally regulated expression of soybean proline-rich cell wall protein genes. Plant Cell 1:937-43.

Hong JC, Nagao RT, Key JL (1990) Characterization of a proline-rich cell wall protein gene family of soybean. A comparative analysis. J Biol Chem 15:2470-5.

Keller B, Lamb CJ (1989) Specific expression of a novel cell wall hydroxyproline-rich glycoprotein gene in lateral root initiation. Genes Dev 3: 1639-46.

Kieliszewski MJ, Lamport DT (1994) Extensin: repetitive motifs, functional sites, post-translational codes, and phylogeny. Plant J 5:157-72.

Kleis-San Francisco SM, Tierney ML (1990) Isolation and characterization of a proline-rich cell wall protein from soybean seedlings. Plant Physiol 94:1897-1902.

Lindstrom JT, Vodkin LO (1991) A soybean cell wall protein is affected by seed color genotype. Plant Cell 3:561-71.

Malamy JE, Benfey PN (1997) Organization and cell differentiation in lateral roots of Arabidopsis thaliana. Development 124: 33-44.

Menke U, Renault N, Mueller-Roeber B (2000) StGCPRP, a potato gene strongly expressed in stomatal guard cells, defines a novel type of repetitive proline-rich proteins. Plant Physiol 122: 677-86.

Sheng J, D'Ovidio R, Mehdy MC (1991) Negative and positive regulation of a novel proline-rich protein mRNA by fungal elicitor and wounding. Plant J 1: 345-54.

Smart LB, Cameron KD, Bennett AB (2000) Isolation of genes predominantly expressed in guard cells and epidermal cells of Nicotiana glauca. Plant Mol Biol 42: 857-69.

Suzuki H, Fowler TJ, Tierney ML (1993) Deletion analysis of SbPRP1, a soybean cell wall protein gene, in roots of transgenic tobacco and cowpea. Plant Mol Biol 21:109-19.

Wyatt RE, Nagao RT, Key JL (1991) Patterns of soybean proline-rich protein gene expression. Plant Cell 4:99-110.

Ye ZH, Song YR, Marcus A, Varner JE (1991) Comparative localization of three classes of cell wall proteins. Plant J 1: 175-83.

Wall-to-Wall Biochemistry: A Personal Perspective

Stephen C. Fry

A fondness of Nature

All natural science, whatever its focus, involves observing Nature and then explaining what we see. Why would anyone devote himself to this goal with plant cell wall biochemistry as its focus? Cell walls are certainly not the corner of Nature which the average student is attracted to first.

Like many biochemists, I entered Science via 'the birds and the bees'. As a boy, I spent much time, indeed whole Easter and summer holidays, catching frogs, newts, butterflies and moths (and playing cricket). When it was too dark or cold to catch butterflies, I spent the time reading about them. A book that I have owned since 1962 and still enjoy reading, for the sheer beauty of its detail, clarity and conciseness, is Sandars (1939).

It was my career aspiration to become a naturalist. I had no idea who might pay naturalists a salary, but I felt sure that someone would. In a sense, I was right, and my current work is a form of 'nature study'. Butterflies led to school biology, then to physics and chemistry. I was inspired by a physics teacher (Michael Pearson) who, with nothing more than 'chalk and talk', could get right to the central point of his lessons, making complex ideas seem intuitively obvious. He made us consider only one variable at a time, never fogging the issue with unnecessary detail. In my experience, more is achieved from intuition than from solving equations.

Facts

Intuition in the absence of facts is just logic; but I wanted to be a biologist not a logician. Between the ages of 18 and 21 I absorbed as many facts as possible while I worked for a BSc in Biology at Leicester University. I was particularly attracted to genetics and botany, mainly because of the lecturers in these subjects. Particularly inspiring were Bob Pritchard and Clive Roberts, who taught fungal genetics, and Bill Cockburn, who introduced plant biochemistry: these three teachers made a lasting impression on me by virtue of constantly drawing on the simple sequence: hypothesis → experiment → observation → deduction. In addition, Bill let us perform a 'Calvin' experiment which involved feeding radioactive CO_2 to *Chlorella* cells and identifying the ^{14}C-photosynthates by paper chromatography. The elegance and power of this simple approach had a lasting impact; few of my publications since 1979 have not involved *in-vivo* radiolabelling and paper chromatography.

In my final year as an undergraduate, I did a lab project supervised by Elisabeth Wangermann on the basipetal auxin transport system. The hypothesis had been published that this transport system was absent from the embryos of seeds and developed a few days after germination. Some simple experiments with radiolabelled IAA disproved this: even embryos dissected from developing (pre-desiccation) seeds transported auxin polarly. This modest success - re-writing a one-line 'fact' in our plant physiology textbook, and producing my first publication (Fry and Wangermann 1976) - using simple bench-top techniques, strengthened my liking for plant science, and I have never looked back (although I still enjoy butterfly-spotting when time permits).

The author of our textbook (though not the originator of the hypothesis) was Professor H.E. Street, head of Botany at Leicester. When I asked him for advice about where to do a PhD, the answer was simple. 'Stay at Leicester', he said; and I did.

Facts seem to have become unpopular. I wish I had £1 for every time an undergraduate student has asked me, after one of my lectures, 'Do we need to *know* this? We can always look it up in a textbook.' Nevertheless, collecting facts is an essential part of a scientist's training. Facts prepare the mind for the interpretation of observations, especially when the latter are unexpected ones.

Daft ideas

Facts are necessary but not sufficient. Dreaming up hypotheses (disparagingly known as 'daft ideas') is also essential to make Science satisfying.

My PhD project initially had nothing to do with cell walls. I wanted to discover how gibberellic acid controls transcription, no less. This is simple to ask but, as I learnt, difficult to achieve - especially using the molecular biological techniques of 1975 [relying on methylated albumen-kieselguhr (MAK) to fractionate RNAs]. My project was to involve isolating protoplasts from cell-cultures, isolating intact nuclei by gentle protoplast lysis (Blaschek et al. 1974), treating these with gibberellic acid, and using [^{3}H]UTP to assay transcription. Unfortunately, each of these steps ran into the ugly realities of what was feasible. Protoplast isolation was tough and led me to realise that cell walls are not just cellulose that can be digested with cellulase (or even 'Driselase'). The original intention was to use Prof. Street's favourite cell-culture, *Acer pseudoplatanus* (Stuart and Street 1971). [I think it was 2 years before I discovered that 'Prof' had a Christian name, Herbert, though this was apparently never used by his colleagues, let alone his students.] However, *Acer* turned out to have indestructible cell walls, so I adopted alternative cultures with more readily digestible walls (spinach and rose - cultures which we still maintain).

Releasing nuclei from protoplasts was another headache. I quantified the nuclei by measuring their DNA using a diphenylamine assay for deoxyribose, supposedly optimised for plants (Giles and Myers 1965). However, this assay was unsatisfactory because other deoxy-sugars, probably the polysaccharide components rhamnose and fucose, interfered: another gentle push for me towards the cell wall.

The moral of the story was clear: plants invest much more effort in cell walls than in nuclei. On the basis that 'if you can't beat them, join them', I decided that cell wall biochemistry was where the action would be. I therefore monitored the effects of gibberellic acid on wall polysaccharide composition, and began some simple pulse-chase experiments to see if gibberellic acid caused wall turnover. Also, on the assumption that in cell-cultures the liquid medium is essentially an extension of the primary wall, I analysed spent medium by spectrophotometry. This revealed that gibberellic acid prevented the secretion of something that absorbed at 402 nm, and promoted the secretion of 280-nm-absorbing material. The former turned out to be peroxidase; the latter was phenolic compounds (substrates of the peroxidase). An excellent, simple, methods book for beginners (Harborne 1973) guided me through the phenolic studies, mainly using paper chromatography. A 'daft idea' was born: apoplastic phenolics (e.g. ferulic acid) get oxidatively cross-linked by peroxidase, thus tightening the cell wall and restraining growth; gibberellic acid blocks peroxidase secretion and thereby prevents the imposition of this restraint. In other words, gibberellins don't loosen the wall, but prevent its tightening. Fortunately, the *Planta* referees didn't deem this idea completely crazy (Fry 1979).

'Prof' had a very enlightened 'hands-off' attitude to PhD students who showed enthusiasm for designing their own experiments. He let us conduct experiments that we were not obliged to report to him if unsuccessful. He clearly enjoyed seeing his students creating and testing crazy ideas. Perhaps it is fortunate that he never knew exactly how crazy some of them were. Tragically, 'Prof' died during my final year as a PhD student.

What's in a name?

My time with 'Prof', studying gibberellin action, had aroused in me an interest in plant cell walls, so I applied for funding to enable me to do Post-Doctoral work on cell walls. The Science Research Council (now Biotechnology and Biological Sciences Research Council) then had an excellent scheme under which students with their own ('daft') ideas could obtain Fellowship funding to work in a lab of their choice. I discussed with 'Prof' where I might hold such a Fellowship, and he enthusiastically endorsed my idea to approach Professor D.H. Northcote at Cambridge University. 'Don' (as he asked to be called, even at our first meeting) happily agreed and I joined his lab in October 1978, aged 24, as a post-doc.

I continued to have daft ideas about phenolics (Don, unlike the editor of *Planta*, was more frank about their daftness). Driselase digestion, radiolabelling and paper chromatography - simple and effective techniques - were put together to isolate radiolabelled feruloyl-oligosaccharides, whose structures revealed the first chemically defined ferulate-polysaccharide linkages (Fry 1982a).

The hypothesis was that wall phenolics undergo oxidative coupling *in vivo* to form cross-linking dimers e.g. diferulate. I wondered if this might also apply to tyrosine, the phenolic amino acid, which in insect structural proteins dimerises to form dityrosine (Figure 1a; Andersen 1964). I therefore acid-hydrolysed cell walls and looked for dityrosine by paper chromatography and paper electrophoresis. I was delighted to find dityrosine (apparently) when the papers were stained with ninhydrin. However, the chromatographic spot lacked the necessary fluorescence to be dityrosine, and the delight turned to disappointment. It took a few weeks of mental rumination before it struck me that there were other ways to dimerise tyrosine than via the C-C biphenyl linkage in dityrosine. An alternative structure that would fit all the data was the isomer shown in Figure. 1b, which, however, was not a known substance. Nevertheless, the additional tests by paper electrophoresis supported this structure, and the disappointment reverted to delight: the downs and ups of Science.

When writing up the tyrosine work for publication, I had to invent a name for the new dimer. The obvious, but dull, name for an isomer of dityrosine was isodityrosine. My girlfriend at the time (now wife), Vreni, has never quite forgiven me for not calling it vrenine. That would certainly have been a prettier name than isodityrosine, but I wasn't brave enough to try this on the editors of the *Biochemical Journal* (Fry 1982b). In retrospect, I needn't have worried: when Jeff Brady later discovered an apparently new trimer of tyrosine in cell walls (trityrosine and isotrityrosine already having been bagged), we proposed to call it 'neotrityrosine' only to find (after correcting the proofs for *Phytochemistry*) that the compound had already been discovered (in egg shells of the sea urchin, *Hemicentrotus pulcherrimus*) and that it had been given the 'pretty' name, pulcherosine. A phone call to Jeffrey Harborne, the editor, was just

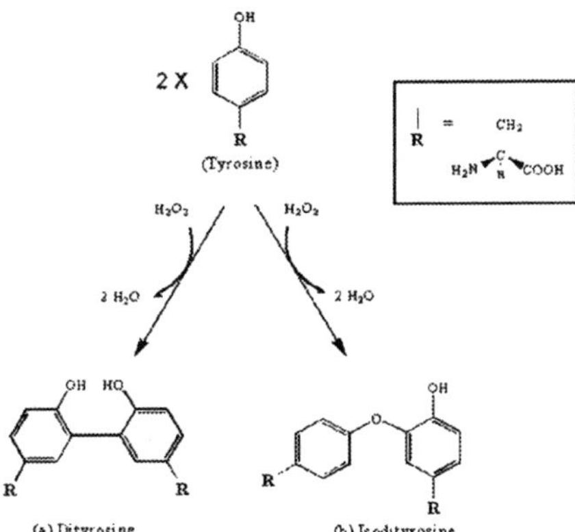

Fig. 1 The oxidative cross-linking of a pair of tyrosine molecules as observed in insect structural proteins (a) and in plant cell wall glycoproteins (b). **The reactions are proposed to be catalysed by peroxidase.**

in time to get our title changed before the paper went to press (Brady et al. 1997).

I picked up another lasting love in Cambridge: high-voltage paper electrophoresis. This analytical technique is so simple and conceptually elegant that it always amazes me that more people don't use it. The run-time is typically 30 - 60 minutes, and a compound's mobility provides an extremely useful piece of information - its charge: mass ratio at the pH chosen for the buffer (Offord 1966). Some people may be disturbed by the Frankensteinesque appearance of the apparatus, but the method is exceedingly versatile (Fry 2000). We are using it today to separate and characterise the apoplastic metabolites of [^{14}C]ascorbate, among many other applications - and the editors of *Nature* have recently blessed this venerable methodology

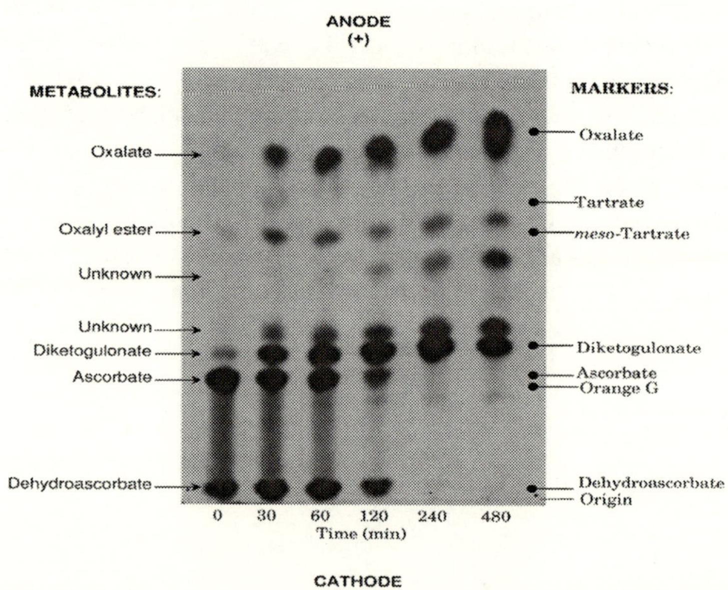

Fig. 2 Autoradiogram of a high-voltage paper electrophoretogram showing the apoplastic metabolism of radiolabelled ascorbate by cultured rose cells after 30–480 minutes). **At least seven degradation products are accumulated in the medium, some of which have been characterised. The positions of marker compounds (right hand margin) serve to define the positions of the unknown ^{14}C-metabolites. Ascorbate metabolites, formed by cell wall enzymes, may serve interesting new roles in cell wall physiology. The radioactive spot marked 'oxalyl ester' has very recently been identified as 4-*O*-oxalyl-L-threonate (Green and Fry, 2005).**

(Green and Fry, 2005). A typical fractionation is shown in Figure 2.

Don Northcote, who died in January 2004, was extremely generous in allowing me to perform whatever daft experiments I fancied, even though he did not agree with all my hypotheses. He introduced me to numerous techniques (Fry and Northcote 1983, Fry 2000) that still form the basis of my research group's activities. His lab was also a most stimulating place to work; colleagues at the time included Elias Baydoun, Giuseppe Dalessandro, Steve Read and Ray Owen - all great blokes for bouncing daft ideas off.

Don supported my application for a Royal Society 'Rosenheim Research Fellowship'. The Rosenheim allowed me to travel - first to Zürich to spend three months working on feruloyl esters with Hans Neukom (discoverer of diferulate) [and getting to know a little of my fiancée's native Switzerland], and then to Peter Albersheim's lab at Boulder, Colorado.

Oligosaccharins

Peter's lab was clearly a sensible place for any cell-wally to move to, and when Peter invited me to join his group in 1982, this set me off on another lasting venture - studying biologically-active oligosaccharides (oligosaccharins). Eighteen months in Peter's lab was another invaluable period of my career, albeit utterly different from Leicester and Cambridge. Peter introduced me to the realities of research grants: his large and astoundingly productive group (>20 people) needed funding, and part of a post-doc's work was to help with grant applications. This taught me much about financing a research lab, whereas in Cambridge the research money (but not much) had just somehow magically 'appeared'. This, and the organisation of regular group meetings, was excellent training for my first independent job when I moved back to Britain.

The Colorado lab focused on oligosaccharin research: I became aware of numerous new methods and developed a particular interest in one specific polysaccharide - xyloglucan. Other important events in Colorado included the arrival of our first daughter, Liz, very patriotically on 4 July.

Edinburgh

Arriving as a 'New Blood' lecturer at Edinburgh in 1983, I had thus assembled an unusual repertoire of methods and interests which I was free to pursue in whatever direction seemed fun (and fundable). The repertoire included chromatography, electrophoresis, radiolabelling, Driselase, phenolics, peroxidase, xyloglucan, NDP-sugars and oligosaccharins. Each of these strands is still running here, 21 years later. It would be very painful to abandon any of one's 'babies'. It is, however, nice to acquire new ones. Two of these (Cath and Helen) were sisters for Liz. The others were transglycosylation, hydroxyl radicals, and wall evolution.

Transglycosylation studies began with a chance observation, originating in oligosaccharins. In 1988, my friend Elias Baydoun was visiting Edinburgh on a mini-sabbatical from Beirut. We were testing the idea that a xyloglucan-derived oligosaccharin (now known as XXFG) gets inactivated *in vivo* by enzymic hydrolysis (a common fate for signalling molecules: read the 'message', then eat it). It seemed disappointing at the time, but our 10^{-8} M [^{3}H]XXFG was not hydrolysed; instead it was unexpectedly converted into high-molecular-weight material. This was the first evidence for *in-vivo* transglycosylation involving xyloglucan (Baydoun and Fry 1989). Later studies elucidated the relationship between donor and acceptor substrates (Smith and Fry 1991) and demonstrated the existence of an enzyme activity, xyloglucan endotransglycosylase (XET; Fry et al. 1992), now called xyloglucan endotransglucosylase. The last four authors of the latter paper were undergraduate project students. Thus, XET started with a 'chance' observation, but perhaps, as Pasteur believed, 'chance favours the prepared mind'. [A 'prepared' mind contains <u>facts</u> - see above.] The moral is not to be overly willing to disregard results that disagree with one's hypothesis. Intriguingly, Kazuhiko Nishitani, whose career has paralleled mine in many respects for 25 years, independently discovered the same enzyme activity and called it endo-xyloglucan transferase (Nishitani and Tominaga 1992). We have now agreed that proteins exhibiting this activity should be termed XTHs (Rose et al. 2002) and it is hoped that this terminology will be widely adopted.

A chance observation also triggered the hydroxyl radical story. In August 1994, we were conducting viscometric assays of cellulase using xyloglucan as substrate. I wanted to use ascorbate as an anti-oxidant to maintain enzyme activity, so I dissolved the xyloglucan in buffer containing 10 mM ascorbate, and stored it in the fridge for use the next day. It was frustrating at the time, but by the next morning the xyloglucan solution had gone runny, losing all hint of its normally high viscosity and rendering it useless as a substrate. However, it prompted some tentative experiments to explore why the xyloglucan's viscosity fell. Simply adding ascorbate to xyloglucan solution caused a pronounced loss of viscosity. A similar effect was observed with all other polysaccharides tested, and gel-permeation chromatography showed that the loss of viscosity was due to polysaccharide chain cleavage. The 'frustration' suddenly turned to excitement: the conclusion was that vitamin C (ascorbate) can cleave polysaccharides, which seemed crazy enough to be interesting. The mechanism turned out to be the production of hydroxyl radicals ($^{\bullet}OH$) by the action of ascorbate on O_2 and tiny traces of Cu^{2+}, followed by the Fenton reaction (Fry 1998). A current working hypothesis is that apoplastic hydroxyl radicals, formed in this way, can cause non-enzymic polysaccharide scission and hence wall loosening (Fry et al. 2001). An offshoot of this work is my lab's current and growing interest in ascorbate metabolism in the cell wall.

The third recent addition to my lab's programme is 'wall evolution', which was introduced to me by a self-propelled PhD student, Zoë Popper. [Self-propelled students (and post-docs) are highly prized!] Almost all plant cell wall analyses had previously been conducted on

Fig. 3 The Edinburgh Cell Wall Group, summer 2003.
Back row, left to right: Craig Mather, Malcolm Bain, Antonio Encina, Steve Fry; front row, left to right: Zoë Popper, Janice Miller, Shona Lindsay, Sandra Sharples.

angiosperm tissues, e.g. pea stems and maize coleoptiles. Zoë convinced me that it was of interest to investigate the changes in primary cell wall chemistry that have occurred over the past few hundred million years since plants colonised the land. She collected cell walls from all the major extant land plant taxa (bryophytes, lycopodiophytes, psilotophytes, equisetophytes, eusporangiate and leptosporangiate ferns, gymnosperms and angiosperms) as well as their probable algal ancestors, the charophytes. Analysis of each wall prep showed that several major steps in plant evolution [e.g. terrestrialisation (charophytes → bryophytes), vascularisation (bryophytes → lycopodiophytes), the acquisition of heterospory, becoming leptosporangiate, and the invention of seeds] were accompanied by remarkable changes in wall composition (Popper and Fry 2003). With the benefit of hindsight, this might have been expected because the cell wall serves so many important functions in the life of the plant, and the demands placed on the wall will have altered as plants changed their life-styles since the Ordovician. However, it takes an enquiring mind to think of testing this in the first place.

The future

Can we predict what will be discovered in the cell wall during the coming decade? I am sorry to disappoint readers, but I think the answer is 'no'. Existing hypotheses can of course be tested and either strengthened or rejected. In particular, the contribution of hydroxyl radicals to wall loosening, and the possibility that transglycosylation affects polysaccharides other than xyloglucan, both need to be evaluated. Perhaps the biochemical functions of those enigmatic polymers the arabinogalactan-proteins will even be glimpsed. And it is certain that *in-vivo* radiolabelling will be an essential tool for observing the behaviour of wall polymers. However, the most fundamental developments will be the result of fortuitous observations of Nature, and - I hope - of intuition enabling chance to favour prepared minds.

Acknowledgements

I am grateful to all my teachers, even though only a few are named in this article. I also thank all my past and present PhD students, post-docs and technicians (Janice Miller in particular, who has ably assisted me for 20 years) for making the Edinburgh Cell Wall Group (Figure 3) an exciting and productive place to work. We are all fortunate to work in a supportive and stimulating global cell wall community, represented at the triennial Cell Wall Meetings (all of which I have attended except the 1st). Most of all, I thank my parents, wife and daughters for tolerating, even encouraging, cell wall studies.

References

Andersen SO (1964) The cross-links in resilin identified as dityrosine and trityrosine. Biochim Biophys Acta 93: 213–215.

Baydoun EA-H, Fry SC (1989) *In-vivo* degradation and extracellular polymer-binding of xyloglucan nonasaccharide, a natural anti-auxin. J Plant Physiol 134: 453–459.

Blaschek W, Hess D, Hoffmann F (1974) Transkription in aus Protoplasten isolierten Zellkernen von *Nicotiana* und *Petunia*. Z Pflanzenphysiol 72: 262–271.

Brady JD, Sadler IH, Fry SC (1997) Pulcherosine, an oxidatively coupled trimer of tyrosine in plant cell walls: its role in cross-link formation. Phytochemistry 47: 349–353.

Fry SC (1979) Phenolic components of the primary cell wall and their possible rôle in the hormonal regulation of growth. Planta 146: 343–351.

Fry SC (1982a) Phenolic components of the primary cell wall: feruloylated disaccharides of D-galactose and L-arabinose from spinach polysaccharide. Biochem J 203: 493–504.

Fry SC (1982b) Isodityrosine, a new cross-linking amino acid from plant cell- wall glycoprotein. Biochem J 204: 449–455.

Fry SC (1998) Oxidative scission of plant cell wall polysaccharides by ascorbate-induced hydroxyl radicals. Biochem J 332: 507–515.

Fry SC (2000) 'The Growing Plant Cell Wall: Chemical and Metabolic Analysis'. Reprint Edition, The Blackburn Press, Caldwell, New Jersey [ISBN 1-930665-08-3].

Fry SC, Northcote DH (1983) Sugar-nucleotide precursors of the arabinofuranosyl, arabinopyranosyl and xylopyranosyl residues of spinach polysaccharides. Plant Physiol 73: 1055–1061.

Fry SC, Wangermann E (1976) Polar transport of auxin through embryos. New Phytol 77: 313–317.

Fry SC, Dumville JC, Miller JG (2001) Fingerprinting of polysaccharides attacked by hydroxyl radicals *in vitro* and in the cell walls of ripening pear fruit. Biochem J 357: 729–735.

Fry SC, Smith RC, Renwick KF, Martin DJ, Hodge SK, Matthews KJ (1992) Xyloglucan endotransglycosylase, a new wall-loosening enzyme activity from plants. Biochem J 282: 821–828.

Giles KW, Myers A (1965) An improved diphenylamine method for the estimation of deoxyribonucleic acid 206: 93.

Green MA, Fry SC (2005) Vitamin C degradation in plant cells via enzymatic hydrolysis of 4-*O*-oxalyl-L-threonate. Nature 433: 83–88.

Harborne, JB (1973) Phytochemical methods: a guide to the modern techniques of plant analysis. Chapman and Hall, London.

Nishitani K, Tominaga R (1992) Endo-xyloglucan transferase, a novel class of glycosyltransferase that catalyzes transfer of a segment of xyloglucan molecule to another xyloglucan molecule. J Biol Chem 267: 21058–21064.

Offord RE (1966) Electrophoretic mobilities of peptides on paper and their use in the determination of amide groups. Nature 211: 591–593.

Popper ZA, Fry SC (2003) Primary cell wall composition of bryophytes and charophytes. Ann Bot 91: 1–12.

Rose JKC, Braam J, Fry SC, Nishitani K (2002) The XTH family of enzymes involved in xyloglucan endotransglucosylation and endohydrolysis: current perspectives and a new unifying nomenclature. Plant Cell Physiol 43: 1421–1435.

Sandars E (1939) A Butterfly Book for the Pocket. Oxford University Press, London.

Smith RC, Fry SC (1991) Endotransglycosylation of xyloglucans in plant cell suspension cultures. Biochem J 279: 529–535.

Stuart R, Street HE (1971) Studies on the growth in culture of plant cells. X. Further studies on the conditioning of culture media by suspensions of *Acer pseudoplatanus* L. cells. J Exp Bot 22: 96–106.

Topography of Polysaccharide Biosynthesis in the Golgi Apparatus.

Ariel Orellana

Hemicelluloses and pectins are synthesized by glycosyltransferases, a group of enzymes that utilize nucleotide sugars as substrates. At the cellular level the biosynthesis of these polysaccharides takes place in the Golgi apparatus, a complex organelle formed by a series of physically separated stacked membranous cisternae. Thus, each cisternae constitutes a unique compartment, separated from the cytosol. Immunolocalization studies showed that newly synthesized xyloglucan and homogalacturonan are found in the lumen of different Golgi cisternae (Zhang and Staehelin 1992). Thus, the following questions arise: What is the topography of the glycosyltransferases in the Golgi apparatus? How do the glycosyltransferases gain access to the nucleotide sugars? Are there other proteins or enzymes, that could overcome the topological constraints associated with the biosynthesis of polysaccharides? The answers to these questions remained largely unknown until a few years ago, when different laboratories began to gain some insights into different aspects related to the biosynthesis of polysaccharides in the Golgi apparatus. Thus, although we are far from knowing the details, we are beginning to understand how the biosynthesis of polysaccharides takes place in the Golgi apparatus. Here, I will describe what we currently know about the topography of biosynthesis of polysaccharides in the Golgi apparatus and I will discuss some ideas that may help us to plan our future research.

Glycosyltransferases

All glycosyltransferases characterized to date use nucleotide sugars as substrates and transfer the sugar moiety to the nascent glycan. Only a few glycosyltransferases have been biochemically characterized and some of them have been cloned (Reiter 2002; Madson et al. 2003). From a biochemical point of view, one of the problems in identifying and characterizing a larger number of glycosyltransferases, is the lack of enzymatic assays to measure their activities. Thus, a challenge for the present and future is to develop assays that can help us to characterize putative glycosyltransferases.

Topography of the glycosyltransferases

Biochemical studies showed that both the pea xyloglucan fucosyltransferase and the homogalacturonan galacturonosyltransferase I have their catalytic domain located in the lumen of Golgi derived vesicles (Sterling et al. 2001; Wulff et al. 2000). The cloning of the *Arabidopsis thaliana* and *Pisum sativum* xyloglucan fucosyltransferase showed that a type II membrane bound protein can be predicted from the protein sequence, suggesting that the xyloglucan fucosyltransferase is a membrane bound protein containing a single transmembrane domain, a small N-terminus cytosolic tail and the majority of the protein facing the lumen of the Golgi cisternae (Perrin et al. 1999). All Golgi localized glycosyltransferases cloned so far also predict type II membrane bound proteins suggesting that glycosyltransferases involved in the biosynthesis of polysaccharides have their catalytic domain facing the lumen of the Golgi cisternae. The topological arrangement of these glycosyltransferases resembles the orientation of glycosyltransferases involved in the biosynthesis of polysaccharides and glycoconjugates in the Golgi apparatus of yeast and mammalian cells, suggesting that this topography is conserved throughout evolution.

Many hemicelluloses and pectins are branched with sugars covalently linked to a backbone. Many of these backbones are linear structures, and the 1,4-β-glucan backbone of xyloglucan resembles the structure of cellulose. Based on this structural similarity, it can be hypothesized that the backbone of xyloglucan is synthesized in the same way as cellulose. In recent years a number of genes encoding for putative cellulose synthases have been identified. These genes have been classified in different families, although some members of this family, named cellulose synthase like (*Csl*) genes, may not encode for enzymes involved in the synthesis of cellulose, and so could encode for glycosyltransferases involved in the biosynthesis of other similar polysaccharides. A quick view of the structure of hemicelluloses shows that in addition to xyloglucan, xylans, mannans and galactans also have linear sugar backbones linked by 1,4-β-linkages. Thus, it is tempting to postulate that the CSL proteins would correspond to glycosyltransferases responsible for the biosynthesis of the backbone of some hemicelluloses with type II glycosyltransferases involved in the addition of sugar sidechains.

The structure of the CSLs predicts multitransmembrane domain proteins, although the topography and location of the catalytic domain of these proteins has not been demonstrated experimentally. However, by comparison to the mechanism postulated for cellulose biosynthesis at the plasma membrane, it can be hypothesized that in all Csls involved in the synthesis of polysaccharides in the Golgi , the nucleotide sugar would be utilized in the cytosolic compartment, and the elongated polymer deposited in the lumen of the Golgi cisternae. Although no experimental data supporting this idea has yet been provided, this is an interesting hypothesis that needs to be tested since it represents a completely new mechanism concerning biosynthesis of polysaccharides in the Golgi apparatus.

Nucleotide sugars

The nucleotide sugars are the activated form of the sugars utilized by the glycosyltransferases. The exact number and classes of all nucleotide sugars utilized during the biosynthesis is not clear since we do not know the actual substrate requirements of many glycosyltransferases. Both the identity of the sugar and the identity of the nucleotide moiety are important. For instance, the glucosyltransferases for xyloglucan biosynthesis use UDP-glucose whereas other glucosyltransferases may use GDP-glucose as substrate for glucomannan biosynthesis.

Synthesis of nucleotide sugars has been detected in cytosolic fractions, leading to the notion that nucleotide sugars interconverting enzymes involved in their synthesis were located mostly in this compartment. Thus, the pool of nucleotide sugars required by the glycosyltransferases in the Golgi apparatus would be available in the cytosol. However, different lines of evidence indicate that some of these enzymes are associated with the Golgi apparatus (UDP-glucose 4-epimerase, UDP-glucuronate carboxy-lyase, UDP-xylose 4-epimerase). Among those nucleotide sugars interconverting enzymes reported to be associated with the Golgi membranes, it is not yet known whether their catalytic domain faces the cytosol or the Golgi lumen. The cDNAs for some of these enzymes have been cloned. Therefore the identification of targeting sequences in the protein can help us to predict the localization of these enzymes. However, many of these proteins lack a targeting signal suggesting that they are cytosolic proteins and so the respective nucleotide sugar is synthesized in that compartment. This does not prevent the possibility that a cytosolic enzyme may associate peripherally with the Golgi membrane as has been observed in the case of the UDP-glucose 4-epimerase (Seifert et al. 2002). Then, the nucleotide sugar (UDP-galactose) would be synthesized on the cytosolic face of the Golgi membrane, with this lipid bilayer separating the substrate from the catalytic site of the glycosyltransferases.

In contrast, a signal sequence can be found in the predicted protein of some nucleotide sugars interconverting enzymes. Thus a type II membrane bound protein, of similar topography to the glycosyltransferases, can be predicted in enzymes involved in the synthesis of UDP-xylose and UDP-arabinose, suggesting that the synthesis of these substrates may occur in the lumen of the Golgi cisternae. In addition, biochemical evidence has also shown that part of the enzymatic activity involved in the biosynthesis of UDP-xylose is located in the Golgi, supporting the idea that some UDP-xylose is indeed synthesized in the lumen of the Golgi apparatus.

Nucleotide sugar transporters

As mentioned above, our current notion is that the majority of the nucleotide sugars are synthesized in the cytosol. This poses a topological problem in the biosynthesis of polysaccharides since the catalytic site of the glycosyltransferases is physically separated by the Golgi membrane from the site of synthesis of their substrates. To solve this problem, nucleotide sugar transporters are present in the Golgi membrane, allowing the substrates to gain access to the lumen of the Golgi cisternae. Biochemical evidence supporting the idea that different nucleotide sugar transporters activities exist has been provided in plants (Muñoz et al. 1996, Wulff et al. 2000), although we still know little about the substrate specificity of the nucleotide sugar transporters. Recently, two cDNAs from *Arabidopsis thaliana* encoding for nucleotide sugar transporters have been characterized. GONST1 corresponds to a Golgi-localized nucleotide sugar transporter that transports GDP-mannose (Baldwin et al. 2001). Whether GONST1 transports other nucleotide sugars is not yet known. On the other hand, AtUTr1 is an nucleotide sugar transporters that can transport both UDP-glucose and UDP-galactose, whereas other UDP- or GDP-sugars are not transported by AtUTr1 Norambuena et al. 2002). These results indicate that nucleotide sugar transporters have a certain degree of specificity, and we can speculate that a number of nucleotide sugar transporters should be required to fulfill the needs of the glycosyltransferases. A comparison of the nucleotide sugar transporter sequences with the *Arabidopsis thaliana* database shows that the genome contains other genes that encode for putative nucleotide sugar transporters. Unfortunately, based on sequence information alone we can not predict the substrate specificity of these putative nucleotide sugar transporters. In any case, the evidence suggests that several distinct nucleotide sugar transporters, playing a role in the biosynthesis of polysaccharides, may be located in the Golgi membranes.

Other Golgi proteins that may be involved in polysaccharide biosynthesis

UDPase**:** Different studies have shown a strong correlation between the synthesis of polysaccharides and the activity of this enzyme (Dauwalder et al. 1969, Ray et al. 1969). Analysis of its localization and topography showed that it is a membrane bound protein with its catalytic site facing the lumen of the Golgi (Orellana et al. 1997). Since the nucleoside diphosphates released upon the transfer of the sugars can inhibit glycosyltransferases, the UDPase would eliminate a potential inhibitor of the addition of new sugars. In addition, the hydrolysis of UDP or GDP produces UMP or GMP, metabolites that are exchanged by nucleotide sugar transporters with cytosolic nucleotide sugars.

Phosphate transporter**:** Studies using P-radiolabelled UDP-glucose have shown that inorganic phosphate (P_i) is produced and released to the cytosol during the synthesis of polysaccharides in the Golgi (Neckelmann and Orellana 1998). Thus the presence of a P_i transporter has been hypothesized. Whether this would be an specific P_i transporter or a more general anionic transporter remains to be tested.

Reversible glycosylated protein (RGP): This self-glycosylating enzyme has been proposed to play a role in the channeling of UDP-sugars into growing polysaccharides, although no direct evidence for this role has yet been provided. Immunoelectronmicroscopy showed that RGP is present in the Golgi apparatus (Dhugga et al. 1997). However, the protein´s primary sequence shows no information for a signal sequence nor for potential transmembrane domains. Thus, it is likely that RGP is bound to Golgi membrane on the cytosolic face. Therefore, from the topological point of view, it is difficult to envision how RGP could be directly involved in the channeling of sugars from UDP-sugars into polysaccharides.

Future directions: The channeling of sugars from cytosolic nucleotide sugars into lumenal elongating polymers plus the formation of a number of well defined glycosidic bonds may involve the concerted action of all players involved in the biosynthetic process (Figure 1). The formation of protein complexes among the glycosyltransferases, nucleotide sugar transporters, UDPase and nucleotide sugars interconverting enzymes may be part of the arrangement required to fulfill the biosynthetic needs. In addition, the localization of these biosynthetic complexes in different Golgi cisternae would be required to explain the compartmentalization from cis to trans that is observed in the synthesis of polysaccharides. Studies to address these ideas should help us in improving our understanding of the biosynthesis of polysaccharides in the Golgi apparatus.

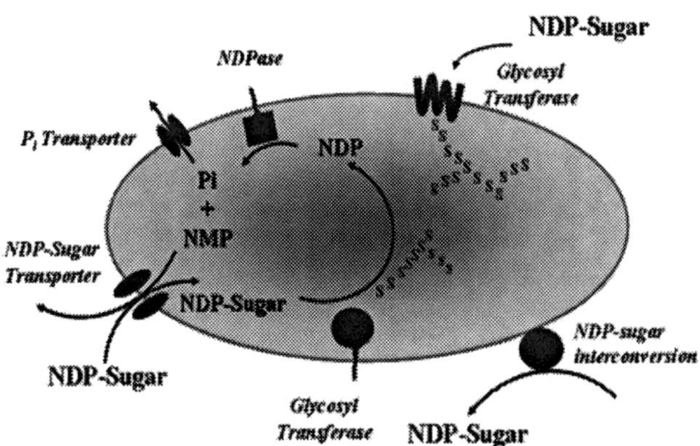

Fig. 1 Hypothetical model of the biosynthesis of polysaccharides in the Golgi apparatus. **A type II membrane bound glycosyltransferase and a cellulose synthase like glycosyltransferase are depicted as involved in sugar transfer. Most of the different elements mentioned in the text are shown in the model. NDP, nucleoside diphosphate; NMP, nucleoside monophosphate**

Acknowledgments

Thanks to current and previous members of the lab for discussion. To Michael Handford for critical reading of the manuscript. To Fondecyt-Chile for continuous grant support.

References:

Baldwin TC, Handford MG, Yuseff M, Orellana A, Dupree P (2001) Identification and characterization of GONST1, a Golgi-localized GDP-mannose transporter in Arabidopsis. Plant Cell 13: 2283-2295.

Madson M, Dunand C, Li X, Verma R, Vanzin GF, Caplan J, Shoue DA, Carpita NC, Reiter WD. (2003) The MUR3 gene of *Arabidopsis* encodes a xyloglucan galactosyltransferase that is evolutionarily related to animal exostosins. Plant Cell 15:1662-1670.

Muñoz P, Norambuena L, Orellana A (1996) UDP-glucose transport into the lumen of Golgi derived vesicles from pea stems (*Pisum sativum*) is required for polysaccharide biosynthesis. Plant Physiol 112: 1585-1594.

Neckelmann G, Orellana A (1998) Metabolism of UDP-glucose in Golgi vesicles from pea stems (*Pisum sativum*). Plant Physiol 117: 1007-1014.

Norambuena L, Marchant L, Berninsone P, Hirschberg CB, Silva H, Orellana A (2002) Transport of UDP-galactose in plants: Identification and functional characterization of AtUTr1, an *Arabidopsis thaliana* UDP-galactose/UDP-glucose transporter. J Biol Chem 277: 32923-32929.

Orellana A, Neckelmann G, Muñoz P, Norambuena L (1997) Topography and function of Golgi UDPase from pea stems (*Pisum sativum*). Plant Physiol 114: 99-107.

Perrin P, DeRocher A, Bar-Peled M, Zeng W, Norambuena L, Orellana A, Raikhel N, Keegstra K (1999) Xyloglucan fucosyltransferase, an enzyme involved in plant cell wall biosynthesis. Science 284: 1976-1979.

Seifert GJ, Barber C, Wells B, Dolan L, Roberts K (2002) Galactose biosynthesis in *Arabidopsis*: genetic evidence for substrate channeling from UDP-D-galactose into cell wall polymers. Curr Biol 12:1840-1845.

Sterling J, Quigley HF, Orellana A, Mohnen D (2001) The catalytic site of the pectin biosynthetic enzyme α-1,4-galacturonosyltransferase (GalAT) is located in the lumen of the Golgi. Plant Physiol 127: 360-371.

Wulff C, Norambuena L, Orellana (2000) A GDP-Fucose incorporation into the Golgi during xyloglucan biosynthesis requires the activity of a transporter-like protein other than the UDP-Glucose transporter. Plant Physiol 122: 867-878.

Zhang GF, Staehelin LA (1992) Functional compartmentation of the Golgi apparatus of plant cells. Plant Physiol 99: 1070-1083.

Xyloglucan and Xyloglucan Endotransglycosylase (XET) during Fruit Cell Wall Degradation

Eva Miedes and Ester P. Lorences

A few years ago, when my PhD supervisor suggested that, for my thesis, I could investigate the theme of the cell wall loosening mechanisms on gymnosperms, the first thing I did was to go to the library to explore what he was talking about. A couple of days later I was still lost between all those unpronounceable polysaccharide names and how they fitted together like jigsaw pieces into the cell wall. I have to admit that it took me a while to realise that cell walls were related with so many different physiological processes, and some of them could be really fascinating. A few years later, when I settled in Valencia, on the Spanish Mediterranean coast, in a region where the production of fruits and vegetables is exported worldwide, I discovered the persimmon fruit. One of the varieties of the fruit, the tastiest one, has a shelf life of only a few days, so the production is just for the local market. I started wondering about the softening process, and a few months later, I suggested to one of my PhD students about characterising the ripening of these fruits. It just so happened that shortly before that, I had spent a year at Dr Fry's lab, where XET was firstly described. The idea that, in addition to all the hydrolytic enzymes, the existence of a restoring enzyme that could reinforce the cell wall, was fascinating. I am still working in cell walls, still interested in the fruit softening process, and starting to gain knowledge of the fruit degradation produced during fungus infection. With the idea that both the processes are irreversible, since a large battery of hydrolytic enzymes is acting, the question at the back of my mind is if both processes could be delayed, and how the XET-mediated wall modification could contribute to the structural integrity of the fruits.

Xyloglucans are the major hemicellulosic matrix polysaccharides in the primary cell walls of plants. Xyloglucans are hydrogen-bonded to the cellulose and are thought to be a load bearing structure because they form cross-links between cellulose microfibres, and the consequence is that xyloglucan chains play an important architectural role in plant cell walls (Fry 1989). We would like to place particular emphasis on three examples: 1) where the breaking and rejoining of xyloglucan chains has been proposed as a mechanism for cell wall loosening that allows cell expansion (Fry 1989), 2) changes in hemicelluloses and in xyloglucan structure that are likely to be important determinants of the fruit softening process (Cutillas-Iturralde et al. 1994), and 3) since the plant cell wall is the major barrier that protects the cell from pathogen infections, the case of degradation of hemicelluloses and xyloglucan during the infection which would allow the fungus to attack the cell (Carpita and Gibeaut 1993). In addition to the structural role of xyloglucan, we also have also to bear in mind a more dynamic role, since xyloglucan may act as a source of signalling molecules. Oligosaccharides derived from xyloglucan have been found to be formed *in vivo*, and they have been shown to possess growth regulating activities in higher plants (Aldington and Fry 1993).

As a consequence of this structural and functional role of xyloglucan, enzymes that can modify the structure of xyloglucan chains are of great interest towards our understanding the mechanism and control of the above mentioned physiological processes. Among all the battery of the xyloglucan modifying enzymes, such as glucanases and glycosylases, we are particularly interested in the endotransglycosylase enzymes (Fry et al. 1992, Nishitani and Tominaga 1992), re-defined as xyloglucan endotransglucosylase/hydrolases (XTH) to unifying nomenclature (Rose et al. 2002). XTHs proteins encoded by the XTH family of genes exhibit different enzyme activities: xyloglucan endotransglycosylase (XET), Xyloglucan endohydrolase (XEH) or both of these activities. The consequence of this broad range of catalytic activity is that XTH family enzymes are playing multifunctional roles in plant growth and development. Over the last few years, a large number of XTH related genes have been cloned from a range of plant species, and XTH encoding genes have been shown to undergo up or down regulated expression in various developmental processes (Rose et al. 2002). As we mentioned above, we are particularly interested in the endotransglycosylase enzymes. XET can cut and re-join intermicrofibrillar xyloglucan chains, causing cell wall loosening, and may also perform the function of incorporating nascent xyloglucan into walls. XET can use xyloglucan oligosaccharides or polymeric xyloglucan as acceptors for the endotransglycosylation reactions, acting in this last option as a xyloglucan endohydrolase enzyme. We wanted to focus the possible role of XET in two physiological processes, previously mentioned, that involve cell wall (and xyloglucan) degradation: fruit ripening and fruit fungus infection and colonisation. Both processes have in common that both involve pectic and hemicellulosic degradation, but the mechanisms, the regulation and the physiological significance are different. In the fruit ripening process, the fruit itself is controlling the cell wall disruption and the final objective is seed-dispersion oriented. In the fruit degradation during fungus infection, a host-pathogen interaction happens, and the final objective for the plant is to trigger the defence mechanisms, meanwhile for the fungus, the objective is tissue colonization.

Xyloglucan structure and XET activity during fruit infection

Although the plant cell wall is an extremely effective physical barrier against attack by pathogens, most phytopathogenic microorganisms produce enzymes that are capable of degrading cell wall polymers such as pectins and hemicelluloses. The breakdown of the bonds holding this structure together will lead to loosening of the stability of this network, which may result in the solubilisation and/or degradation of the cell wall components (Cooper 1983). In our results, we found that in both apple and tomato fruits, the hemicellulosic polysaccharides and xyloglucan were solubilised from the cell wall during fruit infection with *Penicillium expansum*. At the same time, we observed a reduction in the molecular mass of hemicelluloses in infected fruit compared to the control fruits, this reduction being particularly important in xyloglucan. In tomato fruits, the most dramatic changes were observed in the hemicelluloses extracted with 24% KOH, where we found a reduction of 39% in the molecular mass of xyloglucan, between infected and control fruits. In apple fruit, xyloglucan depolymerisation during infection was even higher, 49 and 68%, at 4 and 24% hemicellulosic fractions respectively.

Since we found such a large change in the xyloglucan structure during fungal infection, we studied the XET activity, since this enzyme had been reportedly involved in the modification of the cellulose-xyloglucan network, acting cooperatively with endoglucanases and glycosidases. We found this enzyme present in both apple and tomato fruits, however our results showed that the activity decreased drastically during the infection process in both fruits. We found a decrease of 75% in apple fruit and a decrease of an 86% in XET specific activity during fungus infection. One possibility that might explain such a reduction could be that the decrease in XET activity could be due to proteolytic breakdown of endogenous enzymes by the invading pathogen. It is obviously a possibility, so we did an experiment to further explore this avenue. We measured XET activity of pooled equal amounts of the protein extracts obtained from control and infected fruits, firstly apple and then tomato. We also measured XET activity, where proteins were extracted from pooled infected and control samples of apple fruits. The results suggested that that the decrease in XET activity in the infected fruits was not a consequence of the presence of an XET inhibitor in the infected fruit or in the protein extract, and that the XET activity was specifically reduced during infection. Our suggestion is that the decrease in XET activity during infection could be specifically produced by the fungus. This XET inhibition, in the context of the plant-microbe interaction, could be explained as follows: The decrease in XET activity caused by the fungus reflects a lower xyloglucan endotransglycosylation which, together with the increase in endoglucanases, facilitates the colonization of the host cell wall. During infection, the increase in fungal β-glucanases increased the xyloglucan hydrolysis. This then lowered the degree of polymerization of xyloglucan and increased the concentration of soluble xyloglucan. In consequence, the donors for the transglycosylation reaction catalysed by XET increased and the enzyme could again integrate xyloglucan into the cell wall. However, if the XET activity is inhibited during infection, this possible re-structuring role of XET is not possible and the hemicellulose degradation continues.

Fruit texture, xyloglucan structure and XET activity during fruit ripening

Tissue softening is one of the most significant changes which occur during the development of a fleshy fruit. This process may be due to changes in the cell wall carbohydrate metabolism and to the action of specific cell wall hydrolases (Fisher and Benner 1981). It seems that depending on the fruit, ripening-associated changes in pectin or changes in the hemicellulose structure are likely to be important determinants of the textural changes in fruits (Cutilas-Iturralde et al.

1994). This means that the specific cell wall hydrolases that might also have a more outstanding role as candidates for promoting or for retarding tissue softening, could also vary between the different fruits.

In tomato fruit, our results confirm previous reports. Fruit texture had decreased during development, and mainly after the turning stage of ripening. At the same time, we studied the changes in the average molecular mass of hemicelluloses and xyloglucan. We found that, during growth, the degree of xyloglucan polymerisation did not change much. However, as the fruit ripened and become softer, a decrease in xyloglucan molecular mass was observed. These results indeed show the possible implication of xyloglucan metabolism in maintaining fruit texture. Since we were interested in the role of XET in xyloglucan metabolism and hence fruit ripening, XET specific activity was measured during fruit development. We found that through the different stages of development and ripening, XET specific activity decreased, measuring the lowest values at the red stage of ripening. Different cDNA XTHs have been reported to be differentially expressed in tomato, and some of them during growth and fruit ripening, so we were aware that by measuring the XET specific activity, we were measuring the whole XET "pool" without discriminating whether they were growth or ripening-related. However, we have also to consider that the measured XET activity was present in fruit cells, and there was indeed an alteration in the xyloglucan structure. To some extent, our results were indicating a possible relationship between tomato fruit texture and XET, although no definitive conclusions could be reached.

Fruit texture, xyloglucan structure during fruit ripening in fruits with increased XET activity

We wanted to confirm/discard our suggestion that XET activity could be related with fruit texture by altering the xyloglucan structure. One approach carried out was increasing XET activity in transgenic fruits with a constitutive overexpression of a xyloglucan endotransglycosylase. The specific activity of XET was increased up to five times during fruit ripening in the transgenic lines, and we then studied the hemicellulose and xyloglucan structure. Results showed that in transgenic lines, xyloglucan depolymerisation during ripening was also found, although the xyloglucan molecular weight was higher than in the control lines through all the stages of fruit development and ripening. During ripening, we also found an increase in softening in the transgenic lines. However, the decrease in fruit firmness was lower than in the control fruits, so the fruit texture in the transgenic lines was higher than in the control lines. Our results with the transgenic lines lead us to suggest that during fruit development, xyloglucan structure is an important determinant of fruit texture, and that XET activity is contributing to the maintenance of the structure. As the fruits ripen and soften, XET activity decreases, this decrease being partly responsible for xyloglucan degradation, and thus contributing to the fruit softening processes.

Conclusions

We suggest that the role of xyloglucan in maintaining the structure of cell walls can be modulated by the XET activity, and hence it is contributing to the maintenance of the cell wall integrity. For that reason, in the physiological processes that involve cell wall degradation, such as fruit fungus infection and colonisation and fruit ripening, the decrease in XET activity might be responsible to some extend for loosening the cell wall network. During fruit infection, the decrease in XET activity could be related to an increase in the access to the microfibrill surface of the cellulase complex, increasing the efficiency of cellulose hydrolysis and facilitating the progress of the fungal infection. During fruit ripening, the decrease in XET could be related to an increase in the fruit softening process, allowing fruit maceration and seed dispersion. We still have to investigate if the increase in the average molecular mass of xyloglucan, and the increase in firmness in the transgenic fruits lead to higher stability of the fruit cell wall and how the plant-pathogen interaction are affected.

Acknowledgements

This work was funded by MCYT-BOS2001-2755 and GV-GRUPOS 03/102.

References

Aldington S, Fry SC (1993) Oligosaccharins. Advances in Botanical Research 19: 1-101.

Carpita NC, Gibeaut DM (1993) Structural models of primary cell walls in flowering plants: consistency of molecular structure with the physical properties of the walls during growth. Plant J 3: 1-30.

Cooper RM (1983) The mechanisms and significance of enzymic degradation of host cell walls by parasites. In Biochemical Plant Pathology, Wiley, Chinchester, pp 101-137.

Cutillas-Iturralde A, Zarra I, Lorences EP (1994) Implication of persimmon fruit hemicellulose metabolism in the softening process. Importance of xyloglucan endotransglycoxylase. Physiol Plant 91: 169-176.

Fischer R L, Bennet AB (1991) Role of cell wall hydrolases in fruit ripening. Annu Rev Plant Physiol Plant Mol Biol 42: 675-703.

Fry SC (1989) Cellulases, hemicelluloses and auxin-stimulated growth: a possible relationship. Physiol Plant 75: 532-536.

Fry SC, Smith RC, Renwick KF, Martin DJ, Hodge SK, Matthews KJ (1992) Xyloglucan endotransglycosylase, a new wall-loosening enzyme activity from plants. Biochem J 282: 821-828.

Nishitani K, Tominaga R (1992) Endo-xyloglucan transferase, a novel class of glycosyltransferase that catalyzes transfer of a segment of xyloglucan molecule to another xyloglucan molecule. J Biol Chem 267: 21058-21064.

Rose JKC, Braam J, Fry SC, Nishitani K (2002) The XTH family of enzymes involved in xyloglucan endotransglucosylation and endohydrolysis: current perspectives and a new unifying nomenclature. Plant Cell Physiol 43: 1421-1435.

Thoughts on the Molecular Basis of Plant Cell Wall Extensibility

Simon J. McQueen-Mason

I came to science late, not starting my PhD until I was 29 and having spent several years working as a fisherman. I impressed my committee of advisors from the start, when at our first meeting I was asked what areas of research interested me. I answered that I found plants fascinating but that within that context I believed I would be happy to work on any area, as I saw the process of research fascinating in itself and as a means of self-development. This took off like a lead balloon and was followed up with a question regarding my future career plans, which I answered more or less by saying that no one can tell what the future will hold, and for now I was happy just to put everything into my studies. This answer was greeted with a similar deathly silence. Subsequently, I came to realise that you had to have a clear focus and set goals if you are to make it in an academic setting. The focus eventually came of itself, although goals have never been clear to me, and working both in fishing and science have taught me that just staying afloat is often enough to keep you fully occupied.

The plant cell wall is a beautiful enigma, an intelligent material that appears well adapted to wide-ranging requirements without compromising its performance. A seemingly paradoxical requirement from the growing cell wall is that of providing a material that is sufficiently cohesive to resist the several thousand bar of tension imposed on it by turgor pressure along with the need for controlled expansion during cell growth. To satisfy these requirements, plants have evolved a composite material that combines remarkable strength with dynamic flexibility. The material of the wall is strong enough to contain turgor pressures, yet flexible enough to allow growing cells to increase their volumes several hundred fold. Plant cells may increase their volumes at rates of more than 10% per hour, representing a very substantial increase in surface area. The molecular details of how this is achieved in a closely regulated manner whilst under potentially explosive pressure have preoccupied my research career.

A friend once told me that the biggest break you get in life is where you are born. This is undoubtedly true and it holds well, I think, for where your career as a scientist begins. I began my PhD studentship in a lab just down the corridor from Dan Cosgrove's at Penn State University. Dan was on my PhD committee and invited me to join his lab group for their weekly meetings. I was always in awe of Dan's rigorous approach to thinking and to science, and doubly so as his group took a largely biophysical approach to understanding growth, using language and thinking largely alien to me. At the time, Dan was taking growth biophysics by the scruff of the neck, bringing new techniques and refined thinking to water relations and cell wall biophysics. He had built substantially on the groundbreaking work by Cleland and co-workers in the 1970's who had characterised the phenomenon of acid-induced wall extension that was specifically associated with growing plant tissues and apparently underlay the process of cell wall expansion (Rayle et al. 1970). Early work had shown that acid-induced extension was sensitive to heat and proteolysis (Rayle and Cleland 1972). Dan Cosgrove carried the extensometer studies initiated by Cleland *et al.* into cucumber hypocotyls and showed convincingly that acid-induced extension must be a protein mediated process (Cosgrove 1989) and had been trying, without success, to identify the responsible agent. Fortunately for me, Dan's lab at that time had little experience of working with proteins.

In contrast, I had received a year of excellent training in the art of plant protein isolation under the tutelage of Robert Hamilton, with whom I had worked on aminotransferases in pea (McQueen-Mason and Hamilton 1989). After some pleading on my part, Dan agreed to let me loose in his lab on a rotation project and put me on the trail of the elusive extension-inducing proteins. Within a couple of weeks, I had managed to restore acid-induced extension to heat-inactivated cucumber hypocotyl epidermal strips. I spent the next week trying to convince Dan Durachko (technician and lynchpin of the Cosgrove lab) that I really had it working, and another 3 weeks convincing Dan Cosgrove himself. Eventually, Dan took me on as a PhD student and I spent the next 5 years firstly identifying the proteins and then trying to figure out what they did to cause cell wall extension.

Protein purification can be a painful process, particularly when working with plants and a large proportion of my time as a PhD student was spent in this frustrating pursuit. Eventually, we identified two similar proteins that in isolation could both independently restore acid-induced extension (McQueen-Mason *et al.* 1992) and these were named expansins.

If purifying expansins was a pain, trying to figure out how they worked proved an even greater challenge and one that still hasn't been fully met. Our first best bets were that expansins would be hydrolases or perhaps transglycosidases but all attempts to find such activities associated with the proteins proved negative (McQueen-Mason and Cosgrove 1995). Back in the 70s Peter Albersheim had proposed that acid-induced extension resulted from weakened interactions between wall polymers at low pH. It was subsequently shown that the hydrogen bonds between cellulose microfibrils and hemicelluloses were actually more stable at low pH values and the model was generally dropped, but it gradually crept into my mind that the model may be essentially correct with expansins providing the missing link by weakening these non-covalent interactions.

For me, the breakthrough in this work came through the use of stress-relaxation experiments. Dan Cosgrove had built and used a small device for measuring stress-relaxation some years previously to test some of the observations published by Masuda and co-workers who were applying this method to cell walls (Yamamoto and Masuda 1971). The essence of these measurements is that a piece of material is held between two clamps and stretched very rapidly until a predetermined force is generated. At this point the clamps stop moving apart, the specimen is then held at constant strain and the subsequent decay in stress in the material is measured in real time. Under these conditions, the stress is the material gradually decreases as the polymers of the material (cell walls in our case) rearrange themselves in compliance with the generated force. In these measurements stress decays more rapidly in a more compliant material and less rapidly in a stiffer material. In addition, it is generally held that that the earliest (and most rapid) relaxations reflect the movement of small molecular entities whilst the later (slower) relaxations are due to larger molecular movements.

We found that heat-inactivated growing cell walls were far less compliant than walls that were not heat-inactivated (native walls).

More importantly, we found that by adding expansins back to heat-inactivated walls we could completely restore the patterns of stress-relaxation seen in native cell walls (see figure 4 in McQueen-Mason and Cosgrove 1995). The beauty of the stress relaxation measurements is that the data is gathered in a few seconds and essentially provides a snapshot of the compliance of the cell wall at that moment in time. In contrast, the more usual long-term extension measurement that are used in expansin studies extend for 10s of minutes and reflect long-term on-going molecular events. I argued that if expansins acted as an enzyme that progressively chewed up polymers in the cell walls, then their action would lead to progressively weaker (or looser) cell walls. This is indeed the case if you use hydrolytic enzymes such as cellulases or pectinases on cell walls; the longer you incubate the walls with the hydrolytic enzymes the weaker they become. In contrast to this, we found that it made no difference whether you incubated walls for a few minutes or a few hours with expansins, the magnitude and rate of stress relaxation was exactly the same (McQueen-Mason and Cosgrove 1995). For me, these experiments more or less clinched the argument that expansins were working by disrupting non-covalent bonding in the walls- bonds that in essence can reform between polymers in new positions, so that no matter how long walls are incubated with expansins, the net amount of bonding between polymers in the walls would remain constant, and it is only once the walls are under an imposed load that the effects of the proteins are of major consequence- leading to enhanced creep.

This notion, that expansins operate by disrupting non-covalent interactions between polymers, was subsequently given strong support by the fact that the proteins weaken paper without hydrolysing the component fibres of the material (McQueen-Mason and Cosgrove 1994). These experiments, which utilised cellulosic filter paper, also indicated that expansins were most likely active in disrupting hydrogen bonds between glucan polymers. The hypothesis that that expansins work through hydrogen bond disruption is supported by the observation that expansin action is synergised by chaotropic agents such as 2 M urea, which destabilise H bonds, and is slower in walls where H_2O has been replaced by D_2O leading to more stable H bonds (McQueen-Mason and Cosgrove 1994). It was found that expansins bind weakly to crystalline cellulose, and not detectably to soluble glucans such as xyloglucan, but bound strongly to composites produced by binding xyloglucans (or other hemicelluloses) to crystalline cellulose, and also bound to cellulose that had been swollen by alkali treatment (McQueen-Mason and Cosgrove 1995). This suggests that expansins bind at the point where glucan chains meet, at the border between disordered and stable H-bonded junctions and work by destabilising hydrogen bonds at such junctions. This action would cause the glucans to peel apart, along with expansin affinity for the junction, causes them to progressively move with the opening junction.

Genome sequencing has revealed the extent of the α-expansin gene family, which numbers 27 members in Arabidopsis (Li *et al*. 2002) and 30 in rice (Li *et al*. 2003). There is clear substrate preference amongst α-expansins as shown by the ability of cucumber α-expansin 1 to induce extension in composites of cellulose and xyloglucan, but not in composites made with other hemicelluloses such as glucomanans (Whitney *et al*. 2000). Whether this specificity extends to all α-expansins, or if different expansins have specificity for different polymers remains to be seen. Certainly the fact that β-expansins from maize pollen show little or no activity on dicot cell walls but are active on grass cell walls indicates a clear difference in specificity between these two groups, even though the substrate specificity of the pollen allergen β-expansins remains undefined.

The exact mechanism of expansin action remains undemonstrated and the model described above is in part a product of our current concepts regarding cell wall structure. I fear that a true understanding of expansin function will depend on both a better understanding of cell wall structure and a means of directly testing the hydrogen bond disrupting hypothesis. The current model of expansin action builds from the idea of hemicelluloses acting as cross-linking tethers that bind cellulose microfibrils to one another, and from the assumption that these tethers play a pivotal role in determining cell wall mechanical properties.

A postulated tethering role for xyloglucans has been around for many years and drew major support from microscopic studies that revealed apparent interconnecting crosslinks between microfibrils, along with calculations showing that the average size of a wall xyloglucan was sufficiently long to span the spaces measured between microfibrils (McCann and Roberts 1990). Because of technical dificulties, this tethered model of the wall has not yet been unequivocally demonstrated to match reality. In this regard some interesting observations have been made in model systems that may reflect some aspects of cell wall architecture.In particular recent studies using composites produced from *Acetobacter*, bacteria that secrete cellulose ribbons into their culture media to form tightly matted structures have led to some intriguing observations. Mike Gidley's group showed that if *Acetobacter* cultures are grown in the presence of relatively high concentrations of cell wall matrix polysaccharides that the resulting composite materials showed radically different mechanical properties to the pure cellulose mats. Essentially, composites made with hemicelluloses (Whitney *et al.* 1999), or it turns out with pectins (Chanliaud and Gidley 1999) are weaker and much more extensible than the cellulosic mats and this may reflect a key function of the matrix polysaccharides in making cellulosic cell walls extensible.

Interestingly, we found that cucumber α-expansin 1 could induce extension in composites made with xyloglucans in a very similar fashion to that seen on plant cell walls but not on composites made with other hemicelluloses such as glucomannans (Whitney *et al.* 2000), nor on composites made with pectins (unpublished data). We suggested that the expansin effect was due to disrupting the bonds between cellulose microfibrils and xyloglucans, a model that assumes that the tethers are therefore of mechanical importance to the composite and have a retarding effect on extension. A more recent publication from the Gidley group presents data that appear to contradict this supposition. Chanliaud *et al.* (2004) showed that removal of the xyloglucan from composites using xyloglucanase digestion leads to increased strength and stiffness in the material, although extensibility (strain before failure) was equal to the undigested composite. These data tend to suggest that removal of the "crosslinking" glycans actually makes the composite stiffer and stronger rather than the opposite that would be expected if tethers between microfibrils serve to reinforce the material.

A possibe explanation for these seemingly paradoxical observations is that the the xyloglucans in the composite are acting as both as plasticisers and tethers of the cellulose network. One effect is that the presence of matrix polysaccharides increases alignment of the microfibrils making a less entangled, and therefore more extensible, network. The presence of the polysaccharides may also act to lubricate the mobility of microfibrils relative to one another. This plasticising effect of matrix polysaccharides may simply work by virtue of their functioning to keep the microfibrils apart from one another. If this were the case matrix polysaccharide hydrolysis might decrease wall extensibility rather than increase it, but this may all depend on the extent of hydrolysis and the resulting changes in matrix structure. It is similarly possible that in a more extensible hemicellulose-containing composite, interfibrillar tethers, or entanglements between polysaccharide coats have a significant retarding effect on extension, and these are the targets of expansin action.

The idea of hemicelluloses as lubricants is not new and indeed, McCann and Roberts, proposed that hemicelluloses might act rather like pit props to separate microfibrils from one another as well as functioning as tethers holding them together (McCann and Roberts 1990). This idea also shares similarities with the multi-coat model proposed by Talbott and Ray (1992) and has recently been promoted by Dan Cosgrove (Cosgrove 2001), where microfibrils are kept apart by the matrix polysaccharides rather than being tethered by them. In the plasticising model proposed here, matrix polysaccharide hydrolysis might lead to stiffer cell walls, at least up until the point where matrix hydrolysis destroyed wall cohesiveness in general. As pointed out by Dan Cosgrove, in this model, pectins might be expected to play a key role in maintaining wall extensibility through their greater space-filling properties and higher molecular mobility in the wall as suggested from NMR studies. In this case, partial depolymerisation of some matrix polymers might increase wall extensibility, by increasing matrix mobility without decreasing substantially the space filling aspect of polymer sizes. For example, partial depolymerisation of the backbone of a branched polymer might have such an effect, whereas removal of branches would be more likely to decrease extensibility. However, at some point matrix polysaccharide hydrolysis will decrease the space-filling properties of the polymers and likely lead to a stiffer, less extensible wall. Indeed, as already mentioned, this is what was seen in the case of the action of xyloglucan hydrolase on the *Acetobacter* composites, which led to stiffer stronger materials

(Chanliaud et al. 2004). Thus, it can be anticipated that matrix polymers in the walls of expanding tissues may be more highly branched than those in older tissues.

In this context, recent data from my group has suggested such a role for side chains of pectins, in particular arabinans, in governing the flexibility of stomatal guard cell walls (Jones *et al.* 2003). Pectins are acidic polysaccharides characterised by the presence of galacturonic acid residues and can be classified into 3 types of polymers all of which may exist as domains within a contiguous network (Willats *et al.* 2001). Homogalacturonan (HGA) consists of linear chains of $(1\rightarrow4)$-α-galacturonic acid and these can associate to form rigid gel structures. The carboxyl groups of galacturonosyl residues in HGA are often substituted with an esterified methyl group and the degree of methyl esterification of the polymer influences its ability to form gels. Other pectic polymers or domains are less well understood and this is particularly the case for rhamnogalacturonan-I (RG-I). RG-I is a highly branched polymer consisting of a backbone of alternating rhamnosyl and galacturonosyl residues, which may be contiguous with regions of HGA. The rhamnosyl residues of RG-I can be decorated with extensive side chains of arabinan, galactan or arabinogalactan. We demonstrated that degradation of guard cell arabinan by a purified fungal arabinanase could prevent closed guard cells from opening and open guard cells from closing. These effects could be specifically blocked by pretreating epidermal strips with LM6, a monoclonal antibody recognizing 1,5-α-L-arabinans. On the basis of our experiments, we have proposed a model whereby arabinans control pectin fluidity in cell walls by serving as a steric hindrance prohibiting homogalacturonan chain associations and maintaining wall flexibility.

Recently, there have been some reports that expansins may not be necessary for growth in all cell walls. For example, Reidy et al. (2001) showed that there was little expansin activity associated with the region of maximum expansion in leaf blades of the grass *Festuca pratensis*, whilst considerable expansin activity, protein and transcript were associated with older tissues in the leaf, which had ceased growing. Schopfer (Schopfer 2001) showed that growth and wall expansion in maize coleoptiles were dependent of reactive oxygen species (ROS) and Rodriguez et al. (2002) showed that the growth of maize leaves was ROS driven suggesting an expansin-independent mechanism underlying wall expansion. These experiments appear to have vindicated suggestions by Steve Fry (1998) who first drew attention to ROS as a potential wall loosening factor.

It may prove notable that, so far, all reports regarding expansin-independent wall expansion come from grass leaves. There are well known differences in composition between walls of grasses and other angiosperm plants, and these differences in structure may require or permit different mechanisms of wall extension. The situation is, however, less clear than this because α-expansins are involved in growth in maize roots (Wu et al. 1996) and rice internodes (Cho and Kende 1997) and the Kende group recently showed that α-expansin overexpression could lead to increased growth in transgenic rice plants (Choi et al. 2003).

At the present time expansins have entered the textbooks as providing a general mechanism of cell wall expansion in plants. A number of things related to expansins seem clear and some undeniable. Expansins are largely responsible for the acid-induced extension of plant cell walls that is associated with growth in many plants and tissues. Topical applications of expansin protein (Fleming et al. 1997), or ectopic expansin gene expression (Cho and Cosgrove 2000, Choi *et al.* 2003, Lee *et al.* 2003, Pien *et al.* 2001) can all lead to increases in growth of cells and tissues in plants. However, there are few examples (only 2 that I can think of) where reduced expansin expression has been shown to impact on the growth of plant organs (Cho and Cosgrove 2000, Zenoni *et al.* 2004). Indeed, insertion mutations in many of the Arabidopsis α-expansin genes that we have examined have no obvious impact on growth and development (unpublished data). This may suggest that either expansins are not generally essential for growth or that there is considerable functional redundancy amongst the Arabidopsis expansin gene family. This is a situation that seems likely given the size of the family, and we will have to await the results of experiments to stack mutations in a number of different family members to improve our understanding of this phenomenon.

It has become clear that there is considerable heterogeneity in wall composition between different plant species but also between the walls of different cells within a species. It has also become clear that a number of different factors including proteins such as expansins, XETs , and yieldins, or non protein factors such as ROS can have effects on wall mechanical properties. My current impression of the cell wall is of a material that is tailored by the plant to suit specific situations and one in which a range of different factors may influence extensibility with different aspects dominating extensibility in different situations. Whilst we are slowly getting a better picture of the components of the cell wall, there is still a long way to go before we understand how they fit together and function as a cohesive whole.

References

Chanliaud E, De Silva J, Strongitharm B, Jeronimidis G, Gidley MJ (2004) Mechanical effects of plant cell wall enzymes on cellulose/xyloglucan composites. Plant J 38: 27-37.

Chanliaud E, Gidley MJ (1999) *In vitro* synthesis and properties of pectin/*Acetobacter xylinus* cellulose composites. Plant J 20: 25-35.

Cho HT, Cosgrove DJ (2000) Altered expression of expansin modulates leaf growth and pedicel abscission in Arabidopsis thaliana. Proc Natl Acad Sci USA 97: 9783-9788.

Cho HT, Kende H (1997) Expansins and internodal growth of deepwater rice. Plant Physiol 113: 1145-1151.

Choi DS, Lee Y, Cho HT, Kende H (2003) Regulation of expansin gene expression affects growth and development in transgenic rice plants. Plant Cell 15: 1386-1398.

Cosgrove DJ (1989) Characterization of long-term extension of isolated cell walls from growing cucumber hypocotyls. Planta 177: 121-130.

Cosgrove DJ (2001) Wall structure and wall loosening. A look backwards and forwards. Plant Physiol 125: 131-134.

Fleming AJ, McQueenMason S, Mandel T, Kuhlemeier C (1997) Induction of leaf primordia by the cell wall protein expansion. Science 276: 1415-1418.

Fry SC (1998) Oxidative scission of plant cell wall polysaccharides by ascorbate-induced hydroxyl radicals. Biochem J 332: 507-515.

Jones, L, Milne, JL, Ashford DA, McQueen-Mason, S (2003) Cell Wall Arabinan is Essential for Guard Cell Function. Proc Natl Acad Sci USA 100:11783-11788.

Lee DK, Ahn JH, Song SK, Choi YD, Lee JS (2003) Expression of an expansin gene is correlated with root elongation in soybean. Plant Physiol 131: 985-997.

Li Y, Darley CP, Ongaro V, Fleming A, Schipper O, Baldauf SL, McQueen-Mason SJ (2002) Plant expansins are a complex multigene family with an ancient evolutionary origin. Plant Physiol 128: 854-864.

Li Y, Jones L, McQueen-Mason SJ (2003) The role of expansins and cell growth, Curr Op Plant Biol 6: 603-10.

McCann MC, Wells B, Roberts K (1990) Direct visualisation of cross-links in the primary plant cell wall. J Cell Sci 96: 323-34.

McQueen-Mason S, Cosgrove DJ (1994) Disruption of hydrogen bonding between wall polymers by proteins that induce plant wall extension. Proc Natl Acad Sci USA 91: 6574-6578.

McQueen-Mason SJ, Cosgrove DJ (1995) Expansin mode of action on cell walls: analysis of wall hydrolysis, stress relaxation and binding. Plant Physiol 107: 87-100

McQueen-Mason SJ, Durachko DM, Cosgrove DJ (1992) Two endogenous proteins that induce cell wall extension in plants. Plant Cell 4: 1425-1433

McQueen-Mason SJ, Hamilton RH (1989) The biosynthesis of indole-3-acetic acid from D-tryptophan in Alaska pea plastids. Plant Cell Physiol 30: 999-1005

Pien S, Wyrzykowska J, McQueen-Mason S, Smart C, Fleming A (2001) Local expression of expansin induces the entire process of leaf development and modifies leaf shape. Proc Natl Acad Sci USA 98: 11812-11817

Rayle DL, Haughton PM, Cleland R (1970) An *in vitro* system that simulates plant cell extension growth. Proc Nat Acad Sci USA **67,** 1814-1817

Rayle DL, Cleland R (1972) The *in-vitro* acid growth response: relation to *in-vivo* growth responses and auxin. Planta 104: 282-296

Reidy B, McQueen-Mason S, Nosberger J, Fleming A (2001) Differential expression of alpha- and beta-expansin genes in the elongating leaf of *Festuca pratensis*. Plant Mol Biol 46: 491-504.

Rodriguez AA, Grunberg KA, Taleisnik EL (2002)Reactive oxygen species in the elongation zone of maize leaves are necessary for leaf extension. Plant Physiol 129: 1627-1632.

Schopfer P (2001) Hydroxyl radical-induced cell-wall loosening in vitro and in vivo: implications for the control of elongation growth. Plant J 28: 679-688.

Talbott LD, Ray PM (1992) Changes in molecular size of previously deposited and newly synthesized pea cell wall matrix polysaccharides. Plant Physiol 98: 369-379.

Whitney SEC, Gothard MGE, Mitchell JT, Gidley MJ (1999) Roles of cellulose and xyloglucan in determining the mechanical properties of primary plant cell walls. Plant Physiol 121: 657-663.

Whitney SEC, Gidley MJ, McQueen-Mason SJ (2000) Probing expansin action using cellulose/hemicellulose composites. Plant J 22: 327-334.

Willats WGT, McCartney L, Mackie W, Knox JP (2001)Pectin: cell biology and prospects for functional analysis. Plant Mol Biol 47: 9-27.

Wu YJ, Sharp RE, Durachko DM, Cosgrove DJ (1995) Growth maintenance of the maize primary root at low water potentials involves increases in cell-wall extension properties, expansin activity, and wall susceptibility to expansins. Plant Physiol 111: 765-772.

Yamamoto R, Masuda Y (1971) Stress relaxation properties of the *Avena* coleoptile cell wall. Physiol Plant 25: 330-335.

Zenoni S, Reale L, Tornielli GB, Lanfaloni L, Porceddu A, Ferrarini A, Moretti C, Zamboni A, Speghini S, Ferranti F, Pezzotti M (2004) Downregulation of the Petunia hybrida alpha-expansin gene PhEXP1 reduces the amount of crystalline cellulose in cell walls and leads to phenotypic changes in petal limbs. Plant Cell 16: 295-308.

Does Xyloglucan Endotransglucosylase Promote Cell Elongation?

Takahisa Hayashi and Takumi Takeda

A plant cell that absorbs water due to osmotic pressure maintains its shape by maintaining the wall pressure higher or equal to the osmotic pressure (wall pressure **e"** osmotic pressure). In the growing plant cells, the decrease in the mechanical property of the walls, due to wall loosening, induces water absorption because the wall pressure becomes lower than the osmotic pressure (wall pressure < osmotic pressure). Therefore, water absorption is the driving force of cell enlargement. Auxin (indole-3-acetic acid) functions to promote the cell expansion by loosening the cell wall. Although the mechanism of wall loosening has not been clarified at the molecular level, xyloglucan degradation and solubilization have been proved to be accompanied by auxin-induced cell enlargement (Labavitch 1981). Xyloglucan possesses a 1,4-β-glucan backbone with 1,6-α-xylosyl residues, the backbone of which can bind specifically to cellulose microfibrils by hydrogen bondings (Hayashi et al. 1994) and extends the *in vivo* xyloglucan-cellulose network (Baba et al. 1994).

We have been conducting extensive studies focusing on xyloglucan metabolism, and have demonstrated the removal and cleavage of xyloglucan tethers by cellulase and xyloglucanase, respectively. Evidently, the expression of these enzymes accelerated cell expansion (Park et al. 2003; Park et al. 2004). The phenomenon is similar to that of the expression of expansin in Arabidopsis (Cho and Cosgrove 2000). Other groups have also demonstrated the acceleration by overexpression of cellulase and cellulose-binding domain (Shani et al. 2004; Shpigel et al. 1998). Another interesting enzyme has been proposed as a wall loosening enzyme. It is known as either xyloglucan endotransglycosylase (XET) or endoxyloglucan transferase (EXGT), and is defined as "xyloglucan: xyloglucosyltransferase (EC 2.4.1.207)" (Fry et al. 1992; Nishitani and Tominaga 1992). This enzyme catalyzes cleavage of xyloglucan and transfer of the reducing end of the cleaved xyloglucan to the non-reducing end of another xyloglucan. No evidence has been reported for the involvement of

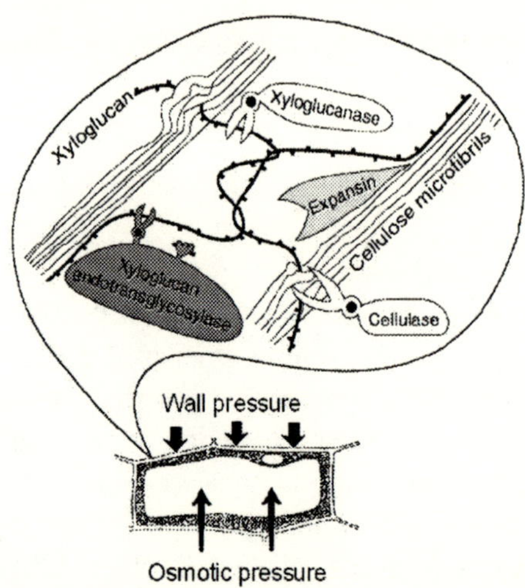

Fig. 1 Wall-loosening enzymes in plant cell
Xyloglucanase cleaves internal linkages adjacent to unsubstituted glucose unit at random, cellulase cleaves 1,4-β-glucan in the non-crystalline site of cellulose microfibrils where xyloglucan is intercalated, expansin cleaves hydrogen bonds between xyloglucan and cellulose, and xyloglucan endotransglycosylase cross-endotransglycosylates xyloglucans in cell walls.

this enzyme in wall loosening. We need to know whether XET promotes cell elongation by generating the integration of xyloglucans into the cell walls (Figure 1).

Xyloglucan is an important component of primary cell walls, as previously shown by the universality of its occurrence in the primary walls of higher plants and the notable change in its molecular size during growth (Hayashi 1989). To examine the function of XET in growing cells, we have tried feeding substrates such as xyloglucan polymer and its oligosaccharide to intact pea stem segments where endogenous XET exist, but the waxy cutin appeared to have prevented these substrates from penetrating the wall. Then, we used 10 mm segments of third internodes of pea partially split longitudinally into halves (split stem segments), which were used for the split test (Van Overbeek and Went 1937). The split segments incubated in 2 mM MES/KOH buffer (pH 6.2) rapidly bent outward to form a "V" shape (Figure 2), due to the higher growth rate of the inner tissue than the outer tissue. When the V-shaped split segments were incubated in

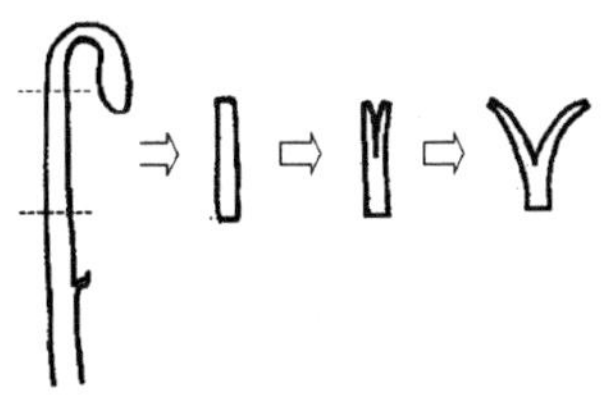

Fig. 2 Split stem segments from pea third internode

auxin, the epidermis grew more rapidly than the inner tissue reducing the bending angle. Thus, the segments straightened about 6 hr later, and bent after a long incubation. We used this method to see the effects of exogenously added substrates for wall-bound XET, because the substrates could be integrated easily at the cut surface of the split segments where XET existed. Incubation of the stem segments in auxin plus xyloglucan oligosaccharide (XXXG), blocked the auxin-mediated straightening (Figure 3). The effect could be interpreted by hypothesizing that XXXG was incorporated by the action of XET into the inner tissue from the cut surface more easily than into the epidermis. The incorporation of XXXG cleaved the xyloglucan tether, resulting in wall loosening and subsequent cell growth. In fact, stiffness of the

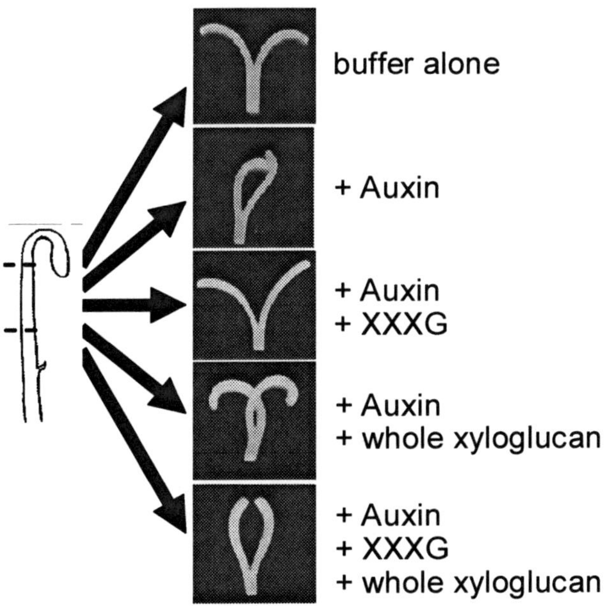

Fig. 3 Effects of auxin, xyloglucan oligosaccharide (XXXG), and whole xyloglucan in the pea split test.

cell wall was decreased by the treatment with XXXG, meaning occurrence of wall loosening.

The xyloglucan polymer is more likely the natural substrate than its oligosaccharide XXXG, because xyloglucan is synthesized in Golgi and secreted as a polymer to the cell walls during growth. The auxin-mediated straightening of the segments was enhanced by incubating in auxin plus whole xyloglucan (around 50 kDa in molecular weight). In fact, the split segments further bent inward at 6 h and formed a hooked cross at 18 h (Figure 3). The effect of whole xyloglucan can be interpreted by hypothesizing that the integration of whole xyloglucan into the wall at the cut surface caused the wall to stiffen. The response of the segments to whole xyloglucan + XXXG + auxin was intermediate between whole xyloglucan + auxin and XXXG + auxin, indicating the integration of XXXG might be competitive with that of whole xyloglucan.

To confirm that the physiological phenomena observed in the split segments treated with xyloglucan and XXXG were caused by the action of pea XETs, we used antibody against pea XET in the split test. To make an antibody, we cloned *PsXTH* cDNA (a gene for pea XET) from the RNAs of pea stems and constructed a high expression vector pET system. *PsXTH* cDNA was expressed in *E. coli* and homogenate of the *E. coli* cells was subjected to SDS-PAGE. The gene product (32 kDa) was isolated and injected into a rabbit to make an antibody against pea XET. The antibody thus obtained inhibited the total XET activity in the crude enzyme preparation of pea stem. We added the purified antibody to the buffer containing auxin and whole xyloglucan or its oligosaccharide, and incubated the split segments in it. The physiological phenomena brought about by the incorporation of xyloglucan and XXXG were partially inhibited by the presence of the antibody, showing that the changes in morphology were caused by the action of XET that incorporated exogenous whole xyloglucan or XXXG into the wall of pea segments (Takeda et al. 2004). The physiological phenomena were observed directly with the naked eyes without a microscope. This is an exciting finding that is almost as epoch-making as when Went discovered auxin.

If the natural substrate for XET is xyloglucan polymer, XET should function to suppress cell elongation during plant growth. However, the level of total XET activity in pea stems was low in the elongating region of the third internode although it was high in the elongated region where cell elongation had ceased and the level of cellulose deposition was high to form the secondary wall. Catala et al. (1997) also found that *LeEXT* mRNA accumulated after the elongation of hypocotyls in tomato, and the treatment with auxin lowered the total XET activity in plant tissues. The cross bridges of xyloglucan between cellulose microfibrils were observed in the non-elongating region of pea stem (Fujino et al. 2000). There is a possibility that XET functions to inhibit cell elongation by generating new xyloglucan tether rather than to accelerate the elongation by cleaving tether. However, we need to define the levels of xyloglucan and its fragment oligosaccharides in the apoplastic spaces of tissues during elongation because the availability of xyloglucan and its oligosaccharide determines the growth rate of plant cells.

The fact that xyloglucan polymer is integrated into the wall shows

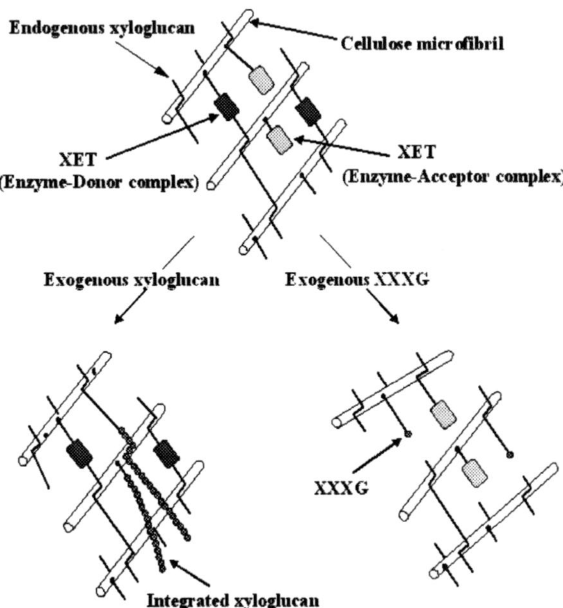

Fig. 4 Possible mechanisms of xyloglucan integration by the action of XET

that the polymer as a donor substrate combined with the enzyme binds to the acceptor molecule of xyloglucan in the wall. XET has a acceptor-binding site for the non-reducing end of xyloglucan (Johansson et al. 2004) and acts as a transferase rather than a hydrolase, although the hydrolase-type XET in nasturtium seeds showed a ping-pong system (Baran et al. 2000). If an XET binds to an internal region of xyloglucan (forming an enzyme-donor complex) in the cell wall and catches a non-reducing end of exogenous xyloglucan, XET cleaves the endogenous one and transfer this newly generated reducing end to the non-reducing end of XXXG. If an XET binds to a non-reducing end of xyloglucan (forming an enzyme-acceptor complex) in the wall and catches the internal region of exogenous xyloglucan, XET cleaves exogenous xyloglucan and transfer it to the endogenous xyloglucan. Thus, xyloglucan in the wall becomes a larger molecule by xyloglucan endotransglycosylation to tighten plant cell walls as shown in Figure 4. The distance between the adjacent cellulose microfibrils was found to be 20 to 40 nm using the shadowed replicas of the sample prepared by the rapid frozen deep etched method (McCann et al. 1990). The next question is how the xyloglucan tether is generated between cellulose microfibrils by integration of the xyloglucan polymer at the molecular level. We are continuing our studies to answer the questions that arise one after another.

Acknowledgement

This work was supported in part by Program for Promotion of Basic Research Activities for Innovation Biosiences (PROBRAIN).

References

Baba K, Sone Y, Misaki A, Hayashi T (1994) Localization of xyloglucan in the macromolecular complex composed of xyloglucan and cellulose in pea stems. Plant Cell Physiol 35: 439-444.

Baran R, Sulova Z, Stratilova E, Farkas V (2000) Ping-pong character of nasturtium-seed xyloglucan endotransglycosylase (XET) reaction. Gen Physiol Biophys 19: 427-440.

Catala C, Rose JKC, Bennett AB (1997) Auxin regulation and spatial localization of an endo-1,4-β-glucanase and a xyloglucan endotransglycosylase in expanding tomato hypocotyls. Plant J 12: 417-426.

Cho HT, Cosgrove DJ (2000) Altered expression of expansin modulates leaf growth and pedicel abcission in *Arabidopsis thaliana*, Proc Natl Aca Sci USA 97: 9783-9788.

Fry SC, Smith RC, Renwick KF, Martin DJ, Hodge SK, Matthews KJ (1992) Xyloglucan endotransglycosylase, a new wall-loosening enzyme activity from plants. Biochem J 282: 821-828.

Fujino T, Sone Y, Mitsuishi Y, Itoh T (2000) Characterization of cross-links between cellulose microfibrils, and their occurrence during elongation growth in pea epicotyl. Plant Cell Physiol 41: 486-494.

Hayashi T (1989) Xyloglucans in the primary cell wall. Annu Rev Plant Physiol Plant Mol Biol 40: 139-68.

Hayashi T, Ogawa K, Mitsuishi Y (1994) Chraracterization of the adsorption of xyloglucan to cellulose. Plant Cell Physiol 35: 1199-1205

Hayashi T (2000) Xyloglucan oligosaccharides in plant cell enlargement. http://www.glycoforum.gr.jp/science/word/saccharide/SA-A04E.html.

Johansson P, Brumer III H, Baumann MJ, Kallas AM, Henriksson H, Denman SE, Teeri TT, Jones TA (2004) Crystal structures of a poplar xyloglucan endotransglycosylase reveal details of transglycosylation acceptor binding. Plant Cell 16: 874-886.

Labavitch JM (1981) Cell wall turnover in plant development. Annu Rev Plant Physiol 32: 385-406.

McCann MC, Wells B, Roberts K (1990) Direct visualization of cross-links in the primary plant cell wall. J Cell Sci 96: 323-334.

Nishitani K, Tominaga R (1992) Endo-xyloglucan transferase, a novel class of glycosyltransferase that catalyzed transfer of a segment of xyloglucan molecule to another xyloglucan molecule. J Biol Chem 267: 1058-21064.

Park YW, Tominaga R, Sugiyama J, Furuta Y, Tanimoto E, Samejima M, Sakai F, Hayashi T (2003) Enhancement of growth by expression of poplar cellulase in *Arabidopsis thaliana*. Plant J 33: 111-118.

Park YW, Baba K, Furuta Y, Iida I, Sameshima K, Arai M, Hayashi T (2004) Enhancement of growth and cellulose accumulation by overexpression of xyloglucanase in poplar. FEBS Letters 564: 183-187.

Popper ZA, Fry SC (2003) Primary cell wall composition of bryophytes and charophytes. Ann. Bot (Lond) 91: 1-12.

Shani Z, Dekel M, Tzbary G, Goren R, Shoseyov O (2004) Growth enhancement of transgenic poplar plants by overexpression of *Arabidopsis thaliana* endo-1,4-β-glucanase (*cel1*). Molecular Breeding 14: 321–330.

Shpigel E, Roiz L, Goren R, Shoseyov O (1998) Bacterial cellulose-binding domain modulates *in vitro* elongation of different plant cells. Plant Physiol 117: 1185-1194.

Takeda T, Furuta Y, Awano T, Mizuno K, Mitsuishi Y, Hayashi T (2002) Suppression and acceleration of cell elongation by integration of xyloglucans in pea stem segments. Proc Natl Aca Sci USA 99: 9055-9060.

Takeda T, Mituishi Y, Sakai F, Hayashi T (1996) Xyloglucan endotransglycosylation in suspension-cultured poplar cells. Biosci Biotech Biochem 60: 1950-1955.

Van Overbeek J, Went FW (1937) Mechanism and quantitative application of the pea test. Bot Gazette 99: 22-41.

A Cell Wall, Autolytic β-Galactosidase that Degrades Pectins during Cell Wall Loosening and Elongation

Emilia Labrador and Berta Dopico

Autolysis of cell walls: the enzymes and sugars involved

Under suitable conditions, cell walls release a series of sugars into the medium by a process called autolysis or autohydrolysis. It was previously thought that this enzyme-mediated process might reflect modifications occurring in the cell wall during growth. The specificity of this process is apparent in monocots, in which only glucose from (1,3;1,4)-ß-glucan is released (Huber and Nevins 1979; Labrador and Nicolás 1982). Release of this compound during autolysis is important due to its involvement in the processes of cell wall relaxation that occur in response to auxins (Loescher and Nevins 1972).

In 1982, during my postdoctoral studies at Iowa State University (USA) with Prof. Nevins, we confirmed that one of the enzymes responsible for autolysis with exoglucanase activity was able to increase elongation in segments of corn coleoptiles, in both the presence and absence of indole acetic acid (IAA) (Labrador and Nevins 1983 and 1989). This linked the autolysis processes with growth.

When I returned to the University of Salamanca, we also examined the autolytic process in dicots. Together with Ph.D. student B. Dopico, I attempted to identify the polysaccharides and enzymes responsible for autolysis in dicots, using chickpeas (*Cicer arietinum*), an important food crop in the Mediterranean region. In dicot cell walls, we found autolysis to be more complex than in monocots. Glucose was not the sole compound released; instead, galactose and arabinose were released jointly, together with small amounts of glucose and xylose (Dopico *et al.* 1989 and 1990b).

The enzyme responsible for this process was identified by analyzing the action of different hydrolytic enzymes, present in the cell wall, on the walls themselves. Only one protein fraction, with β-galactosidase activity (named βIII-galactosidase), was able to release sugars at a proportion and composition similar to those solubilized in the autolytic process of active cell walls (Dopico *et al.* 1989, Dopico *et al.* 1990a) (Figure 1). Antibodies against βIII-galactosidase clearly suppressed autolytic reactions of isolated walls, supporting the idea that it is the primary enzyme responsible for chickpea autolysis (Valero and Labrador 1993).

At that time, β-galactosidase was known to be involved in the breakage of bonds between polysaccharides that occurs during cell wall loosening (Konno *et al.* 1986). An increase in the activity of β-galactosidase after induction with auxins in *Pisum* (Tanimoto and Igari 1976; Tanimoto 1985) was also reported.

In both chickpea autolysis and hydrolysis by βIII-galactosidase, 70% of the sugars released are monosaccharides, primarily galactose (Dopico *et al.* 1990b), indicating that this enzyme mainly acts in its exo-form, although the release of a polysaccharide component composed of arabinose and galactose indicates that the enzyme may also act in an endo-hydrolytic way.

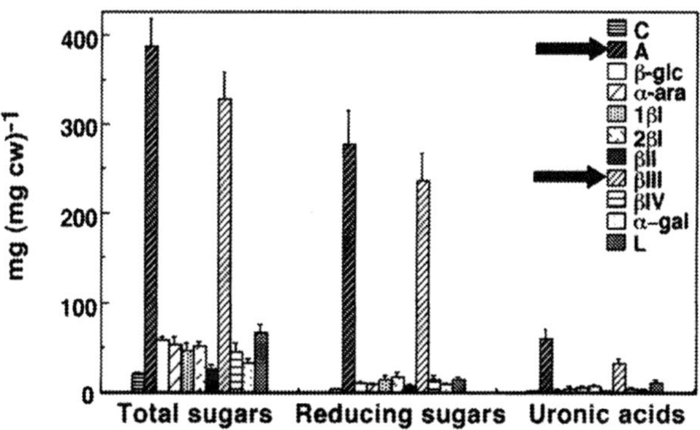

Fig. 1 Total and reducing sugars and uronic acids released during cell wall autolysis (A) and in cell wall hydrolysis mediated by cell wall protein fractions obtained after several purification steps of chickpea cell wall proteins, extracted in 3 M LiCl. Results are expressed as μg quantities of sugars released from 1 mg of cell wall by 100 μg proteins. Control (C) represents μg quantities of sugars released from 1 mg of inactivated cell wall. The protein fractions exhibit the following enzymatic activities: 1βI, 2βI, βII, βIII, and βIV (β-galactosidase); β-glc (β-D-glucosidase); α-ara (α-arabinosidase); α-gal (α-galactosidase) and L (activity against laminarin: 1,3-β-glucan). The data are means of three replicates of each of four independent isolations ± SE (vertical bars).

In 1991, during a stay at the University of California, Davis, in the laboratory of Prof. Nevins and with the aid of Prof. Labavitch's group, I conducted an analysis of autolytic substrates in studies involving methylation and magnetic resonance. The results indicated that the autolytic substrate and the substrate of βIII-galactosidase is a type I arabinogalactan: 1,4-β-galactan chains with short chains of 1,5-α-arabinose at the O-3 of the galactose units, joined to a rhamnogalacturonan I (RG-I). This structure explained why the action of the enzyme in its exo-form releases large amounts of galactose as a monosaccharide, but due to its endo-activity, it is also able to release small polysaccharides, as we had demonstrated.

These pectic arabinogalactan side chains bound to RG-I probably participate in the formation of inter-polymeric cross-links and, conceivably, side-chain hydrolysis may bring about significant effects in upsetting the structural integrity and cohesiveness of the wall. As early as 1980, Nishitani and Masuda linked the modifications in arabinogalactans with the pectin degradation that may occur during elongation, which could connect our studies of the autolytic process to cell wall extension during epicotyl growth.

Relationship between autolysis, βIII-galactosidase, and the growth process

The autolytic process in epicotyl cell walls of chickpeas and the hydrolysis of heat-inactivated cell walls mediated by βIII-galactosidase are maximal on the fourth day of germination, coinciding with maximum growth capacity. The processes decrease over the following days, during which the growth rate diminishes (Dopico *et al.* 1990c) (Figure 2). This relationship between autolysis and growth is further supported by the observation that IAA, which induces epicotyl growth, also elicits an increase in the autolytic release of sugars (Seara *et al.* 1988). Moreover, the stronger growth in the apical segments of epicotyls is correlated with higher levels of autolysis, with both growth rate and autolysis decreasing in medial and basal epicotyl segments (Seara *et al.* 1988).

The βIII-galactosidase protein is nearly absent from 3-day-old epicotyls, coinciding with the scant autolysis appreciable at that age. However, both the quantity of the protein and its specific β-galactosidase activity increase alongside growth, which is in disagreement with the decrease in autolysis observed from the fifth day. If βIII-galactosidase is responsible for autolysis, but both the activity and amount of the enzyme continued to increase, why does autolysis decrease and the enzyme cease to work in the wall?

To address these issues, we explored the hypothesis that modifications might occur in the substrate and the structure of the wall, which might prevent βIII-galactosidase from acting in older epicotyls. Indeed, we observed that a decrease occurred in the autolytic substrate along with growth (Dopico *et al.* 1991), which explains why βIII-galactosidase is unable to maintain function despite its levels and activity. These results led us to theorize that the variations in autolysis and hydrolysis caused by βIII-galactosidase during growth processes are due to structural modifications in the cell walls; this prevents the release of galactose even though the enzyme is present. Thus, we confirm that the βIII-galactosidase from 4-day-old epicotyls is unable to hydrolyse the aged epicotyl cell walls to the same extent as young walls. Accordingly, the βIII-galactosidase from aged epicotyl cell walls hydrolysed young walls more efficiently than cell walls from aged material (Dopico *et al.* 1991).

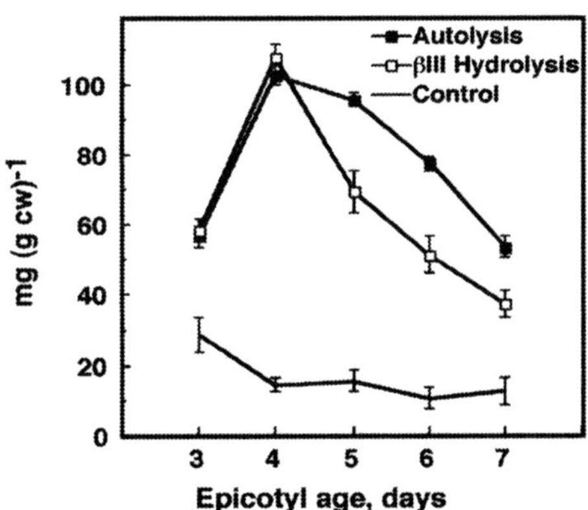

Fig. 2 Time course of cell wall autolytic activity and hydrolytic capacity shown by the protein fraction βIII-galactosidase extracted from epicotyls of varying ages. **The results are expressed as mg quantities of sugar released per g of cell wall. In hydrolysis by βIII-galactosidase, 100 µg protein was used for each incubation. Heat-inactivated cell walls were used as a control. Data are means of three replicates in each of three independent isolations ± SE (vertical bars).**

Physiological significance of pectin degradation by βIII-galactosidase during cellular elongation

Until that time, our studies had indicated a relationship between cellular elongation and pectin degradation carried out by βIII-galactosidase. Next, we question what role pectin degradation might play in the elongation processes. We believed that changes in pectins may lead to a loss of rigidity (loosening) of the wall, even though other changes in the cellulose/xyloglucan network that would later lead to extension and elongation are necessary. The problem was, however, that whenever elongation is studied under natural conditions, the wall loosening and extension occur simultaneously.

To test our hypothesis, we needed to separate the processes of loosening and extension. Fortunately, we found a system in which cell walls loosened but the cells did not elongate. Under condition of PEG-induced osmotic stress, the epicotyls did not elongate but were subject to the loosening process. Our results indicated that the changes observed in the pectic fractions during growth - namely, a decrease in galactose - are related to cell wall loosening, which explains why these changes took place even though the epicotyls did not elongate during PEG-treatment (Muñoz *et al.* 1993). This involvement of the pectin fraction in cell wall loosening supports the idea that cell wall autolysis is an *in vitro* process representing *in vivo* cell wall loosening. It therefore seems clear that the physiological function of βIII-galactosidase is to produce the changes taking place in the loss of cell wall rigidity that precede and accompany growth under normal conditions, even though the changes elicited by other enzymes are further required for elongation.

Confirming the participation of βIII-galactosidase in the elongation process, we observed in chickpea seedlings that this protein appeared only in cell walls of elongating zones, such as in epicotyls and roots, but could not be detected either in cotyledons (reserve organs) or in meristematic hooks (Esteban *et al.* 2003). This allowed us to conclude that βIII-galactosidase acts in the degradation of pectins of the primary cell wall of organs undergoing elongation.

Cloning of the chickpea βIII-galactosidase gene

During a later stage of our inquiries, we wished to identify the cDNA encoding the cell wall βIII-galactosidase of *C. arietinum*. To this end, we used a cDNA library constructed from poly $(A)^+$ RNA extracted from epicotyls of 5-day-old etiolated chickpea seedlings grown in water. After several screenings of the library, we collected four different cDNAs encoding β-galactosidases (Esteban *et al.*, 2003). Identification of the specific cDNA clone encoding the cell wall pectin-degrading βIII-galactosidase (named CanBGal-3) was carried out by comparing the deduced amino acid sequences of the four isolated chickpea β-galactosidase clones to the purified βIII-galactosidase peptide sequence.

The expression pattern of the βIII-galactosidase gene was in concordance with the fluctuations of the enzyme in different seedling organs, specifically present in elongating organs such as epicotyls and roots. Thus, βIII-galactosidase and its corresponding gene expression are associated only with elongating organs, in support of the involvement of chickpea βIII-galactosidase in cell wall elongation.

To confirm that the βIII-galactosidase gene encodes a pectin-degrading enzyme, we constructed transgenic plants by introducing the βIII-galactosidase cDNA (CanBGal-3) into the *Solanum tuberosum* genome and studied its *in vivo* expression in potato tubers. The cell wall proteins extracted from several potato transformants showed hydrolytic activity against inactivated chickpea cell walls, to an extent that correlated with the transgene expression level in the transformants (Esteban *et al.* 2003). Cell wall proteins from wild-type potato tubers were able to release sugars only in low amounts, similar to the amounts released by a control of inactivated cell wall. The composition of the sugars hydrolysed by proteins from transgenic lines supports the notion that the *in vivo* expressed βIII-galactosidase increases the amount of galactose released from chickpea cell walls.

Our studies of a β-galactosidase able to hydrolyse cell wall pectins helps us understand the function of pectin in the cell wall. β-galactosidase activity on cell wall pectins could significantly affect pectin solubility and cell wall porosity (De Veau *et al.* 1993; Baron-Epel *et al.*, 1988). Several reports regarding RG-I-associated galactan have suggested that this specific pectic polysaccharide may be involved in the regulation of cell wall porosity (Fenwick *et al.* 1999; Brummell and Harpster 2001). Kakegawa (2000) also suggested that neutral side chains may be responsible for cell wall rigidity. Thus, the action of β-galactosidase on the neutral side chains of the pectic backbone could have major implications for the matrix of the cell wall, by rendering substrates susceptible to hydrolysis by other cell wall-degrading enzymes during extension.

Most studies of β-galactosidase genes refer to their expression in ripening fruits, in which the degradation of galactan is associated with the ripening process (reviewed in Brummell and Harpster 2001). To our knowledge, few studies have examined β-galactosidase genes in relation to vegetative organs. Our studies of βIII-galactosidase gene expression point to an important role for this enzyme in pectin degradation in elongating chickpea organs (Esteban *et al.* 2003), as was previously suggested on the basis of the enzymatic activity of this chickpea galactosidase (Dopico *et al.* 1989, 1990b).

Current studies by our group regarding the immunolocalization of βIII-galactosidase protein should afford deeper insight into the activity of this enzyme and its involvement in cell growth. Likewise, we are studying the other β-galactosidase cDNAs identified in the chickpea, hoping to attribute a specific function to each of the enzymes (Esteban *et al.* 2005). The future construction of transgenic plants with these clones should allow us to unequivocally establish their activities in the cell wall.

Acknowledgments

The authors wish to thank Profs. Nevins, Nicolás, and Zarra for their help during the early stages of our research. The authors also wish to thank the students who, over the years, contributed to the studies of β-galactosidases: P. Valero, J. Seara, F.J. Muñoz, S. Romo, R. Esteban, and I. Martín. The research reported here was supported by grants from the European Community, the Ministerio de Ciencia y Tecnología (Spain), and the Junta de Castilla y León (Spain).

References

Baron-Epel O, Gharyal PK, Schindler M (1988) Pectins as mediators of wall porosity in soybean cells. Planta 175: 389-395.

Brummell DA, Harpster MH (2001) Cell wall metabolism in fruit softening and quality and its manipulation in transgenic plants. Plant Mol Biol 47: 311-340.

De Veau EJI, Gross KC, Huber DJ, Watada AE (1993) Degradation and solubilization of pectin by ß-galactosidases purified from avocado mesocarp. Physiol Plant 87: 279-285.

Dopico B, Nicolás G, Labrador E (1989) Partial purification of cell wall β-galactosidases from *Cicer arietinum* epicotyls. Relationship with cell wall autolytic proceses. Physiol Plant 75: 458-464.

Dopico B, Nicolás G, Labrador E (1990a) Characterization of a cell wall β-galactosidase of *Cicer arietinum* epicotyls involved in cell wall autolysis. Physiol Plant 80: 629-635.

Dopico B, Nicolás G, Labrador E (1990b) Cell wall localization of the natural substrate of a β-galactosidase, the main enzyme responsible for the autolytic process of *Cicer arietinum* epicotyl cell walls. Physiol Plant 80: 636-641.

Dopico B, Nicolás G, Labrador E (1990c) Changes during epicotyl growth of an autolysis-related β-galactosidase from the cell wall of *Cicer arietinum*. Plant Science 72: 45-51.

Dopico B, Muñoz FJ, Nicolás G, Labrador E (1991) Cell wall structure regulates the autolytic process throughout growth of *Cicer arietinum* epicotyls. Physiol Plant 83: 659-663.

Esteban R, Dopico B, Muñoz FJ, Romo S, Martín I, Labrador E (2003) Cloning of a *Cicer arietinum* β-Galactosidase with pectin-degrading function. Plant Cell Physiol 44: 718-725.

Esteban R., Labrador, E., Dopico, B. (2005) A family of β-galactosidase cDNAs related to development of vegetative tissue in *Cicer arietinum*. Plant Science 168: 457-466.

Fenwick KM, Apperley DC, Cosgrove DJ, Jarvis MC (1999) Polymer mobility in cell walls of cucumber hypocotyls. Phytochem 51: 17-22.

Huber DJ, Nevins DJ (1979) Autolysis of the cell wall β-D-glucan in corn coleoptiles. Plant Cell Physiol 20: 201-212.

Kakegawa K, Edashige Y, Ishii T (2000) Metabolism of cell wall polysaccharides in cell suspension cultures of *Populus alba* in relation to cell growth. Physiol Plant 108: 420-425.

Konno H, Yamasaki Y, Katoh K (1986) Characteristics of β-galactosidase purified from suspensión cultures of carrot. Physiol Plant 68: 46-52.

Labrador E, Nevins DJ (1983) Cell wall proteins: Their role in autohydrolysis and cell elongation. Plant Physiol 72: 73.

Labrador E, Nevins DJ (1989) An exo-β-D-glucanase derived from *Zea* coleoptile walls with a capacity to elicit cell elongation. Physiol Plant 77: 479-486.

Labrador E, Nicolás G (1982) Autolytic activities of the cell wall in rice coleoptiles. Effects of nojirimycin. Physiol Plant 55: 345-350.

Loescher W, Nevins DJ (1972) Auxin-induced changes in *Avena* coleoptile cell wall composition. Plant Physiol 50: 556-563.

Muñoz FJ, Labrador E, Dopico B (1993) Effect of osmotic stress on the growth of epicotyls of *Cicer arietinum* in relation to changes in the cell wall composition. Physiol Plant 87: 552-560.

Nishitani K, Masuda Y (1980) Modifications of cell wall polysaccharides during auxin-induced growth in azuki bean epicotyl segments. Plant Cell Physiol 21: 169-181.

Seara J, Nicolás G, Labrador E (1988) Autolysis of the cell wall. Its posible role in endogenous and IAA-induced growth in epicotyls of *Cicer arietinum*. Physiol Plant 72: 769-774.

Tanimoto E, Igari M (1976) Correlation between β-galactosidase and auxin-induced elongation growth in etiolated pea stems. Plant Cell Physiol 17: 673-682.

Tanimoto E (1985) Axial distribution of glycosidases in relation to cellular growth and ageing in *Pisum sativum* root. J Exp Bot 36: 1267-1274.

Valero P, Labrador E (1993) Inhibition of cell wall autolysis and auxin-induced elongation of *Cicer arietinum* epicotyls by β-galactosidase antibodies. Physiol Plant 89: 199-203.

Genetic Analysis of Cell Walls

Chris Somerville

My interest in plant cell walls began during a conversation with Peter Albersheim in 1985. We were invited to a small workshop on Fraser Island Australia that was organized by Jim Peacock to encourage dialog between US and Australian Scientists. The meeting was held in a very isolated but beautiful setting on the ocean that was conducive to long open-ended conversations. As Peter described the many challenges associated with understanding how cell walls were made and how they functioned, I became intrigued by the idea that the field would greatly benefit by a genetic approach. I had recently had some success in identifying a series of mutants that was useful in resolving certain difficult problems in photorespiration (Somerville 2001) and, in collaboration with John Browse and Peter McCourt, had also had some successes in isolating a large series of mutants that would ultimately prove useful in working out various aspects of lipid metabolism (Browse et al. 1985; Somerville and Browse 1991). I was, and am, of the opinion that mutations could be identified in essentially any process by screening directly for variation in heavily mutagenized populations. In the case of lipid metabolism, we had isolated mutants by simply screening large numbers of plants for alterations in fatty acid composition by gas chromatography. I assumed that the same method could work for finding variation in cell wall composition.

For the next five years I tried repeatedly without success to interest any of my students or postdocs in working on cell wall mutants. My lab attracted people who were interested in using Arabidopsis to study a wide variety of topics but the cell wall was not seen as a suitably fashionable topic. The idea of looking for mutants that affected the wall was seen as tedious and difficult at best. In 1990 Clint Chapple and Wolf-Dieter Reiter joined my lab as postdocs. To my surprise, they both expressed interest in the project and so we decided that they should undertake the project as collaboration. Clint had just finished a degree in chemistry and was comfortable with analytical methods. Dieter had done a degree in microbial molecular genetics and his previous experience was, therefore, very complementary to

Clint's.

Our basic idea was to test the degree to which we could modify the polysaccharide composition of the cell wall by single mutations. We decided to take an unbiased approach and simply screen every plant in a chemically mutagenized population. We reasoned that if we inactivated a particular enzyme with a mutation, and if the corresponding gene was not redundant to another gene, and if the mutation was not lethal, the mutation should cause a change in the overall sugar composition of cell walls. One concern we had about this scheme was the possibility that many mutations with effects on growth and development might cause non-specific changes in cell wall composition that could obscure our ability to identify mutants with defects in biosynthetic enzymes. We tested this by making measurements on a variety of ecotypes and developmental mutants (Reiter et al. 1997). The results indicated that none of the mutants or ecotypes that we tested had significant effects on total cell wall sugar composition. Therefore, we decided to proceed.

We grew about 10,000 M_2 plants from an ethylmethanesulfonate-mutagenized population. We marked each plant with a small numbered flag (a piece of tape on a toothpick) and then took one leaf, extracted the soluble sugars, then hydrolyzed the leaf in sulfuric acid. The resulting sugars were converted to alditol acetates and measured by gas chromatography. Because uronic acids were degraded by the procedures, the screen was not designed to identify mutants with certain defects in pectin biosynthesis. Approximately three days a week, Clint and Dieter would produce a total of 300 samples and during the remainder of the week they would analyze the results and retest any line that seemed promising. The stack of paper chromatograms from this mutant screen filled twelve bankers boxes and made a stack about 13 feet in height (Figure 1). It took about a year to do the screen but it was very productive. We identified twenty-three unambiguous mutants that were subsequently found to represent eleven different genes (Reiter et al. 1997) and fifteen additional putative mutations in which the phenotypes were not pronounced enough to support additional genetic studies. The mutants included eight lines lacking fucose (*mur1-1* to *mur1-8*), three lines with approximately 50% reduction in fucose (*mur2* and *mur3-1, 3-2*), five lines with reduced arabinose (*mur4-1, 4-2, mur5, mur6, mur7*) one line with reduced rhamnose (*mur8*) one line with reduced fucose and xylose (*mur9*) and three lines with more complex changes (*mur10-, 10-2, mur11*).

The class of mutants that most attracted our attention initially were the *mur1* mutants, which were essentially lacking fucose in the aerial parts of the plant (Reiter et al. 1993). Since a significant proportion of fucose is found in xyloglucan, one implication of the mutant was that it had a strongly altered xyloglucan in which the repeating nonasaccharide XXFG had become XXLG. This was particularly interesting because previous studies from Will York, Alan Darvill, Steven Fry, Peter Albersheim and colleagues had suggested that XXFG might have an in vivo function corresponding to the observation that exogenous application of XXFG inhibited auxin-induced elongation growth (York et al. 1984). Since McDougall and Fry (1989) had shown that the terminal fucose was required for oligosaccharin activity, we thought that the relatively normal growth and development

Fig. 1 The mutant screen finally ends. **The boxes in the background hold the gas chromatograms from the mutant screen. From left to right: Clint Chapple, Chris Somerville, Wolf-Dieter Reiter.**

of the *mur1* mutants was evidence that the role of XXFG as an oligosaccharin was less important than proposed. Zablackis et al (1996) subsequently showed that the missing fucose in xyloglucan of the *mur1* mutant was partially replaced by L-galactose, and that the galactose-containing homolog of XXFG had comparable oligosaccharin activity to the fucose-containing variant. However, characterization of the fucose-deficient *mur2* mutant showed that it lacked the fucosyltransferase that adds the terminal fucose to xyloglucan and, as a result, the *mur2* mutant lacks XXFG but does not show any major effects on growth and development (Vanzin *et al.* 2002). Thus, several of the mutants were ultimately very useful in critically testing the in vivo functions of oligosaccharins. During the original characterization of the *mur1* mutant we had observed that there was a large decrease in the amount of force required to break cell walls (Reiter *et al.* 1993). Because fucose is a quantitatively minor component of cell walls, and because cellulose is thought to be the main load-bearing component of cell walls, we were surprised by the effects of the *mur1* mutation on tensile strength. The *mur2* mutant did not have a similar decrease in tensile strength, suggesting that a factor other than xyloglucan was responsible for the effect. Subsequently, the crosslinking of RGII by borate diesters was discovered (Kobayashi *et al.* 1996; ONeill *et al.* 1996). O'Neil and colleagues then showed that in the *mur1* mutant, the fucose residues that are normally present in RGII are replaced by α-L-galactose (O'Neill *et al.*, 2001). Furthermore, they showed that the somewhat reduced growth of the *mur1* mutants could be reversed by providing increased levels of borate, which overcame a reduced rate of borate di-ester formation by the structurally modified RGII molecules in the mutant. This provided a very compelling demonstration that the growth rate of plants depends to a surprising degree on the efficiency of RGII crosslinking in the cell walls.

Ryden *et al.* (2003) tested the effect of borate supplementation on the tensile strength of the *mur1* cell walls, and several other mutants. The tensile properties of the hypocotyls and of the inflorescence stems of *mur1* were rescued by growth in the presence of high concentrations of borate. Thus, the RGII borate di-esters were shown to have a major effect on the tensile strength of the cell walls. From comparison of the mechanical responses of *mur2* and *mur3*, they also concluded that galactose-containing side chains of xyloglucan make a major contribution to overall wall strength, whereas xyloglucan fucosylation plays a comparatively minor role.

Several of the other *mur* mutants have recently been cloned. The *mur3* gene was found to encode a galactosyltransferase involved in xyloglucan synthesis (Madson *et al.* 2003). The corresponding enzyme acts specifically on the third xylose residue within the XXXG core structure of xyloglucan. The *mur4* mutants have a 50 % decrease in the amount of arabinose in leaf polysaccharides. Positional cloning of the corresponding gene revealed that it encodes an epimerase that converts UDP-xylose to UDP-arabinose (Burget and Reiter 1999; Burget *et al.* 2003). The gene was found to be one of four closely related genes in plants, explaining why the mutant was not completely deficient in arabinose. Interestingly, a mutation (*rhd1*) originally isolated in a root hair mutant screen in my lab, was found to be defective in one of the five UDP-glucose epimerases (Seifert *et al.* 2002). The RHD1 isoform is specifically required for the galactosylation of xyloglucan and type II arabinogalactan but is not involved either in galactose detoxification or in galactolipid biosynthesis. Epidermal cell walls in the root expansion zone lack arabinosylated (1→6)-β-galactan and galactosylated xyloglucan. In cortical cells of the *rhd1* mutant, galactosylated xyloglucan is absent, but an arabinosylated (1→6)-β-galactan is present. It was concluded that the flux of galactose from UDP-galactose into different downstream products is compartmentalized at the level of cytosolic UDP-glucose epimerase isoforms. This suggests that substrate channeling plays a role in the regulation of plant cell wall biosynthesis.

Although the mutant screen produced a number of useful mutations, not all of which have yet been characterized at the molecular level, it failed to result in identification of any of the genes for enzymes that make the backbones of the major polymers. Thus, since the screen was quite extensive, it is apparent that such mutants will not be found in mutant screens of relatively normal plants. With the completion of the genome sequence of Arabidopsis, it has become possible to isolate the gene corresponding to any mutation. The use of these methods has greatly facilitated the cloning of some of the genes corresponding to the *mur* mutants. Although it is possible to identify the genes by positional cloning methods, it requires the analysis of several thousand samples by gas chromatography to map the gene with sufficient accuracy. Thus, many people will prefer to try and find a less tedious approach to identifying useful mutations. The availability of large numbers of sequence-indexed mutations in Arabidopsis has made it possible to obtain mutations in most of the probable glycosyltransferases by email. As a result, there are now easier ways to study cell wall biosynthesis and function than by classical genetic methods. Nevertheless, the mutants have been used in a variety of other studies and will, no doubt, continue to be a useful set of biological materials for studies of cell wall structure and function.

Looking back on the experiment it still seems to me that it was a good idea and that perhaps it is time to redo it with some variations that might allow us to expand the range of mutants recovered. It is apparent from the genome sequence of Arabidopsis that there are a large number of genes with characteristics suggestive of a role in wall synthesis. Therefore, it should be possible to identify many additional useful mutants with a more sensitive screen.

References

Browse JP, McCourt P, Somerville CR (1985) A mutant of *Arabidopsis* lacking a chloroplast-specific lipid. Science 227: 763-765.

Burget E, Reiter W (1999)The *mur4* mutant of Arabidopsis is partially defective in the de novo synthesis of uridine diphospho L-arabinose. Plant Physiol 121: 383-389.

Burget E, Verma R, Molhoj M, Reiter W (2003) The biosynthesis of L-arabinose in plants: Molecular cloning and characterization of a Golgi-localized UDP-D-xylose 4-epimerase encoded by the *MUR4* gene of Arabidopsis. Plant Cell 15: 523-531.

Kobayashi M, Matoh T, Azuma J (1996) Two chains of rhamnogalacturonan II are cross-linked by borate-diol ester bonds in higher plant cell walls. Plant Physiol 110: 1017-1020.

Madson M, Dunand C, Li X, Verma R, Vanzin G, Calplan J, Shoue D, Carpita N, Reiter W (2003) The MUR3 gene of Arabidopsis encodes a xyloglucan galactosyltransferase that is evolutionarily related to animal exostosins. Plant Cell 15: 1662-1670.

McDougall GJ, Fry SC (1989) Structure-activity-relationships for xyloglucan oligosaccharides with antiauxin activity. Plant Physiol 89: 883-887.

O'Neill MA, Eberhard S, Albersheim P, Darvill AG (2001) Requirement of borate cross-linking of cell wall rhamnogalacturonan II for Arabidopsis growth. Science 294: 846-849.

ONeill MA, Warrenfeltz D, Kates K, Pellerin P, Doco T, Darvill AG, Albersheim P (1996) Rhamnogalacturonan-II, a pectic polysaccharide in the walls of growing plant cell, forms a dimer that is covalently cross-linked by a borate ester - In vitro conditions for the formation and hydrolysis of the dimer. J Biol Chem 271: 22923-22930.

Reiter WD, Chapple CC, Somerville CR (1993) Altered growth and development in a fucose-deficient cell wall mutant of *Arabidopsis*. Science 261: 1032-1035.

Reiter WD, Chapple CC, Somerville CR (1997) Mutants of *Arabidopsis thaliana* with altered cell wall polysaccharide composition. Plant J 12: 335-345.

Ryden P, Sugimoto-Shirasu K, Smith AC, Findlay K, Reiter WD, McCann MC (2003) Tensile properties of Arabidopsis cell walls depend on both a xyloglucan cross-linked microfibrillar network and rhamnogalacturonan II-borate complexes. Plant Physiol 132: 1033-1040.

Seifert G, Barber C, Wells B, Dolan L, Roberts K (2002) Galactose biosynthesis in Arabidopsis: Genetic evidence for substrate channeling from UDP-D-galactose into cell wall polymers. Curr Biol 12: 1840-1845.

Somerville CR, Browse J (1991) Plant Lipids: Mutants, metabolism and membranes. Science 252: 80-87.

Somerville CR (2001) An early Arabidopsis demonstration. Resolving a few issues concerning photorespiration. Plant Physiol 125: 20-24.

Vanzin GF, Madson M, Carpita NC, Raikhel NV, Keegstra K, Reiter WD (2002) The *mur2* mutant of Arabidopsis thaliana lacks fucosylated xyloglucan because of a lesion in fucosyltransferase AtFUT1. Proc Natl Acad Sci USA 99: 3340-3345.

York WS, Darvill AG, Albersheim P (1984) Inhibition of 2,4-dichlorophenoxyacetic acid-stimulated elongation of pea stem segments by a xyloglucan oligosaccharide. Plant Physiol 75: 295-297.

Zablackis E, York WS, Pauly M, Hantus S, Reiter WD, Chapple CCS, Albersheim P, Darvill A (1996) Substitution of L-fucose by L-galactose in cell walls of Arabidopsis *mur1*. Science 272: 1808-1810.

Cell Wall Formation and Xylem Integrity

Simon Turner

Xylem vessels

The differentiation of xylem vessels is a complex process that involves cell expansion and cell death. This leads to the removal of cell contents and a functioning vessel that is composed only of the remaining cell wall material. Vessels have long been noted for the localised deposition of the secondary cell wall that gives vessels their characteristic patterned appearance. What is also intriguing about the localised cell wall deposition is that it is restricted to the lateral walls of the cell and must be excluded from the end wall where the perforation plate will form. Wilted-dwarf is a tomato mutant in which the secondary cell wall is not excluded from the end wall and the wilting phenotype is presumably a consequence of the perforation plate being partially occluded (Alldridge 1964). It remains an important mutant since it should give some insight into how plants retain the information that allows them to distinguish the lateral walls from the end walls.

Isolation of secondary cell wall mutants

It was obvious to me that once a xylem vessel looses its contents it has no turgor pressure to maintain its cell shape. What is not always clear is that the secondary cell wall is not required to prevent surrounding cells from crushing it, but to resist the negative pressure that is generated inside the vessel during water transport. Two important papers demonstrate the importance of the secondary cell wall. In the presence of AOPP, an inhibitor of phenylalanine ammonia lyase, the xylem vessels of bean seedlings collapse inwards. This is presumably a result of reductions in lignin or other phenylpropanoids in the secondary cell wall (Amrhein et al. 1983; Smart and Amrhein 1985). It was reading these papers that made me realise that this collapse of xylem vessels represented the phenotype of any mutant with a defective secondary cell wall.

Screening for the phenotype was laborious and could only be done by hand sectioning stems of mutated M2 plants and then viewing the toluidine blue stained sections under the microscope. Later efforts involved pre-screening for weak or abnormal looking stems followed by sectioning. Since the phenotype could be caused by any secondary cell wall defect it was surprising that the initial mutants fell into three complementation groups with a specific defect in cellulose and one with a lignin deficient phenotype (Turner and Somerville 1997; Jones et al. 2001). Subsequent screening identified further alleles and an additional cellulose deficient phenotype (Taylor et al. 2003). Despite its abundance and importance, most people would agree that our understanding of how cellulose is synthesised remains poor. Consequently, the isolation of these mutants appeared an excellent opportunity to gain important insight into the synthesis of this very abundant and important plant polymer.

Multiple CESA proteins are required for cellulose synthesis

irx1,3 and *5* are all caused by a mutation in member of the Arabidopsis *CESA* gene family (Taylor et al. 1999; Taylor et al. 2000; Taylor et al. 2003). This family of proteins was first identified as an EST in cotton fibres that exhibited homology to bacterial cellulose synthase genes and the protein product bound to UDP-glucose (Pear et al. 1996). CESA proteins were subsequently demonstrated to be unequivocally required for cellulose synthesis when the cellulose deficient *rsw1* mutants were shown to be caused by a defect in AtCesA1 (Arioli et al. 1998). Both genetic and expression data suggest that three CESA proteins are absolutely required to make cellulose in the secondary cell wall (Taylor et al. 2003). This was very clearly shown to be true in both monocots and dicots when it was demonstrated that three cellulose-deficient *brittle-culm* mutants of rice were caused by a mutation in the rice homologues of irx1, 3 and 5 (Tanaka et al. 2003).

Co-precipitation experiments demonstrate that IRX1, 3 and 5 all form part of the same complex (Taylor et al. 2000; Taylor et al. 2003). The essential question is why multiple CESA proteins are required for cellulose synthesis. Studying mutants in which one of the CESA proteins is missing has given some insight. The *irx5-1* insertion is caused by a transposon and contains no detectable IRX5 protein, in this mutant IRX1 and IRX3 no longer associate. The *irx1-1* allele is caused by a change to an aspartate residue that is conserved in all processive glucosyltransferases that is presumed to be essential for catalysis. The protein is assumed to have no activity but appears to be made at normal levels. In an *irx1-1* mutant all proteins appear to associate normally and form a complex. Our current working hypothesis is that three different CESA proteins are required to make the intersubunit contacts for the very large rosette structures that represent the plant cellulose synthase complex. If this idea is correct then specific interaction between subunits are probably mediated by the two variable regions, an idea that should be testable.

One of the most essential questions about cellulose synthesis in plants is what determines the structure of a basic microfibril. It is likely that the organisation of catalytic sites within the rosette has an essential role in determining microfibril structure. Clearly these questions could be answered by solving the structure of the rosette, but this goal remains a long way off at present. A more realistic short term goal at present is to determine all the essential components required to make cellulose and to determine their function. The problems are typified by the *KORRIGAN* gene (Nicol et al. 1998) that has been identified independently by several groups and demonstrated to cause a defect in cellulose synthesis (Szyjanowicz et al. 2004). How exactly it functions in cellulose synthesis, however, remains unclear. While genetics has contributed to identifying a number of genes that are potentially essential for cellulose synthesis, it is likely that at least some of these genes are involved in more general housekeeping functions within the cell (e.g. Gillmor et al. 2002). This must be borne in mind since it is likely that genomic techniques, that combine proteomics or microarrays with mutants generated via reverse genetic, will identify more genes potentially involved in cellulose synthesis.

Cortical microtubules and secondary cell wall formation

It has been well documented that during vessel development bands of cortical microtubules coincide with areas of secondary cell wall deposition. Alteration in the pattern of cortical microtubules using drugs results in altered patterns of cell wall deposition that exactly matches the pattern of microtubules. The arguments concerning how, or whether, microtubules regulate the orientation of the cellulose microfibrils are discussed briefly below, however, it is clear that the microtubules have a variety of functions in secondary cell wall deposition. One report suggests that secondary cell wall deposition is regulated by self assembly and that cellulose deposition is required for localised deposition of lignin (Taylor et al. 1992). Our work on cell wall mutants does not support this idea. In cellulose deficient mutants or lignin deficient mutants transmission electron microscopy demonstrates that secondary cell wall deposition follows a similar distribution as found in wild type. Even in *irx3/irx4* double mutants that possess a cell wall presumably composed largely of xylan and protein the secondary cell wall is still located in the banded pattern characteristic of the wild type (L. Jones and S. Turner unpublished). How bands of microtubules are themselves targeted to the specific regions of the plasma membranes and whether these band of microtubules play a direct role in the targeting on cell wall components remain intriguing questions.

Localisation of CESA proteins and microtubules in developing xylem has confirmed that banded patterns of microtubules appear before CESA proteins localisation to the plasma membrane and so clearly mark the sites of future cellulose deposition. Given that all three secondary cell wall CESAs form part of a complex, it is perhaps not surprising that they all colocalise with the bands of cortical microtubules. In mutants such as the *irx5-1* however, in which one CEAS protein is absent the two remaining CESA proteins appear to be retained within the cell (Gardiner et al. 2003). The fact that localisation is normal in an *irx1-1* mutant suggest that the presence of the protein and not activity is required for proper organisation. This is consistent with studies on other membrane bound multisubunit enzyme complexes, in that all subunits must be present for the complex to exit the endoplasmic reticulum. The large size and complex organisation of the cellulose synthase rosette that presumably contains in the order of 18-36 CESA proteins means that assembly is likely intricate process that is likely to require specific chaperonins.

Control of cellulose microfibril orientation

As mentioned above many studies on developing xylem have contributed to the debate as to if, or how, microtubules regulate the orientation of cellulose microfibrils deposition. Studies on xylem vessels have contributed to the idea that microtubules constrain and/or direct the movement of the cellulose synthase rosette and hence the orientation of cellulose microfibrils. An excellent comparison of the different models can be found in a review by Heath and Seagull (1982) and a detailed account reviewing a wide range of experiments was published recently by Baskin (2001). Nearly every study to date has relied on examining the correlation between cellulose microfibril orientation and microtubule organisation. Movement of the rosette is inferred from the orientation of the cellulose microfibrils. The availability of antibodies to the CESA proteins has allowed us to visualise the rosette directly and simultaneously visualise microtubule organisation. There is a very good correlation between retention of microtubule banding with localisation of the CESA proteins (Gardiner et al. 2003). Cortical microtubules appear to be continuously required for proper CESA localisation supporting a direct role of microtubules in orientating cellulose microfibrils. Clearly more work is required, but work with CESA:GFP fusions together with similar constructs that monitor microtubules dynamics are likely to shed more light on the dynamics of this interaction.

Conclusions

Despite the recent progress made in our understanding of cellulose synthesis the central questions remain very similar to those that have been studied for at least 30 years. How do microtubules affect movement of the cellulose synthase rosette, what is the composition and organisation of the rosette, what is the role of primers in cellulose synthesis? Many of these questions depend upon being able to isolate large quantities of a purified cellulose synthase complex. With new technology the answers to many of these question should be within our grasp, although happily the debates are likely to continue for quite some time.

Acknowledgements

I would like to thank Jon Pitman and Nicola High for their comments on the manuscript and all my colleagues who have contributed to the work described in this manuscript.

References

Alldridge NA (1964) Anomolous vessel elements in wilty-dwarf plants. Botanical Gazette 40: 877-889.

Amrhein N, Frank G, Lemm G, Luhmann HB (1983) Inhibition of lignin formation by L-alpha-aminooxy-beta-phenylpropionic acid, an inhibitor of phenylalanine ammonia- lyase. Eur J Cell Biol 29:139-144.

Arioli T, Peng LC, Betzner AS, Burn J, Wittke W, Herth W, Camilleri C, Hofte H, Plazinski J, Birch R, Cork A, Glover J, Redmond J, Williamson RE (1998) Molecular analysis of cellulose biosynthesis in Arabidopsis. Science 279: 717-720.

Baskin TI (2001) On the alignment of cellulose microfibrils by cortical microtubules: a review and a model. Protoplasma 215: 150-171.

Gardiner JC, Taylor NG, Turner SR (2003) Control of cellulose synthase complex localization in developing xylem. Plant Cell 15: 1740-1748.

Gillmor CS, Poindexter P, Lorieau J, Palcic MM, Somerville C (2002) alpha-Glucosidase I is required for cellulose biosynthesis and morphogenesis in Arabidopsis. J Cell Biol 156: 1003-1013.

Heath IB, Seagull RW (1982) Orientated cellulose fibrils and the cytoskeleton: a critical comparison of models. In CW Lloyd (ed) The Cytoskeleton in Plant Growth and Development. Academic Press, London, pp 163-184.

Jones L, Ennos AR, Turner, SR (2001) Cloning and characterization of *irregular xylem4* (*irx4*): a severely lignin-deficient mutant of Arabidopsis. Plant J 26: 205-216.

Nicol F, His I, Jauneau A, Vernhettes S, Canut H, Hofte H (1998) A plasma membrane-bound putative endo-1,4-beta-D-glucanase is required for normal wall assembly and cell elongation in Arabidopsis. EMBO J 17: 5563-5576.

Pear JR, Kawagoe Y, Schreckengost WE, Delmer DP, Stalker DM (1996) Higher plants contain homologs of the bacterial celA genes encoding the catalytic subunit of cellulose synthase. Proc Natl Acad Sci USA 93: 12637-12642.

Smart CC, Amrhein N (1985) The influence of lignification on the development of vascular tissue in *Vigna radiata* L. Protoplasma 124: 87-95.

Szyjanowicz PMJ, McKinnon I, Taylor NG, Gardiner J, Jarvis MC, Turner SR (2004) The irregular xylem2 mutant is an allele of korrigan that affects the secondary cell wall of *Arabidopsis thaliana*. Plant J 37: 730-740.

Tanaka K, Murata K, Yamazaki M, Onosato K, Miyao A, Hirochika H (2003) Three distinct rice cellulose synthase catalytic subunit genes required for cellulose synthesis in the secondary wall. Plant Physiol 133: 73-83.

Taylor JG, Owen TP, Koonce LT, Haigler CH (1992) Dispersed lignin in tracheary elements treated with cellulose synthesis inhibitors provides evidence that molecules of the secondary cell-wall mediate wall patterning. Plant J 2: 959-970.

Taylor NG, Scheible WR, Cutler S, Somerville CR, Turner SR (1999) The irregular xylem3 locus of arabidopsis encodes a cellulose synthase required for secondary cell wall synthesis. Plant Cell 11: 769-779.

Taylor NG, Laurie S, Turner, SR (2000) Multiple cellulose synthase catalytic subunits are required for cellulose synthesis in Arabidopsis. Plant Cell 12: 2529-2539.

Taylor NG, Howells RM, Huttly AK, Vickers K, Turner, SR (2003) Interactions among three distinct CesA proteins essential for cellulose synthesis. Proc Natl Acad Sci USA 100: 1450-1455.

Turner SR, Somerville CR (1997) Collapsed xylem phenotype of Arabidopsis identifies mutants deficient in cellulose deposition in the secondary cell wall. Plant Cell 9: 689-701.

Fucose-Deficiency of *Arabidopsis thaliana* Mutant *mur1* Decreases Cell Wall Porosity

María Teresa Herrera, Beatriz Barral, Gloria Revilla and Ignacio Zarra

Primary walls are sufficiently permeable to allow the movement of low relative molecular mass. However, the porosity of cell walls restricts the size of the macromolecules that can diffuse through it, the size of this filter being important for several aspects of cell physiology (Read and Bacic 1996). So, the mobility through the wall and consequently substrate accessibility of enzymes that process incoming nutrients and modify wall structure would be regulated by the wall porosity. Cell wall porosity has been measured using different techniques (ultrastructural, bulk exclusion and functional assays), the range of values for different materials and techniques being 3.5-9.2 nm (see Read and Bacic 1996 for review). However the range obtained using the same technique was narrower. Thus although some part of this variation undoubtedly represents real differences in the porosity of different cell types, the majority come from differences in the methods and their inherent assumptions.

Pectins (Baron-Epel et al. 1988; Ehwald et al. 1992) and more specifically RG II (O'Neill et al. 2001; Fleischer et al. 1999), have been considered to provide the background porosity of the walls, although the participation of other cell wall components as hemicellulosic polysaccharides cannot be discarded. So, the Arabidopsis mutants altered in their wall composition represent a very useful tool to study the relation between cell wall structure and wall porosity. Thus, the *mur1* mutation with a severe fucose deficiency in the cell walls that it has been shown to confer a dwarf phenotype and their cell walls became considerably more fragile than normal type (Reiter et al. 1993), was used for this study.

Suspension-cultured cells generated from basal leaves of both *Arabidopsis thaliana* Col0 and *mur1* mutant lines were grown on MS medium supplemented with 0.5 mg L^{-1} 2,4-dichlorophenoxyacetic acid and 0.05 mg L^{-1} kinetin at 22 °C under a 16 h light and 8 h dark photoperiod and orbital shaking (100 rpm). The absence of fucose in the walls of the mutant was confirmed by GC after alditol-acetate derivatization.

The pore size of cell walls was measured using two different methods based on two different processes, solute exclusion and dextran permeation analysis previously described by Carpita et al. (1979) and Woehlecke and Ehwald (1995), respectively. First, the pore size was measured using the solute exclusion technique (Carpita et al. 1979). The cells were immersed in a hypertonic 0.3 M PEG solution of different Stokes' radii and they were observed under the microscope to check if they underwent plasmolysis or cytorrhysis. Plasmolysis, where the protoplast shrinks and pulls away from the inner surface of the cell wall, occurs if the external solute is able to penetrate through the wall pores and accumulates in the periplastic space around the protoplast. However, if the external solute is not able to penetrate through the wall pore because its Stokes radii is over the pore radius, then water movement out of the cells results that both the protoplast and wall shrink and the whole cell collapses causing cytorrhysis. The size of the largest molecules causing plasmolysis but not cytorrhysis is thus a measure of the size of cell wall pores. Both wild type and *mur1* suspension-cultured cells were plasmolysed and not cytorrhysed when incubated with PEG 1000, but the percent of cytorrhysed cells increased with PEG molecules of larger radii. However, the behavior of the suspension-cultured cells of wild type and the *mur1* mutant when incubated with PEG solutions of different Stokes radii were different at a $p < 0.05$ level (Figure 1). The Stokes' radii of the PEG molecules causing a 50% of cytorrhysis were calculated and considered as the radii of the cell wall pores. The values for wild type and *mur1* wall pores were 2.1 and 1.9 nm, respectively.

The size of wall pores was also measured using a diffusion-based technique (Woehlecke and Ehwald 1995). During incubation of denatured cells with the polydisperse dextran mixture (DPS), selective dilution of permeating dextran fractions took place. This resulted in modified HPLC chromatograms, which were compared with the original ones (Figure 2A). They were significantly different at a $p < 0.001$ level. Column calibration allowed the elution time to be transformed into Stokes' radius, so that the concentration signal of the detector could be related to molecular size. The figure 2B was derived from a curve, giving the quotient Q of detector signals (dilution factor) as a function of Stokes' radius. The Q value depends on the dilution of all dextran fractions by the extracellular liquid and size dependent diffusion into the cell lumina. The transition from complete exclusion to approximated equilibration with the entire intracellular space was indicated by the change from a lower to a higher Q level. As the borders of the transition range depend on the chromatographic resolution, we determined the average pore radius, which may be regarded as the Stokes' radii of the dextran size fractions equilibrating with half of the intracellular volume. The values obtained for the wild type and the *mur1* mutant were 2.6 and 1.7 nm, respectively.

The size of pore walls of the *mur1* line was smaller than the one of the wild type, measured by both methods (Figures 1 and 2). The values obtained for pore radii were in the range of 1.7-2.6 nm which are in the range of values obtained for different plant materials and different techniques (Read and Bacic 1996). Consensus on the most appropriate techniques has thus not yet been attained. Techniques assessing bulk movement of a solute, as solute exclusion method, render different results than those techniques measuring the penetration of only a small proportion of an external tracer, as the diffusion-based method did. That difference suggested the existence of a

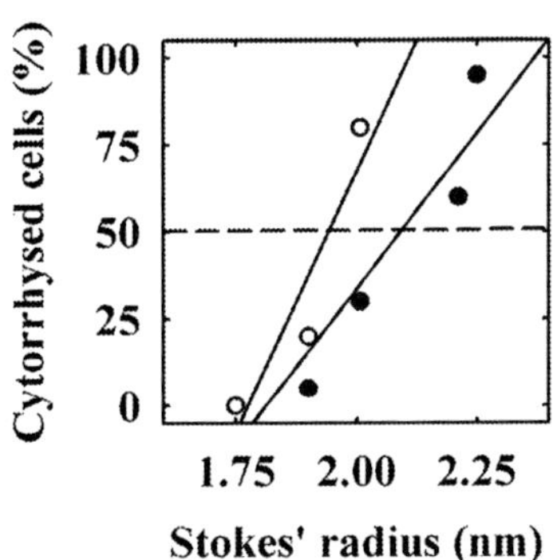

Fig. 1 Determination of cell wall porosity using a solute exclusion technique. **Arabidopsis cells from wild type (•) and *mur1* (o) were incubated with 0.3 M PEG of different Stokes' radii (PEG-1000, PEG-1500, PEG-2000, PEG-3000 and PEG-4000) and after 15 min the per cent of cytorrhysed cells was determined. One hundred cells for each incubation condition were counted and three independent experiments were carried out. The size of the largest PEG molecules causing plasmolysis but not cytorrhysis was considered as a measure of the diameter of the pore cell walls.**

continuous range of pore sizes in the wall in contrast to the concept of single limiting pore radii as it has been postulated by Horn et al. (1992). Nevertheless, although the results obtained using two different methods were clearly different, the pore radius for the fucose-lacking mutant was always lower as compared with the wild type. Thus, it seems clear that the absence of fucose in the cell wall of Arabidopsis *mur1* mutant causes a decrease in the wall porosity.

L-Fucose is a monosaccharide constituent of arabinogalactan proteins (van Hengel and Roberts 2002), various N-linked glycoproteins (Rayon et al. 1999) and cell wall polysaccharides. The most common fucose-containing cell wall polysaccharides are RG II and xyloglucan

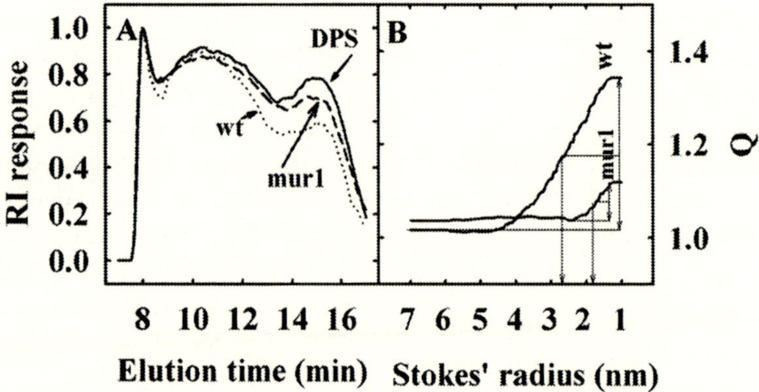

Fig. 2 Determination of cell wall porosity using a dextran permeation technique. **Arabidopsis cells from wild type and *mur1* denaturized with 1% acetic acid in 90% ethanol and stored at 4°C, were rehydrated and incubated for 30 min with a polydisperse dextran probing solution (DPS). The size dependence of dextran partitioning was determined by high-perfomance gel permeation chromatography on a Superdex-HR 200 column using a refractive index detector and 10 mM sodium phophate buffer pH 7.0 containing 100 mM NaCl as eluent at a flow rate of 1 mL min^{-1}. A) Molecular mass distribution of original dextran probing solution (DPS) and after incubation with denaturized cells from Arabidopsis wild type and *mur1*. A representative chromatogram for each condition is shown. B) Dilution quotient (Q) of dextran probing solution after incubation with wild type and mur1 at different Stokes' radii. The mean pore radius obtained at the middle of the step in the Q diagram. Q is calculated as the ratio between the concentration of original and modified dextran probing solution after incubation at different elution times. The Stokes radii corresponding at different elution times were obtained by calibration of the column with authentic dextrans.**

(Carpita and Gibeaut 1993). A substitution of L-fucosyl residues by L-galactosyl residues in *mur1* has been shown in N-linked glycoproteins (Rayon et al. 1999), RG II (O'Neill et al. 2001.), and xyloglucan (Zablackis et al. 1996).

Pectins (Baron-Epel et al. 1988; Ehwald et al. 1992) and more specifically RG II (O'Neill et al. 2001; Fleischer et al. 1999), have been considered to provide the background porosity of the walls. An important reduction in the RG II dimerization in the *mur1* mutant has been found, suggesting that the substitution of L-fucose and 2-*O*-methyl-L-fucose by L-galactose and 2-*O*-methyl-L-galactose, respectively, was not able to replace the borate cross-linking of RGII (O'Neill et al. 2001). Furthermore, the hydrophobicity of *mur1* RG II might also likely to be altered by reducing its 2-*O*-methyl-xylose content (Fleischer et al. 1999). Furthermore, Ryden et al. (2003) have shown that the fucosyl residues of RG II are important for RG II to become load-bearing element in cell walls. Thus, the structural changes in the pectin network of the *mur1* cell walls may be the responsible for the changes in the pore size and may participate in regulating cell wall mechanics and extensibility. Ehwald et al. (2002) have found that the cell walls of the Arabidopsis mutant *mur1* released more borate as well as compared with the wild type when they were incubated at pHs lower than 4.5 for a long period, over 18 h. Thus, the structural integrity of at least some parts of the RG II molecules seems to be essential for formation of borate cross-links. Since, RG II and its borate-ester cross-links have been related with cell wall porosity (see for review O'Neill et al. 2004), it seems reasonable that the substitution of L-fucose and 2-*O*-methyl-L-fucose by L-galactose and 2-*O*-methyl-L-galactose found in the *mur1* mutant may cause changes in the wall porosity. Ehwald et al. (2002) reported an increase in the pore size of the *mur1* mutant when their cells were incubated for a long period at acidic pH, the maximum difference with wild type being found at pH 2.5. Moreover, that increase took place at the same time that borate was released from the treated walls, suggesting that the stability of the dimeric dRG II-B complex against acid pHs is modified by relatively minor changes in the glycosyl residues composition of RG II. However, when the conditions used to measure the pore radii were much milder, polyethylene glycol solutions or water, and shorter incubation periods, 15 and 30 min, for solute exclusion and dextran permeation analyses, the results were very different. So, we report a significant decrease in the wall porosity of the *mur1* mutant as compared with the wild type, measured using both different techniques (Figures 1 and 2). Since, under that mild conditions, it seems unlikely the release of borate as reported for the acidic conditions, might be assumed that the pore state under our experimental conditions was more close to *in vivo* conditions, than in conditions used by Ehwald et al. (2002) that caused instability of the borate ester linkage releasing borate.

The absence of fucose might modify the ability to form borate cross-links leading to a modification, probably a decrease, in the strength supported by the RG II. The decrease in the number of borate cross-links might reduce the entanglement of the pectin network with the others wall polymers causing that the walls collapse reducing their pore size, at least under the experimental conditions, absence of turgor pressure, generally used for porosity measurements. A cell wall collapse has been suggested for *Commellina communis* guard cells to explain the effect of arabinan degradation on stomata mechanism (Jones et al. 2003).

Although, minor structural changes in RG II seems to be the most probably candidate for the changes in pore size of the Arabidopsis *mur1* mutant, it cannot be discarded the participation of other wall polymers. So, xyloglucan is another wall polysaccharide that is affected by the fucose absence in the *mur1* mutant. However, an effect on the mechanical properties of their cell walls has not been found (Ryden et al. 2003), but a reduction in the *O*-acetylation of galactose residues in xyloglucan side chains having an effect on the polymer hydrophobicity has been suggested (Perrin et al. 2003). The absence of L-fucose residues on the N-linked complex glycan side chains of glycoproteins synthesized by *mur1* plants (Rayon et al. 1999) is also unlikely to result in the dwarf phenotype because an *Arabidopsis* mutant (*cgl1*) that synthesizes glycoproteins that lack L-fucose grows normally (von Schaewen et al. 1993).

Acknowledgements

This work was supported by grants from the MCyT (BOS2002-03417) and Xunta de Galicia (PGIDIT03PXIC20004PN). We thank to Prof. N. Sakurai from Hiroshima University for his helpful comments.

References

Baron-Epel O, Gharyal PK, Schindler M (1988) Pectins as mediators of wall porosity in soybean cells. Planta 175: 389-395.

Carpita NC, Gibeaut DM (1993) Structural models of primary cell walls in flowering plants: consistency of molecular structure with the physical properties of the walls during growth. Plant J 3: 1-30.

Carpita NC, Sabularse D, Montezinos D, Delmer DP (1979) Determination of the pore size of cell walls of living plant cells. Science 205: 1144-1147.

Ehwald R, Fleischer A, Schneider H, O'Neill M (2002) Stability of the borate-ester cross-link in rhamnogalacturonan II at low pH and calcium activity in muro and in vivo. In: Goldbach HE, Rerkasem B, Wimmer MA, Brown PH, Thellier M, Bell RW (eds) Boron in Plant and Animal Nutrition. Kluwer Academic/Plenum Publishers, pp. 157-166.

Ehwald R, Woehlecke H, Titel C (1992) Cell wall microcapsules with different porosity from suspension cultured *Chenopodium album.* Phytochemistry 31: 3033-3038.

Fleischer A, O'Neill MA, Ehwald R (1999) The pore size of non-graminaceous plant cell walls is rapidly decreased by borate ester cross-linking of the pectic polysaccharide rhamnogalacturonan II. Plant Physiol 121: 829-838.

Horn MA, Heinstein PF, Low PS (1992) Characterization of parameters influencing receptor-mediated endocytosis in cultured soybean cells. Plant Physiol 98: 673-679.

Jones L, Milne JL, Ashford D, McQueen-Mason SJ (2003) Cell wall arabinan is essential for guard cell function. Proc Natl Acad Sci USA 100: 11783-11788.

O'Neill MA, Eberhard S, Albersheim P, Darvill AG (2001) Cross-linking of cell wall rhamnogalacturonan II for Arabidopsis growth. Science 294: 846-849.

O'Neill MA, Ishii T, Albersheim P, Darvill AG (2004) Rhamnogalacturonan II: Structure and function of a borate cross-linked cell wall pectic polysaccharide. Annu Rev Plant Biol 55: 109-139.

Perrin RM, Jia Z, Wagner TA, O'Neill MA, Sarria R, York WS, Raikhel NV, Keegstra K (2003) Analysis of xyloglucan fucosylation in *Arabidopsis.* Plant Physiol 132: 768-778.

Rayon C, Cabanes-Macheteau M, Loutelier-Bourhis C, Salliot-Maire I, Lemoine J, Reiter WD, Lerouge P, Faye L (1999) Characterization of N-glycans from Arabidopsis. Application to a fucose-deficient mutant. Plant Physiol 119: 725-733.

Read SM, Bacic A (1996) Cell wall porosity and its determination. In: Linskens HF, Jackson JF (eds) Modern Methods of Plant Analysis. Vol. 17. Springer-Verlag, pp 63-80.

Reiter WD, Chapple CCS, Somerville CR (1993) Altered growth and cell walls in a fucose-deficient mutant of *Arabidopsis.* Science 261: 1032-1035.

Ryden P, Sugimoto-Shirasu K, Smith AC, Findlay K, Reiter WD, McCann MC (2003) Tensile properties of Arabidopsis cell walls depend on both a xyloglucan cross-linked microfibrillar network and rhamnogalacturonan II-borate complexes. Plant Physiol 132: 1033-1040.

van Hengel AJ, Roberts K (2002) Fucosylated arabinogalactan-proteins are required for full root cell elongation in Arabidopsis. Plant J 32: 105-113.

von Schaewen A, Sturm A, O'Neill J, Chrispeels MJ (1993) Isolation of a mutant *Arabidopsis* plant that lacks N-acetyl-Glucosaminyl Transferase I and is unable to synthesize Golgi-modified complex N-linked glycans. Plant Physiol 102: 1109-1118.

Woehlecke H, Ehwald R (1995) Characterization of size-permeation limits of cell walls and porous separation materials by high-performance size-exclusion chromatography. J Chromatogr A 708: 263-271.

Zablackis E, York WS, Pauly M, Hantus S, Reiter WD, Chapple CCS, Albersheim P, Darvill A (1996) Substitution of L-fucose by L-galactose in cell walls of *Arabidopsis mur1.* Science 272: 1808-1810.

Molecular and Functional Analysis of Pectins in Intercellular Attachment by T-DNA Tagging Using a Haploid Tobacco Tissue Culture System

Hiroaki Iwai, Tadashi Ishii, Shingo Sakai and Shinobu Satoh

Spatially and temporally controlled intercellular attachment is indispensable for the organized development of higher organisms. In higher plants, intercellular attachment is mediated by pectin, which consists of homogalacturonan (HG), rhamnogalacturonan (RG-I and RG-II) domains (Carpita et al. 1993; Ridley et al. 2001; Willats et al. 2001). Although some of the genes responsible for cell wall polysaccharide synthesis are known (Scheible and Pauly 2004), only a few glycosyltransferase genes involved in pectin biosynthesis have been identified (Iwai et al. 2002). This is partly because plants with mutations that cause defects in matrix polysaccharides have embryonic lethal phenotypes, hindering the biochemical analysis of the cell walls of such mutants.

In tissue culture systems derived from plant species such as carrot, tobacco, and rice, the loss of morphogenetic competence (the capacity for embryogenesis or formation of adventitious buds) in the callus often occurs during long-term culture, with an accompanying reduction in the size of the cell clusters as a result of the loosening of intercellular attachments (Satoh 1998). It seems likely, therefore, that close intercellular attachments are essential for morphogenetic processes, such as embryogenesis and organogenesis. To characterize the mechanisms of intercellular attachment that are involved in morphogenesis in higher plants, we compared the cell wall polysaccharides in carrot embryogenic callus-forming large cell clusters and non-embryogenic callus-forming small cell clusters (Iwai et al. 1999; Iwai et al. 2001; Kikuchi et al. 1995; Kikuchi et al. 1996). From these studies, we proposed possible roles in intercellular attachment for arabinan and xylose in the neutral sugar regions of pectins (Satoh 1998). Although somaclonal variations in the mechanisms involved in cell wall synthesis and functioning may result in weak intercellular attachment in a non-embryogenic callus, it is difficult to more extensively analyze such spontaneous mutants because the variations sometimes occur at the chromosomal level.

We therefore established a novel T-DNA transformation system for the production of nolac (non-organogenic callus with loosely attached cells) mutants, using in vitro cultures of leaf disks of a haploid *Nicotiana plumbaginifolia* line (Figure 1; Iwai et al. 2001; Iwai et al. 2002). The nolac mutants are defective in intercellular attachment, resulting in a failure to form adventitious buds in the presence of exogenous cytokinin. Haploid *N. plumbaginifolia* plants (Marion-Poll et al. 1993) are suitable for the generation and study of such mutants because mutations have a direct effect on the phenotype, and cells with embryo-lethal mutations can be maintained in tissue culture as an unorganized callus, enabling the analysis of mutant cell

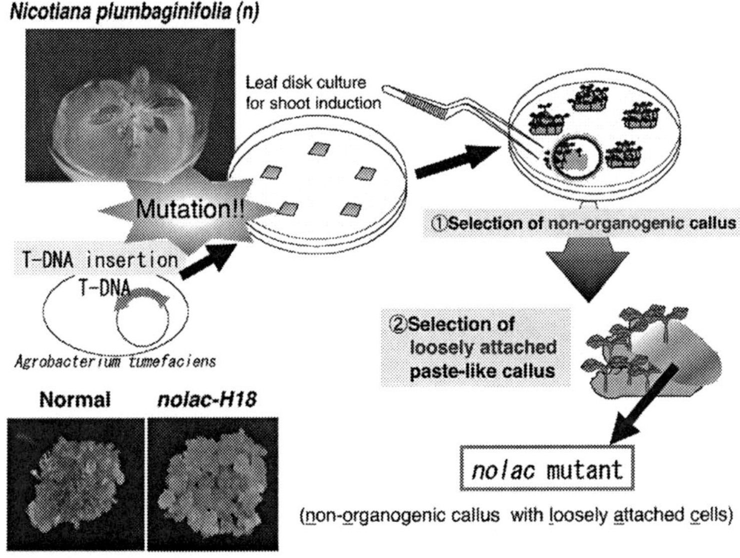

Fig. 1 Schematic illustration of the method used for the production of the *nolac* (*n*on-*o*rganogenic callus with *l*oosely *a*ttached *c*ells) tobacco intercellular attachment mutants. **Leaf disks from a haploid line of *Nicotiana plumbaginifolia* were transformed using *Agrobacterium-tumefaciens*-mediated transformation, and cultured on shoot-inducing medium supplemented with cytokinin. *nolac* mutants appeared on approximately 7% of the leaf disks. In the first round of selection of the mutant callus lines, non-shoot-forming calluses were selected, followed by a second round of selection for soft calluses, as determined by touching with forceps. The mutant *nolac-H18* is defective in a pectin glucuronyltransferase (NpGUT1).**

walls. We initially identified 199 callus-producing lines with loosely attached cells from cultures of 2,970 transformed leaf disks. Although only 25 of the lines continued to grow on the medium, two lines with mutant calluses, *nolac-H14* and *nolac-H18*, had callus growth rates similar to that of normal calluses.

The cells of *nolac-H14* calluses are loosely attached and the cell clusters are of irregular morphology; in normal calluses, however, the cells are tightly attached to one another and form shoot meristems at the surface of the callus. Transmission electron microscopy revealed that the middle lamella and cell wall of *nolac-H14* calluses were barely stained by ruthenium red, even after demethylesterification with NaOH, whereas the entire cell wall and the middle lamella were strongly stained in normal calluses. In cultures of *nolac-H14* calluses, the level of sugar components of pectic polysaccharides was greatly reduced in the hemicellulose fraction and strongly elevated in the culture medium, relative to the levels of normal calluses. These results indicate that pectic polysaccharides are not retained in the cell walls and middle lamellae of *nolac-H14* calluses and, instead, leak into the culture medium.

In *nolac-H14*, the ratio of arabinose to galactose (Ara/Gal) was low compared to the normal line. Glycosyl linkage analysis showed that the mutant contains very short neutral-sugar side chains, composed primarily of linear arabinan. These results suggest that arabinose-rich pectins in the hemicellulose fraction play an important role in intercellular attachment. In particular, long arabinan regions in pectic polysaccharide side chains might be associated with cellulose-hemicellulose complexes (Jarvis et al. 2003; Willats et al. 1999). The gene responsible for the *nolac-H14* mutation might be involved in the formation of long arabinan regions. Characterization of the mutated gene, which encodes a type of membrane protein, is now in progress.

The T-DNA-tagged gene of another mutant cell line, *nolac-H18*, was isolated, and has a morphology similar to that of *nolac-H14*. The gene encodes a protein with significant sequence similarity to the glucuronyltransferase domain of animal exostosins and was therefore named *NpGUT1* (glucuronyltransferase 1; AB08676; Iwai et al. 2002). Transformation of *nolac-H18* with *NpGUT1* complemented the mutation, while in contrast, transformation of normal *Nicotiana tabacum* leaf disks with a 35S-promoter-driven antisense *NpGUT1* gene resulted in transformants with abnormal shoots. In these transformants, the leaf and stem tissues were ragged and brittle, and callus-like cell clusters were observed on the shoot apex. These results indicate that NpGUT1 is responsible for the *nolac-H18* mutation.

A prominent feature in NpGUT1 is the presence of the putative glycosyltransferase catalytic domain, pfam03016, in glycosyltransferase family 47 (GT47). The GT47 family grouping is based on the presence of the β-glucuronyltransferase domain of animal exostosins. MUR3, a xyloglucan β-galactosyltransferase, is also a member of the GT47 family. Numerous plant genes have been shown to encode proteins of this family (http://afmb.cnrs-mrs.fr/CAZY/). For example, the Arabidopsis genome contains 39 GT47 family members, and the rice genome has at least 22 members (Zhong and Ye 2003).

We next examined the cell wall composition of the *nolac-H18* mutant. The glucuronic acid levels in the pectin fraction of the cell wall were dramatically reduced relative to the wild type. Further fractionation and composition analyses demonstrated that the mutation blocks the addition of glucuronic acid to RG-II in pectin, abolishing the terminal galactose-glucuronic acid disaccharide in the side chain. As a result, little RG-II dimerization is inducible by borate in *nolac-H18*. Recent studies have demonstrated that the essential micronutrient boron cross-links two chains of RG-II to form a RG-II dimer in the cell wall, and the formation of this dimer is essential for plant development and growth. These results indicate that the *NpGUT1* gene encodes a glucuronyltransferase that transfers glucuronic acid to the RG-II domain of pectin, and suggest that glucuronic acid in the RG-II domain of pectin plays a significant role in the formation of the pectin-pectin network generated by borate cross-linking of RG-II.

The *NpGUT1* gene is predominantly expressed in the meristem and other young tissues that require particularly strong intercellular attachment for intercellular communication and organization. Disruption of the plasmodesmatal connection between cells may be one of the primary causes of the abnormal shoot apex morphology in both *nolac-H18*- and antisense *NpGUT1*-transformed shoots. These results indicate that the glucuronyltransferase gene is essential for intercellular organization in plant meristems and tissues. The involvement of the gene in the formation of reproductive tissues is currently under study.

Intercellular attachment between plant cells is usually established when daughter cells are formed by cytokinesis; epigenetic adhesion between existing cell walls of mature cells takes place only in limited cases. Plant growth involves the coordinated and directed expansion of adherent cells (Jarvis et al. 2003). In certain developmental stages, cell separation takes place in a controlled manner. As pectin plays a critical role in intercellular attachment, the pectin network is a modification target during developmental processes such as cell wall swelling and softening during fruit ripening, cell separation during leaf and fruit abscission, pod dehiscence, and root cap cell differentiation (Ridley et al. 2001).

During various phases of development in the plant life cycle, each of the three domains of pectin, homogalacturonan, RG-I, and RG-II may have significant and specific roles between and/or within primary cell walls (Figure 2). The presence of divalent calcium ions allows ionic cross-linking between galacturonic acid residues in the homogalacturonan domains, resulting in the formation of pectin gel concentrated in the middle lamella, between the primary walls. This type of bond is abundant and usually irreversible in older tissues, in which differentiation of cells has ceased, since homogalacturonan is originally secreted as the methyl-esterified form and de-esterification occurs during cell maturation (Brett and Waldron 1996). The side chains, composed of arabinan and arabinogalactan in the RG-I domains, are also thought to be involved in intercellular attachment (Iwai et al. 2001). The long arabinan side chain is especially important for the anchoring of pectin to the cellulose-hemicellulose complexes of the primary wall. This type of side chain is rich in young meristems (Willats et al. 2001), in which the calcium-induced cross-linking of homogalacturonan is less abundant. In young meristems, the primary walls are rich in the RG-II domain. RG-II forms dimers by cross-linking with borate to generate complex pectin networks (Kobayashi et al. 1996; Ishii et al. 1999; O'Neill et al. 1996 and 2001). This might explain why a borate deficiency can cause severe malformations

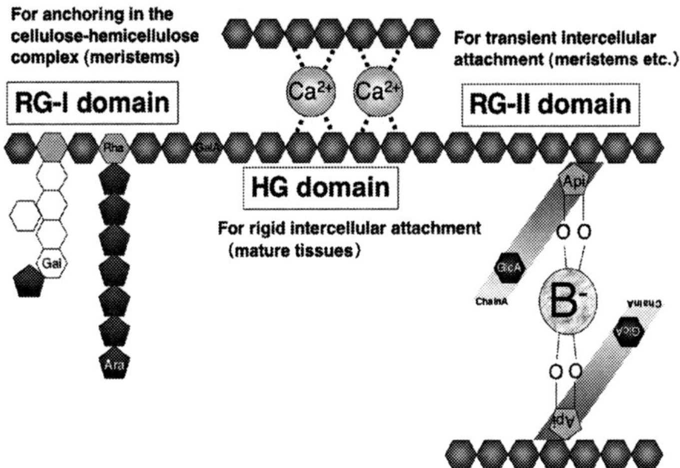

Fig. 2 A simplified model of the roles in intercellular attachment for the three major pectin domains: homogalacturonan (HG), rhamnogalacturonan-I (RG-I), and rhamnogalacturonan II (RG-II). **Both RG-I and RG-II are thought to be covalently linked to the HG domain. Gal, galactose; Ara, arabinose; GalA, galacturonic acid; GlcA, glucuronic acid; Api, apiose; B, borate. The RG-I domain is abundant in meristematic tissue and is involved in the anchoring of pectin into the cellulose-hemicellulose complex in the primary wall. The HG domain is abundant in the middle lamella of mature tissues and has a role in rigid, irreversible intercellular attachments. The RG-II domain is involved in transient and flexible intercellular attachment in meristematic tissues.**

in meristematic tissues of crop plants. Two opposing functions, smooth slippage and sufficient intercellular attachment, are required in the developing meristem and peripheral cells. The intercellular attachment generated by the RG-II domain of pectin via borate might be flexible and reversible. The three pectin domains might be functionally specialized in temporal and spatial manners during plant development.

Acknowledgements

We are grateful to Dr. M. Jullien of the Institut National de la Recherche Agronomique (INRA) Centre de Versailles, France, for the gift of haploid *Nicotiana plumbaginifolia* plants. This work was supported in part by a Grant-in-Aid for Scientific Research on Priority Area (No. 14036204), and by the Program for Promotion of Basic Research Activities for Innovative Biosciences (PROBRAIN).

References

Brett C, Waldron K (1996) Cell wall architecture and the skeletal role of the cell wall, In Physiology and Biochemistry of Plant Cell Walls. Chapman & Hall, London, pp 44-74.

Carpita NC, Gibeaut D (1993) Structural models of primary cell walls in flowering plants: consistency of molecular structure with the physical properties of the walls during growth. Plant J 3: 1-30.

Ishii T, Matsunag T, Pellerin P, O'Neill MA, Darvill A, Albersheim P. (1999) The plant cell wall polysaccharide rhamnogalacturonan II self-assembles into a covalently cross-linked dimer. J Biol Chem 274: 13098-13104.

Iwai H, Kikuchi A, Kobayashi T, Kamada H, Satoh H (1999) High levels of non-methylesterified pectins and low levels of peripherally located pectins in loosely attached non-embryogenic callus of carrot. Plant Cell Rep 18:561-566.

Iwai H, Ishii T, Satoh S (2001) Absence of arabinan in the side chains of the pectic polysaccharides strongly associated with cell walls of *Nicotiana plumbaginifolia* non-organogenic callus with loosely attached constituent cells. Planta 213: 907-915.

Iwai H, Masaoka N, Ishii T, Satoh S (2002) A pectin glucuronyltransferase gene is essential for intercellular attachment in the plant meristem. Proc Natl Acad Sci USA 99: 16319-16324.

Jarvis MC, Briggs SPH, Knox JP (2003) Intercellular adhesion and cell separation in plants. Plant Cell Envir 26: 977-989.

Kikuchi A, Satoh S, Nakamura N and Fujii T (1995) Differences in pectic polysaccharides between carrot embryogenic and non-embryogenic calli. Plant Cell Rep 14:279-284.

Kikuchi A, Edashige Y, Ishii T, Fujii T, Satoh S (1996) Variations in the structure of neutral sugar chains in the pectic polysaccharides of morphologically different carrot calli and correlations with the size of cell clusters. Planta 198: 634-639.

Kobayashi M, Matoh T, Azuma J (1996) Two chains of rhamnogalacturonan II are cross-linked by borate-diol ester bonds in higher plant cell walls. Plant Physiol 110: 1017-1020.

Marion-Poll A, Marin E, Bonnefoy N, Pautot V (1993) Transposition of the maize autonomous element Activator in transgenic *Nicotiana plumbaginifolia* plants. Mol. Gen. Genet. 238: 209-217.

O'Neill MA, Warrenfeltz D, Kates K, Pellerin P, Doco T, Darvill A, Albersheim P (1996) Rhamnogalacturonan-II, a pectic polysaccharide in the walls of growing plant cell, forms a dimer that is covalently cross-linked by a borate ester. In vitro conditions for the formation and hydrolysis of the dimer. J. Biol. Chem. 271: 22923-22930.

O'Neill MA, Eberhard S, Albersheim P, Darvill AG (2001) Requirement of borate cross-linking of cell wall rhamnogalacturonan II for Arabidopsis growth. Science 294: 846-849.

Ridley BL, O'Neill MA, Mohnen D (2001) Pectins: structure, biosynthesis, and oligogalacturonide-related signaling. Phytochemistry 57: 929-967.

Satoh S (1998) Function of the cell wall in the interactions of plant cells: analysis using carrot cultured cells. Plant Cell Physiol 39: 361-368.

Scheible WR and Pauly M (2004) Glycosyltransferase and cell wall biosynthesis: novel players and insights. Curr Opin Plant Biol 7: 1-11.

Willats WG, Steele-King CG, Marcus SE, Knox JP (1999) Side chains of pectic polysaccharides are regulated in relation to cell proliferation and cell differentiation. Plant J 20: 619-628.

Willats WG, McCartney L, Mackie W, Knox JP (2001) Pectin: cell biology and prospects for functional analysis. Plant Mol. Biol. 47: 9-27.

Zhong R and Ye ZH (2003) Unraveling the functions of glycosyltransferase family 47 in plants. Trend Plant Sci 8: 565-568.

Early Insights into the Plant Cell Wall Proteome

Jocelyn K.C. Rose and Sang-Jik Lee

The proteome of the plant cell wall/apoplast is less well characterized than those of other subcellular compartments. This largely reflects the many technical challenges involved in extracting extracellular proteins, many of which resist isolation and identification, and in capturing a population that is both comprehensive and relatively uncontaminated with intracellular proteins. However, a range of disruptive techniques, involving tissue homogenization and subsequent sequential extraction, as well as non-disruptive approaches has been developed. These approaches have been complemented more recently by other genome-scale screens, such as secretion traps, that reveal the genes encoding proteins with N-terminal signal peptides that are targeted to the secretory pathway, many of which are subsequently localized in the wall. While the size and complexity of the wall proteome is still unresolved, the combination of experimental tools and computational prediction is rapidly expanding the catalog of known wall-localized proteins, suggesting the unexpected extracellular localization of other polypeptides and providing the basis for further exploration of plant wall structure and function.

Introduction

Recent research across the spectrum of plant biology has emphasized the importance of the plant cell wall as the interface with the biotic and abiotic environment, in additional to its mechanical role, and is revealing an ever-growing number of developmental processes that are directly or indirectly influenced by wall-localized molecular interactions and signaling pathways. Accordingly, the tightly regulated expression of cell wall proteins has been observed in association with a diverse range of developmental events, metabolic processes and in response to many external stimuli. In the context of this chapter, 'extracellular', 'apoplastic' and 'wall-associated' are used as synonymous terms when describing proteins that are secreted from the protoplast. This implies localization outside the plasma membrane and a physical proximity to the wall, rather than necessarily a direct or biologically significant interaction with the extracellular matrix.

Isolating cell wall proteins: a major obstacle

Substantial progress has been made over the last few years in characterizing a range of plant subcellular proteomes, such as those of chloroplasts and mitochondria (Rose et al. 2004). In comparison, far less is known about the cell wall proteome, which likely reflects some major technical challenges that have to be surmounted. Firstly, and perhaps most importantly, the cell wall compartment is a continuum and not conveniently bounded by a single discrete membrane. An essential goal of subcellular fractionation prior to proteomic analysis is to capture the most comprehensive representation of the protein complement, whilst minimizing contamination with proteins from other subcellular locations. Highly effective protocols have been developed to isolate intracellular organelles, typically involving controlled tissue disruption followed by density gradient centrifugation, and in these cases the membrane acts both to contain the organellar proteins and to exclude extraneous protein contaminants. In contrast, any tissue disruption that compromises the integrity of the plasma membrane instantly leads to contamination of the cell wall fraction with intracellular proteins, many of which bind to the wall matrix with high affinity. The second related technical hurdle is that proteins in the wall/apoplast exhibit a remarkably wide range of biochemical characteristics and affinities with the polysaccharide matrix. Many are highly soluble with no apparent interaction with the wall polysaccharides and are therefore combined with the cytosolic protein fraction during tissue homogenization and effectively 'lost' from the wall extract. However, other wall-localized proteins, such as extensins (Wojtaszek et al. 1995), can be covalently linked into the cell wall matrix and are consequently highly resistant to extraction. In addition, many wall proteins are glycosylated, which can complicate extraction, isolation and identification, while yet other extracellular proteins have domains that anchor them into the plasma membrane (Takos et al. 1997; Borner et al. 2003; Eisenhaber et al. 2003; Oxley and Bacic 1999), or that span the membrane.

In conclusion, no one single protocol can be used to extract all cell wall proteins and so a range of techniques must be employed that collectively will capture a greater proportion of the wall proteome. These may broadly divided into three categories: disruptive, non-disruptive and secretion screens.

Non-disruptive strategies to isolate cell wall proteins

One approach to obtain proteins that are localized in the apoplast/cell wall that minimizes contamination with proteins from within the protoplast involves the use of suspension cultured plant cells. Robertson et al. (1997) used this system in a pioneering study of secreted proteins from five plant species that represented the first attempt to systematically separate and identify large numbers of cell wall-associated proteins. Protein populations were isolated from the suspension cell culture medium and from intact cells that were sequentially washed with a series of solvents that were designed to leave the plasma membrane intact. Numerous families of known cell wall-localized proteins were identified, although a large proportion could not be assigned to a protein functional class based on sequence homology. In many cases, this probably reflected the scarcity of DNA sequence information for some of the species that were studied (e.g. carrot). Approximately 30% of the proteins from Arabidopsis could not be classified but since publication of this paper, the Arabidopsis genome sequence has become available. A post-genome re-evaluation of the published sequences now allows a considerably greater proportion, but not all, of the proteins to be assigned a putative biochemical function (SJ Lee, JKC Rose unpublished data). A more recent analysis of the sequential washing approach identified additional loosely bound wall proteins from Arabidopsis suspension cells, but cautioned that the plasma membrane is easily ruptured, resulting in contamination with cytosolic proteins (Borderies et al. 2003). Great care must therefore be taken, even when using this supposedly non-destructive approach. The authors also noted that individual protein isoforms were not exclusively eluted in one solvent, which further complicates analysis. Contamination with intracellular proteins was also observed in a study of cell wall-associated proteins in the green algae *Haematococcus pluvialis* that also used sequential washing of the cell walls (Wang et al. 2004). A similar strategy has been used to identify new secreted pathogenesis-related proteins (Okushima et al. 2000), examine responses to fungal elicitors (Ndimba et al. 2003) and to identify secreted proteins that are associated with secondary wall formation (Blee et al. 2001; Kärkönen et al. 2002).

While suspension cells provide a convenient source of homogenous plant material that can be rapidly regenerated and from which cell walls and wall proteins may easily be obtained, they represent an artificial biological system. The complement of wall proteins in complex plant tissues is likely to be significantly different, given the associated cellular heterogeneity, and to exhibit substantial spatial and temporal variability. Different tissues and cell types will have distinct subsets of wall proteins with diverse functions and this variability will emerge through careful proteomic analysis of subtypes of plant material. Similarly, little is currently known about the dynamic aspects of wall protein populations during cell growth and differentiation. However, some preliminary results have recently started to emerge, such as an examination of cell wall-associated proteins from the developing xylem of compression and non-compression wood of Sitka spruce (McDougall 2000).

An alternative experimental approach to isolate cell wall/apoplastic proteins from complex tissues, and one that can be adapted to minimize contamination with cytosolic proteins, is to use pressure-rehydration and vacuum infiltration protocols to extract the apoplastic fluid from the target sample. This approach has been used successfully to extract extracellular proteins from several tissue or organ types, including roots, leaves, stems, fruit, tubers, xylem and phloem (Lee et al. 2004). In addition, a range of solutions can be infiltrated into the tissues to release different subsets of proteins, such as buffers containing relatively high salt concentrations that liberate proteins that are ionically bound to the wall. The disadvantages of this technique are that the protein yield is typically low and great care has to be taken to avoid cell lysis. Consequently, throughput is generally slow and many wall-localized proteins cannot be recovered without rupturing the plasma membrane. Assessment of contamination is an important factor that should ideally be assessed to obtain an estimation of purity or enrichment.

Disruptive strategies to isolate cell wall-bound proteins

Many reports describe the isolation of a "cell wall protein fraction" by tissue homogenization in a buffer with a low salt concentration, followed by sequential washing of the wall pellet with a low-salt buffer to remove cytosolic protein contaminants and then with solutions with high concentrations of salt to release proteins that are ionically bound to the wall. However, since the polygalacturonate component of cell wall pectin can act as a polyanionic matrix, positively charged proteins from within the protoplast have the potential to bind to the wall once the plasma membrane has been ruptured, contaminating the sample. In some cases a cytosolic protein can associate so strongly with the wall pellet that a detergent such as SDS, or harsh solvents and chaotropic reagents are subsequently required to re-solubilize the protein (RS Saravanan, JKC Rose unpublished data). Therefore, considerable caution should be used when classifying proteins that are isolated using this disruptive technique as cell wall proteins and ideally other approaches should be used to verify their subcellular localization.

This type of strategy was used in an analysis of wall-associated proteins extracted from *Arabidopsis thaliana* suspension cells (Chivasa et al. 2002), where a cell wall fraction was removed from disrupted cells and sequentially extracted with calcium chloride and urea. The authors used 2-DE followed by mass spectrometry analysis to identify 69 different proteins, which included numerous known wall proteins with well-established biochemical functions, a number of unclassified proteins and several polypeptides whose location in the wall was unexpected. These latter two classes of proteins demonstrate the potential value of this approach for identifying new cell wall/apoplastic proteins and wall-localized biochemical pathways. Such experiments will provide a platform for subsequent functional studies. A recent review highlighted the contamination issue and the authors reaffirmed the need for confirmatory analyses using techniques such as immunolocalization (Slabas et al. 2004). Such identification of supposedly intracellular proteins in wall extracts is not unique (Chivasa et al. 2002; Kärkönen et al. 2002; Watson et al. 2004).

This disruptive "grind and find" approach is clearly effective for many classes of wall proteins, particularly those that are more tightly associated with the extracellular matrix. However, ideally, contamination with non-wall localized proteins should be rigorously assessed, as has been described in analyses of Arabidopsis suspension cells (Chivasa et al. 2002; Bayer et al. 2004). The *in vivo* localization should subsequently be confirmed using other techniques, such as immunolocalization or green fluorescent protein (GFP)-fusion protein analysis.

Additional complementary approaches to define the cell wall proteome

In addition to the above strategies to characterize the cell wall proteome, a number of genome-scale screens have been developed in the last few years that either directly or indirectly reveal populations of secreted proteins. For example, high-throughput screening of plant protein subcellular localization has been described following the fusion of plant cDNA libraries with the DNA sequence encoding green fluorescent protein and screening transformed plants by fluorescence microscopy. A recently reported such screen of a cDNA library from *Nicotiana benthamiana* roots, revealing a number of new extracellular proteins, as well as details of localization of certain proteins in specific microdomains within the wall matrix (Escobar et al. 2003).

A more focused approach to identify extracellular proteins involves "secretion traps" that specifically target secreted and plasma membrane-spanning proteins. Groover et al. (2003) described an ingenious strategy that involved the analysis of Arabidopsis transposon insertion lines expressing β-glucuronidase (GUS):protein fusions. If the fusion proteins are routed through the secretory pathway, they can be identified by growing the plants in the presence or absence of a glycosylation inhibitor, since glycosylation of β-glucuronidase in the ER alters β-glucuronidase enzyme activity. This analysis allowed the identification of a range of both known and new wall-localized proteins and membrane-bound proteins, in addition to revealing their expression patterns in specific tissues.

Another high-throughput screen that targets secreted proteins is termed the yeast signal sequence trap (Tashiro et al. 1993; Kristoffersen et al. 1996). This involves ligating cDNAs in frame at the 5' end of the DNA sequence encoding a truncated yeast invertase gene without the initiator methionine and signal peptide. This library is then transformed into an invertase-deficient yeast mutant, which is grown on a medium with sucrose as the sole carbon source. Any yeast transformant containing a plant-derived cDNA encoding a N-terminal signal peptide (SP) sequence for targeting to the ER and the secretory pathway has the potential to secrete the polypeptide as an invertase fusion protein, resulting in reconstitution of extracellular invertase activity and the rescue of the mutant. This secretion trap has now been used to identify substantial populations of genes encoding secreted plant proteins (Belanger et al. 2003; Goo et al. 1999; Hugot et al. 2004; Yamane et al. 2005) and while it has some limitations, such as redundancy and a low level of false positives, it represents an effective and rapid screen that importantly allows the identification of some wall-localized proteins that are not found by traditional protein extraction techniques.

Concluding Remarks

The complexity and dynamics of the cell wall proteome is now being elucidated using a range of experimental approaches and tools. It is noteworthy that most larger scale studies of cell wall proteins, whether isolating and sequencing native proteins from plant tissues, or performing secretion trap screens, result in the identification of proteins that are not generally recognized as being secreted. In some cases, as described above, this may be attributed to contamination, or an artifact of the screen; however, in other cases these may represent legitimate localization. Another common observation from the wall protein isolation studies and secretion trap screens is that proteins are often identified that are not predicted to be secreted by the web-based programs, such as SignalP (www.cbs.dtu.dk/services/SignalP), that are used to determine the presence of an N-terminal signal sequence, or to suggest subcellular localization. These results may be interpreted to reflect the contamination issue, but another explanation is that there exist currently unknown non-classical secretory signal sequences and pathways, as shown in yeast, animals and bacteria (described in Chivasa et al. 2002) but that have yet to be demonstrated in plants. However, a simpler explanation for the paradoxical localization in many cases is simply that prediction software regularly incorrectly assigns location. Unambiguous demonstration of protein localization is a challenging goal as each technique has specific limitations, whether immunolocalization or GFP-fusion studies.

At present, the combination of complementary screens, protein isolation and *in silico* analysis is generating a more comprehensive catalog of the wall proteome, although it difficult to estimate how many wall proteins remain to be identified. Future milestones will include developing a picture of protein dynamics and localization within individual walls and wall micro-domains, the identification of protein complexes and interaction networks in the wall/apoplast and an assessment of post-translational modification. While these goals are not at present within reach, doubtless new technologies and techniques will incrementally reveal the complexity of the wall proteome, which will in turn provide important new insights into cell wall structure and function.

Acknowledgements

Support in this research area was provided by a grant from KOSEF to S.J.L. and by grants to J.K.C.R. from the National Science Foundation (IBN-009109) and the United States Department of Agriculture (2001-52100-1137). The authors thank the various members of the Rose lab who contributed to the research and Dr. Carmen Catalá for helpful discussion and assistance.

References

Bayer E, Thjomas CL, Maule AJ (2004) Plasmodesmata in *Arabidopsis thaliana* suspension cells. Protoplasma 223: 93-102.

Belanger KD, Wyman AJ, Sudol MN, Sigla-Pareek SL, Quatrano RS (2003) A signal peptide secretion screen in *Fucus distichus* embryos reveals expression of glucanase, EGF domain-containing, and LRR receptor kinase-like polypeptides during asymmetric cell growth. Planta 217: 931-950.

Blee KA, Wheatley ER, Bonham VA, Mitchell GP, Robertson D, Slabas AR, Burrell MM, Wojtaszek P, Bolwell GP (2001) Proteomic analysis reveals a novel set of cell wall proteins in a transformed tobacco cell culture that synthesizes secondary walls as determined by biochemical and morphological parameters. Planta 212: 404-415.

Borderies G, Jamet E, Lafitte C, Rossignol M, Jauneau A, Boudart G, Monsarrat B, Esquerré-Tugayé M-T, Boudet A, Pont-Lezica R (2003) Proteomics of loosely bound cell wall proteins of *Arabidopsis thaliana* cell suspension cultures: A critical analysis, Electrophoresis 24: 3421-3432.

Borner GHH, Lilley KS, Stevens TJ, Dupree P (2003) Identification of glycosylphosphatidylinositol-anchored proteins in Arabidopsis. A proteomic and genomic analysis. Plant Physiol 132: 568-577.

Chivasa S, Ndimba BK, Simon WJ, Robertson D, Yu X-L, Knox JP, Bolwell P, Slabas AR (2002) Proteomic analysis of the *Arabidopsis thaliana* cell wall. Electrophoresis 23: 1754-1765.

Eisenhaber B, Wildpaner M, Schultz CJ, Borner GH, Dupree P, Eisenhaber F (2003) Glycosylphosphatidylinositol lipid anchoring of plant proteins. Sensitive prediction from sequence- and genome-wide studies for Arabidopsis and rice. Plant Physiol 133: 1691-1701.

Escobar NM, S. Haupt S, Thow G, Boevink P, Chapman S and Oparka K (2003) High-throughput viral expression of cDNA–green fluorescent protein fusions reveals novel subcellular addresses and identifies unique proteins that interact with plasmodesmata. Plant Cell 15: 1507-1523.

Goo JH, Park AR, Park WJ, Park OK (1999) Selection of Arabidopsis genes encoding secreted and plasma membrane proteins. Plant Mol Biol 41: 415-423.

Groover AT, Fontana JR, Arroyo JM, Yordan C, McCombie WR, Martienssen RA (2003) Secretion trap tagging of secreted and membrane-spanning proteins using Arabidopsis gene traps. Plant Physiol 132: 698-708.

Kärkönen A, Koutaniemi S, Mustonen M, Syrjänen K, Brunow G, Kilpeläinen I, Teeri TH, Simola LK (2002) Lignification related enzymes in *Picea abies* suspension cultures. Physiol Plant 114: 343-353.

Kristoffersen P, Teichmann T, Stracke R, Palme K (1996) Signal sequence trap to clone cDNA encoding secreted or membrane-associated plant proteins. Anal Biochem 243: 127-132.

Lee SJ, Saravanan RS, Yamane H, Kim BD, Rose JKC (2004) Digging deeper into the plant cell wall proteome. Plant Physiol Biochem 42: 979-988.

McDougall GJ (2000) A comparison of proteins from the developing xylem of compression and non-compression wood of branches of Sitka spruce (*Picea sitchensis*) reveals a differentially expressed laccase. J Exp Bot 51: 1395-1401.

Ndimba BK, Chivasa S, Hamilton JM, Simon WJ, Slabas AR (2003) Proteomic analysis of changes in the extracellular matrix of Arabidopsis cell suspension cultures induced by fungal elicitors. Proteomics 3: 1047-1059.

Okushima Y, Koizumi N, Kusano T, Sano H (2000) Secreted proteins of tobacco cultured BY2 cells: identification of a new member of pathogenesis-related proteins. Plant Mol Biol 42: 479-488.

Oxley D, Bacic A (1999) Structure of the glycosylphosphatidylinositol anchor of an arabinogalactan protein from *Pyrus communis* suspension-cultured cells. Proc Natl Acad Sci USA 96: 14246-14251.

Robertson D, Mitchell GP, Gilroy JS, Gerrish C, Bolwell GP, Slabas AR (1997) Differential extraction and protein sequencing reveals major differences in patterns of primary cell wall proteins from plants. J Biol Chem 1272: 15841-15848.

Rose JKC, Bashir S, Giovannoni JJ, Jahn MM, Saravanan RS (2004) Tackling the plant proteome: Practical approaches, hurdles and experimental tools. Plant J 39: 715-733.

Slabas AR, Ndimba B, Simon HWJ, Chivasa S (2004) Proteomic analysis of the Arabidopsis cell wall reveals unexpected proteins with new cellular locations. Biochem Soc Trans 32: 524-528.

Takos AM, Dry IB, Soole KL (1997) Detection of glycosyl-phosphatidylinositol-anchored proteins on the surface of *Nicotiana tabacum* protoplasts. FEBS Lett 405: 1-4.

Tashiro K, Tada H, Heilker R, Shirozu M, Nakano T, Honjo T (1993) Signal sequence trap: a cloning strategy for secreted proteins and type I membrane proteins. Science 261:600-603.

Wang SB, Hu Q, Sommerfeld M, Chen F (2004) Cell wall proteomics of the green alga *Haematococcus pluvialis* (Chlorophyceae). Proteomics 4: 692-708.

Watson BS, Lei Z, Dixon RA, Sumner LW (2004) Proteomics of *Medicago sativa* cell walls. Phytochem 65: 1709-1720.

Wojtaszek P, Trethowan J, Bolwell GP (1995) Specificity in the immobilisation of cell wall proteins in response to different elicitor molecules in suspension-cultured cells of French bean (*Phaseolus vulgaris* L.). Plant Mol Biol 28: 1075–1087.

Yamane H, Lee SJ, Kim BD, Tao R, Rose JKC (2005) A coupled yeast signal sequence trap and transient plant expression strategy to identify genes encoding secreted proteins from peach pistils. J Exp Bot 56: 2229-2238.

Functional Wall Glycomics through Oligosaccharide Mass Profiling

Nicolai Obel, Veronika Erben and Markus Pauly

With the advent of full genome sequences, bioinformatic tools and numerous genetic resources the stage is set to advance significantly our knowledge in cell wall polysaccharide biosynthesis, metabolism and function. However, to take advantage of this opportunity sensitive methods with a medium to high throughput capacity are necessary to quickly assess the structure of wall components in detail. Techniques that have been established in the last decades to analyse wall polysaccharides such as monosaccharide composition (Albersheim et al. 1967), glycosidic linkage analysis (York et al. 1986), sequential extraction of wall components (Redgwell 1980), and antibody labelling (Willats et al. 2000) can give quantitative and detailed information about the cell wall structure, but they are time-consuming and labour intensive. In order to improve the speed of analysis new techniques have to be developed. A rapid way to analyse cell walls is Fourier Transformed Infra Red (FT-IR) spectroscopy which can distinguish wall mutants, but little structural knowledge of wall polysaccharides is obtained from the spectra (Mouille et al. 2003). A different and very promising technique takes advantage the specificity of glycosylhydrolases to solubilize polysaccharides from wall materials in form of oligosaccharides and those are then analysed facilitating various means (Table I).

One of these methods is OLigosaccharide Mass Profiling (OLIMP), which combines high sensitivity and rapid analysis (Table 1). The method is based on enzymatic solubilization of specific oligosaccharides from wall materials and subsequent mass profiling using a Matrix Assisted Laser Desorption Ionization Time Of Flight (MALDI-TOF) Mass Spectrometer (MS). The resulting ion signals represent the molecular mass of the oligosaccharides and clearly indicate their sugar composition such as number of hexoses, pentoses, deoxyhexoses, uronic acids and other substituents. Due to the mild extraction procedure and lack of a derivatization procedure the oligosaccharides are solubilized in their native state. Therefore, non-carbohydrate substituents such as esters remain on the

Table I Comparison of current oligosaccharide profiling techniques

	Derivatisation	Sensitivity level	Speed of analysis	Quantification	Structural information	Reference
OLIMP	No	10 pmol	2 min/ sample	relative abundance	Molecular mass of oligosaccharides. No separation of isomers	(Wang et al. 1999; Lerouxel et al. 2002)
PACE	Yes	500 fmol	2h/ up to 15 samples in parallel	absolute abundance	Only retention times. Requires characterized standards	(Goubet et al. 2002; Lerouxel et al. 2002)
CE	Yes	40 fmol	20 min/sample	absolute abundance	Only retention times. Requires characterized standards	(Zhang et al. 1996; An et al. 2003)
HPLC	Yes	2 pmol	40 min/ sample	absolute abundance	Only retention times. Requires characterized standards	(Kakita et al. 2002)
HPAEC	No	100 pmol	60 min/ sample	absolute abundance	Only retention times. Requires characterized standards	(Lerouxel et al. 2002; van der Hoeven et al. 1998)
NMR	No	5 nmol	5 min/ sample	absolute abundance	Chemical shifts of the anomeric centres	(Duus et al. 2000; Perrin et al. 2003)

oligosaccharides and can be identified using OLIMP. However, information about the precise nature and configuration of e.g. a hexose is not revealed as is not its glycosidic linkage or sequence within the oligosaccharide and structural isomers cannot be distinguished by this technique. Nevertheless, advanced MS-techniques have been described to distinguish between oligosaccharide isomers (Fernandez et al. 2004). If the specificity of the enzyme is well established and its products have been analysed before in detail by other techniques such as NMR-spectroscopy, a putative structure can be assigned to each ion signal. Furthermore, the relative abundance of the oligosaccharides present in the mixture can be assessed by integrating the ion signals. The generated mass spectrum represents thus a structural semi-quantitative fingerprint of the polymer in the wall allowing e.g. a rapid confirmation of wall mutants (Lerouxel et al. 2002). Here, we report on an application of OLIMP that highlights the speed and sensitivity of the method: The assessment of the natural variation of xyloglucan, the major hemicellulose present in non-gramineceous higher plants, in an Arabidopsis ecotype collection.

Improvement of OLIMP

OLIMP has already been demonstrated to be a very powerful tool for the confirmation of wall mutants (Lerouxel et al. 2002). Advantage was taken of the sensitivity of mass spectrometry, and an optimization of the solubilization procedure lead to the generation of reproducible spectra from wall material derived from a single etiolated Arabidopsis hypocotyl (fresh weight ~200 µg, Figure 1).

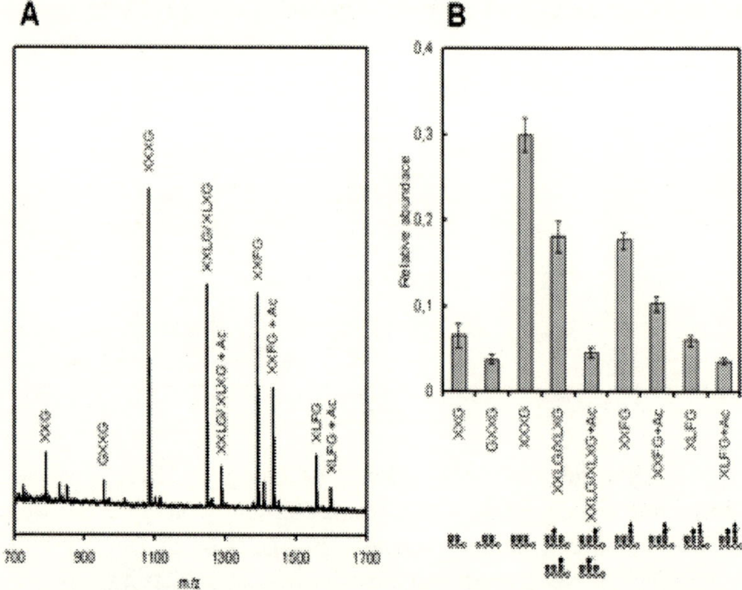

Fig. 1 Oligosaccharide profiling of a single etiolated Arabidopsis hypocotyl. **A MALDI-TOF mass spectrum of released xyloglucan oligosaccharides from a single hypocotyl. The structures of the various ion signals are shown using the nomenclature described by Fry et al. 1993 and a graphical representation of the various structures is shown. (O) - glucose, (■) - xylose, (●)- galactose, (▲) - fucose; (Ⓘ) - O-acetyl-substituent. B Relative abundance of the various oligosaccharides, n = 6.**

In such a typical optimized experiment Arabidopsis thaliana seedlings were grown on agar plates, containing 8.8 g/L Murashige and Skoog basal medium in 0.8% agar. Sterilized seeds were placed on the plates and incubated for 4 days at 22°C in the dark. One hypocotyl was placed in a single well of a 96 well sample preparation plate. The hypocotyls were frozen in liquid nitrogen and grinded using a 96 well retsch-mill. The residue was washed with 70 % aqueous ethanol, and pelleted through centrifugation. The pellet was washed with a chloroform-methanol mixture (1:1 v:v). After centrifugation, decanting, and evaporation of the solvent, 100 µl of 100 mM ammoniumformate, pH 4.5, containing 0.02 units of a purified recombinant xyloglucan specific endoglucanase (EC 3.2.1.151, Pauly et al. 1999) was added and incubated for 17h at 37°C. After digestion the suspension was centrifuged and 50 µl of the supernatant containing the solubilized oligosaccharides was transferred to another 96-well plate. The entire supernatant was dried in a speed-Figure concentrator and resuspended in 5 µl water. Ion exchange resin, ~ 10 beads, was added to remove buffer salts. After an eight minute incubation the supernatant (1 µl) was spotted onto a MALDI target plate that contained already vacuum dried crystallized 2,5-dihydroxybenzoic acid (10 µg). Mass spectra were obtained using a VoyagerDE-Pro MALDI-TOF MS in positive mode with an accelerating voltage of 20,000 V and a delay time of 350 ns. The spectra were recorded automatically and output-data files containing a list of the ion signals and their area were generated and saved automatically. The recorded spectra were analysed by an in house developed PERL program (Lerouxel et al. 2002), which calculated the relative area of the xyloglucan ion signals of each individual spectra and performed a pair wise comparison and statistical assessment by a student's t-test.

This improvement allows the use of plant material, which has been grown under controlled standardized conditions on agar plates in just four days, thus avoiding the extensive time and space requirements when investigating large plant populations.

Ecotype analysis

Seeds from 77 different Arabidopsis ecotypes (Torjek et al. 2003) were germinated and grown under standardized conditions. After four days the seedlings were harvested and their xyloglucan oligosaccharide profile was determined. All seedlings of each ecotype were grown, harvested and analyzed independently three times, and each time at least three replicates were made, thus allowing an assessment of the natural variation of xyloglucan structure. For each ecotype the length of the seedling was determined with a ruler, thus enabling a correlation between seedling growth and xyloglucan structure.

The clustering of ecotypes containing similar xyloglucan structures was achieved by hierarchical clustering analysis (data not shown) and Principal Component Analysis using the program Pirouette 2.6. The Data was mean centred and the Euclidian distance was used for calculating the distance matrix. The analyses were performed based on the mean of the relative abundance of single xyloglucan oligosaccharide structures from each independently harvested ecotype (Figure 2). Principal components one and two accounted for more than 97 % of the variance in the sample set. Component one was mainly representing the abundance of the structure XXXG, while component two was mainly representing the abundance of structures XXLG/XLXG and XXFG. As seen in Figure 2, the ecotypes did not show any distinct grouping and this was confirmed in the hierarchical clustering analysis. However, 15 of the 77 ecotypes were reproducibly located in distinct regions of the principal component analysis, as shown for the ecotypes Je54 and Yo in Figure 2. These two ecotypes were separated along principal component 1, indicating a difference in the abundance of XXXG. Indeed, a pair wise comparison of the spectra of these two ecotypes revealed that the abundance of the XXXG structure in the ecotype Je54 was 50% higher when compared to Yo, and a student's t-test comparison of the two ecotypes showed that the abundance of several oligosaccharides were significantly altered (see Figure 2 inlet). It appeared that the higher abundance of the XXXG fragment in ecotype Je54 was mainly compensated by a

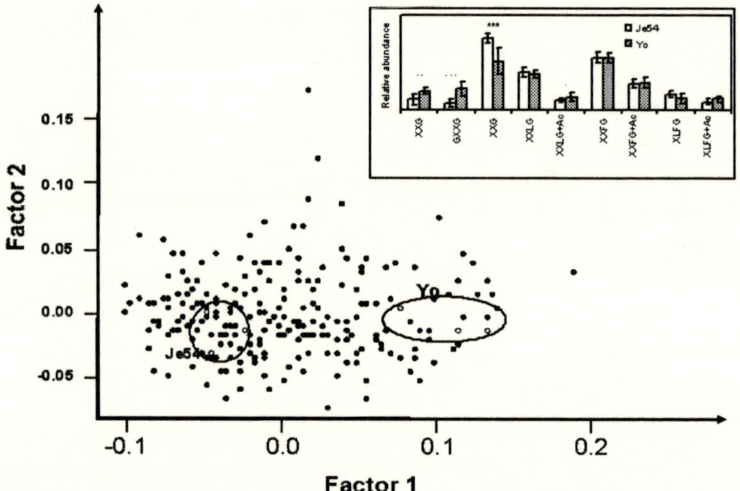

Fig. 2 Principal component analysis of xyloglucan oligosaccharide profiles from the analyzed ecotypes. **Each point represents the position of xyloglucan oligosaccharide spectra from three independent harvests. The results from each harvest of the two ecotypes, Yo and Je54, is highlighted with open circles. The inlet shows the relative abundance of the different xyloglucan oligosaccharides from the mass spectra of Yo and Je54. Stars indicate structures, which are significantly different (n=13), * $p<0.05$, ** $p<0.01$; *** $p<0.001$**

higher abundance of the smaller fragments, XXG and GXXG in the ecotype Yo, while most of the galactosylated and fucosylated xyloglucan oligosaccharides had an almost equal abundance in both ecotypes.

Previously, geographically different ecotypes have been analyzed by amplified fragment length polymorphism markers and correlations between geographic origin and the abundance of specific markers was observed for a number of ecotypes. Accordingly, ecotypes were found in central Europe, Scandinavia and the Iberian Peninsula, which showed distinct genetic variations (Sharbel et al. 2000). However, here no correlation between geographic origin of the ecotype and xyloglucan structure was observed.

The sizes of the seedlings after 4 days varied among the ecotypes between 0.8 and 1.5 cm (data not shown). Previous work has demonstrated the importance of the xyloglucan structure for plant growth and development probably due to xyloglucan metabolism (Hayashi et al. 1987). In peas, changes in the two oligosaccharides XXG and XXXG between elongating and non-elongating tissue has previously been reported (Pauly et al. 2001), and it has been shown that oligosaccharides containing galactose are important for the incorporation of xyloglucan into the existing wall (Pena et al. 2004). Furthermore, fucosylated oligosaccharins have been shown to inhibit or enhance 2,4-dichlorophenoxyacetic acid induced growth (Albersheim et al. 1992; Takeda et al. 2002). However, in the generated dataset no correlation between specific xyloglucan oligosaccharides and length of the hypocotyl was observed (data not shown), thus not substantiating further any of these hypotheses. The lack of correlation might be due to the nature of the plant material used here, as it contained a mixture of both elongating and non-elongating material. Furthermore, only enzyme soluble xyloglucan is characterized with OLIMP, not all xyloglucan present in the wall.

Extension of OLIMP to pectic homogalacturonan

The oligosaccharide mass profiling technique can be extended to other cell wall polysaccharides when specific hydrolytic enzymes are available. In the case of homogalacturonan it was possible to obtain an oligosaccharide profile using a combination of pectinmethylesterase 0.08 units (1 unit releases 1 μmol of methanol per min) and endopolygalacturonase. The released fragments corresponded to oligogalacturonides substituted with methylester- and acetyl-groups (Figure 3A). Ions signals representing potentially methylesterified or O-acetylated oligosaccarides shifted in mass upon base treatment (data not shown) clearly demonstrating the presence of ester-substituents. Similar to the xyloglucan profile the relative abundance of the various oligogalacturonides fragments is reproducible when analyzing hypocotyls (Figure 3B) thus providing a good basis for a routine medium to high-throughput analysis of homogalacturonan.

OLIMP, originally established for plant cell wall derived xyloglucan, can easily be extended to other classes of wall polysaccharides as shown here for the pectic polysaccharide homogalacturonan. The condition is the availability of an appropriate enzyme that specifically solubilizes a particular wall component. Hence, there is no obvious reason why oligosaccharide mass profiling cannot be extended also

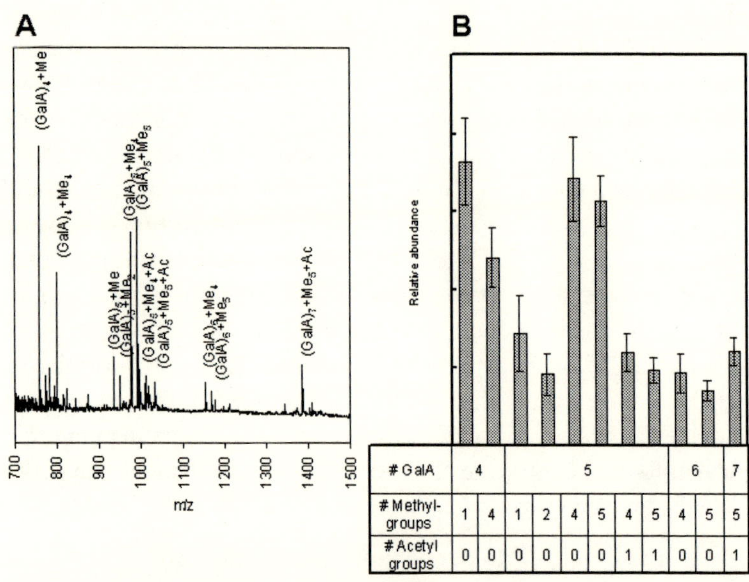

Fig. 3 Oligosaccharide profiling of homogalacturonan. **A, Mass spectrum of the oligosaccharides generated by digestion of wall material derived from five hypocotyls with endopolygalacturonase and pectin methylesterase. B, The relative abundance of various homogalacturonan fragments, In the table below the number of galacturonic acids, methyl groups and acetyl substituents in each signal are indicated, n=6.**

to wall glycoproteins. The more information is known about the hydrolase-specificity and/or the released products, the more detailed conclusion can be drawn from the resulting mass spectrum.

OLIMP is sensitive, only minute amounts of plant tissue are necessary. Currently, OLIMP is the most rapid method to generate an oligosaccharide profile (see Table 1). Together, these attributes allow a medium to high throughput polysaccharide analysis of samples, making this technique ideal for functional wall glycomics. Future applications include reverse, forward, and association genetic approaches to gain further insights into wall biosynthesis, metabolism and function.

Acknowledgement

We thank Dr Kirk Schnorr (Novozymes A/S, Bagsvaerd, Denmark) for the generous gift of xyloglucan endoglucanase and pectin methyl esterase. Financial support of this work was provided by the German GABI project 0312277D.

References

Albersheim P, Darvill A, Augur C, Cheong JJ, Eberhard S, Hahn MG, Marfa V, Mohnen D, O'Neill MA, Spiro MD, York WS (1992) Oligosaccharins - oligosaccharide regulatory molecules. Accounts Chem Res 25: 77-83.

Albersheim P, Nevins DJ, English PD, Karr.A. (1967) A method for the analysis of sugars in plant cell-wall polysaccharides by gas-liquid chromatography. Carbohydr Rcs 5: 340-345.

An HJ, Franz AH, Lebrilla CB (2003) Improved capillary electrophoretic separation and mass spectrometric detection of oligosaccharides. J Chromatogr A 1004: 121-129.

Duus JO, Gotfredsen CH, Bock K (2000) Carbohydrate structural determination by NMR spectroscopy: Modern methods and limitations. Chem Rev 100: 4589-+.

Fernandez LEM, Obel N, Scheller HV, Roepstorff P (2004) Differentiation of isomeric oligosaccharide structures by ESI tandem MS and GC-MS. Carbohyd Res 339: 655-664.

Fry SC, York WS, Albersheim P, Darvill A, Hayashi T, Joseleau JP, Kato Y, Lorences EP, Maclachlan GA, McNeil M, Mort AJ, Reid JSG, Seitz HU, Selvendran RR, Voragen AGJ, White AR (1993) An unambiguous nomenclature for xyloglucan-derived oligosaccharides. Physiol Plantarum 89: 1-3.

Goubet F, Jackson P, Deery MJ, Dupree P (2002) Polysaccharide analysis using carbohydrate gel electrophoresis: A method to study plant cell wall polysaccharides and polysaccharide hydrolases. Anal Biochem 300: 53-68.

Hayashi T, Marsden MPF, Delmer DP (1987) Pea xyloglucan and cellulose V. xyloglucan-cellulose interactions *in vitro* and *in vivo*. Plant Physiol 83: 384-389.

Kakita H, Kamishima H, Komiya K, Kato Y (2002) Simultaneous analysis of monosaccharides and oligosaccharides by high-performance liquid chromatography with postcolumn fluorescence derivatization. J Chromatogr A 961: 77-82.

Lerouxel O, Choo TS, Seveno M, Usadel B, Faye L, Lerouge P, Pauly M (2002) Rapid structural phenotyping of plant cell wall mutants by enzymatic oligosaccharide fingerprinting. Plant Physiol 130: 1754-1763.

Mouille G, Robin S, Lecomte M, Pagant S, Hofte H (2003) Classification and identification of Arabidopsis cell wall mutants using Fourier-Transform InfraRed (FT-IR) microspectroscopy. Plant J 35: 393-404.

Pauly M, Andersen LN, Kauppinen S, Kofod LV, York WS, Albersheim P, Darvill A (1999) A xyloglucan-specific endo-beta-1,4-glucanase from *Aspergillus aculeatus*: expression cloning in yeast, purification and characterization of the recombinant enzyme. Glycobiology 9: 93-100.

Pauly M, Qin Q, Greene H, Albersheim P, Darvill A, York WS (2001) Changes in the structure of xyloglucan during cell elongation. Planta 212: 842-850.

Pena MJ, Ryden P, Madson M, Smith AC, Carpita NC (2004) The galactose residues of xyloglucan are essential to maintain mechanical strength of the primary cell walls in Arabidopsis during growth. Plant Physiol 134: 443-451.

Perrin RM, Jia ZH, Wagner TA, O'Neill MA, Sarria R, York WS, Raikhel NV, Keegstra K (2003) Analysis of xyloglucan fucosylation in Arabidopsis. Plant Physiol 132: 768-778.

Redgwell RJ (1980) Fractionation of plant-extracts using ion-exchange sephadex. Anal Biochem 107: 44-50.

Sharbel TF, Haubold B, Mitchell-Olds T (2000) Genetic isolation by distance in Arabidopsis thaliana: biogeography and postglacial colonization of Europe. Mol Ecol 9: 2109-2118.

Takeda T, Furuta Y, Awano T, Mizuno K, Mitsuishi Y, Hayashi T (2002) Suppression and acceleration of cell elongation by integration of xyloglucans in pea stem segments. Proc Natl Acad Sci USA 99: 9055-9060.

Torjek O, Berger D, Meyer RC, Mussig C, Schmid KJ, Sorensen TR, Weisshaar B, Mitchell-Olds T, Altmann T (2003) Establishment of a high-efficiency SNP-based framework marker set for Arabidopsis. Plant J 36: 122-140.

van der Hoeven RAM, Hofte AJP, Tjaden UR, van der Greef J, Torto N, Gorton L, Marko-Varga G, Bruggink C (1998) Sensitivity improvement in the analysis of oligosaccharides by on-line high-performance anion-exchange chromatography ion spray mass spectrometry. Rapid Commun Mass Sp 12: 69-74.

Wang J, Sporns P, Low NH (1999) Analysis of food oligosaccharides using MALDI-MS: Quantification of fructooligosaccharides. J Agr Food Chem 47: 1549-1557.

Willats WGT, Steele-King CG, McCartney L, Orfila C, Marcus SE, Knox JP (2000) Making and using antibody probes to study plant cell walls. Plant Physiol Bioch 38: 27-36.

York WS, Darvill AG, McNeil M, Stevenson TT, Albersheim P (1986) Isolation and characterization of plant-cell walls and cell-wall components. Method Enzymol 118: 3-40.

Zhang ZQ, Pierce ML, Mort AJ (1996) Detection and differentiation of pectic enzyme activity in vitro and in vivo by capillary electrophoresis of products from fluorescent-labeled substrate. Electrophoresis 17: 372-378.

Xylem Cell Expansion – Lessons from Poplar

Ewa J. Mellerowicz

Xylem cell size and shape determines its function

Wood evolved as a tissue to carry out two essential functions: water conduction from roots to shoots and the mechanical support necessary to elevate shoots to the photosynthetic light. In angiosperm species, such as poplar, these two functions are performed by separate cell types: vessel elements and fibers (Figure 1A). Poplar wood contains in addition parenchyma cells, primarily arranged in radial files forming uniseriate rays. All cell types have strong cell walls, referred to as secondary to indicate that the wall thickened markedly following cell expansion when only primary wall was present, and are distinguished by their size and shape. The fibers are slender and elongated, with the length reaching typically 25 fold the radial diameter while vessel elements are shorter and greatly radially expanded showing the length to diameter ratio of 5 at the average, which can however vary considerably because vessel diameter is regulated by season, water availability and position within the stem. The degree of fiber elongation determines how much support it will provide to the stem while extend of vessel element lateral expansion determines how effective it will be as a conductive vessel and how easy it will fail during drought conditions (Tyree and Zimmermann 2002). Thus, the cell size and shape of both vessel element and fibers determine their respective functions. How poplar regulates the size and shape of its xylem cells? To understand growth processes in a complex tissue such as the wood, one needs to go back to anatomy to describe how mechanistically the growth occurs. This will ensure that the right tissue will be sampled and right questions asked.

Anatomy of xylem cell growth

The xylem cells start growing when they are still within the vascular cambium, but most growth is observed right after the exit from the meristem, in the zone of radial cell expansion (Figure 1B). Vessel elements and fibers both arise from the already-elongated fusiform initials (Figure 2A). Vessel elements do not elongate during differentiation, except possibly for a very limited tail growth, but they greatly radially expand (Larson 1994). This is accomplished by the differential stretching of radial walls and the intrusive displacement of adjacent fibers (Figure 2B). Fibers also expand radially but to a lesser degree and they substantially elongate by the intrusive tip growth (Figure 2A). During elongation, the central part of the fiber remains fixed and growth is limited to the tip. Tip growth might continue even after the deposition of the secondary wall in the central part of the fiber. Ray parenchyma cells, which originate from isodiametric ray initials, elongate in radial direction by a diffuse growth. During growth, ray cell radial walls remain attached to the expanding radial walls of adjacent vessel elements or fibers (Figure 1B). This is an example of symplastic growth.

Thus, within the developing wood tissue different types of cell growth co-occur, including diffuse and tip growth, as well as intrusive and symplastic growth. Wood development involves also a great deal of localized cell expansion. Thus, the anatomical analysis tells us that growth regulating agents must differ for different cell types and in most cases also be localized within the cells.

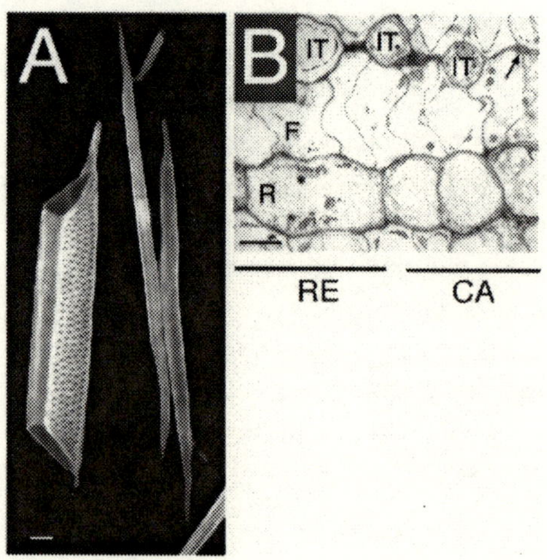

Fig. 1 Poplar wood.

A. Vessel element and fibers separated by maceration in the acetic acid and hydrogen peroxide. Bar=30μm

B. Beginning of an intense cell expansion at the exit from the vascular cambium (CA) as seen in the transverse section. Notice the co-existence of symplastic growth between the developing fiber (F) and the contacting ray cell (R), and the intrusive tip growth (IT) of fibers. Intrusive growth starts from cell separation (arrow). RE – radial expansion zone. Bar=5 μm.

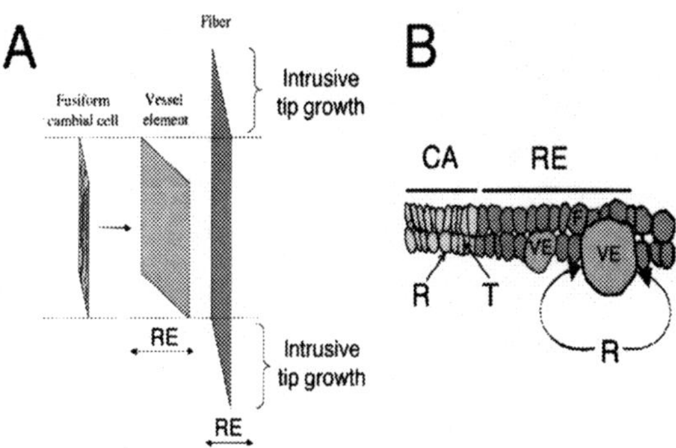

Fig. 2 Contrasting patterns of cell expansion of cambial derivatives.

A. Vessel elements expand radially but do not elongate while fibers expand radially and elongate by intrusive tip growth. Both cell types are formed from fusiform cambial cells. The direction of expansion is determined by the orientation of cellulose microfibrils drawn here for the fusiform cambial cell as an irregular longitudinal network.

B. Wood developmental series showing cambium (CA) and radial expansion zone (RE) as seen in the transverse section. Note that fibers (F) and vessel elements (VE) expand by extension of their radial walls (R) while the tangential walls (T) remain the same.

Direction of growth

Cellulose in plant cells forms crystalline microfibrils that form a skeleton of cell wall. Since the cellulose has a high tensile index, i.e. is not stretchable, the expansion is possible only in the direction perpendicular to the cellulose microfibrils. In the axially elongated developing wood cells at the beginning of expansion, there is a tendency for longitudinal microfibril orientation although the microfibrils are not very well ordered (Funada 2000). This net longitudinal orientation would allow the lateral cell expansion but not the longitudinal expansion by diffuse growth. Thus, the analysis of cellulose microfibril orientation tells us that the fibers cannot elongate by a diffuse growth as the cells deriving from apical meristems do. In contrast, developing fibers elongate by the intrusive tip growth when the new wall is being deposited at the tip. In the ray initials however, the arrangement of cortical microtubules (which reflects the orientation of cellulose microfibrils) is random and very sparse (Chaffey et al. 2002), thus allowing the cell expansion by diffuse growth in any orientation. In fact, ray cells elongate radially, probably as a result of pooling forces through the contacts with expanding fibers and vessel elements (Figure 1B).

Major players of wood expansion

Cell expansion is usually considered as a turgor-driven irreversible cell wall extension that is prerequisite for growth. Turgor pressure might play an important role in extend of vessel element expansion (Langer et al. 2002), but in fibers, the expansion is largely determined by cell wall plasticity. This way, the developing wood cells can grow even at low turgor pressure and can precisely control their size and shape. Cell wall strength depends on the net-like structure of non-stretchable cellulose microfibrils, which are coated and cross-linked by non-stretchable single molecules of hemicelluloses, mainly xyloglucan, to form a framework of the wall. Enzymes affecting wall plasticity are implicated in either cutting the cross linking hemicelluloses (xyloglucan endotransglycosylases or XETs, xyloglucanases, and cellulases), or disrupting their bonding to cellulose (expansins and cellulases). Wall extensibility depends also on other cell wall components, for example arabinogalactan proteins (Shi et al. 2003) or pectins (Ryden et al. 2003), which may interact with cellulose-hemicellulose framework.

We have demonstrated that the cambial region is very abundant in XET activity (Bourquin et al. 2002), and in collaboration with Dr. Simon McQueen Mason from the York University in UK, we found a substantial expansin activity in developing wood (Gray-Mitsumune et al. 2004). Several different pectin methyl esterase (PME) isoforms were found in the cambium and in expanding wood cells (Micheli et al. 2000). These agents certainly have a role in wood cell expansion. The challenging task now is to identify the remaining enzymes involved and to determine what is the exact role and relative contribution of each of these agents. The difficulty of this task lies in the fact that most wall biosynthetic and modifying enzymes and proteins are encoded by multi-gene families whose members are sometimes very similar in their coding sequence. Each of these genes can have either specific or redundant functions. Redundant functions can be assumed if genes are both expressed in the same tissue and if the knock out of each gene's function does not induce considerable effects in plants while the combination of knockouts produces the effect. An abundance of genes encoding wall synthesizing and modulating enzymes and proteins was found in developing wood (Mellerowicz et al. 2001, and unpublished) and many genes exhibited differential expression patterns across wood developmental zones (Hertzberg et al. 2001). Thus, transcriptional regulation appears to play a role in determination of xylem cell size and shape. However, the lack of differential transcription cannot be considered as a proof of non-involvement. For example, many enzymes could be regulated at the activity level by pH, ions and co-factors. Sapwood generally has an acidic pH while heartwood is slightly basic (Kroll et al. 1992). pH environment regulates enzymatic activity of wall plasticizing enzymes and proteins such as expansins and might play a role in wood development. Non-enzymatic mechanisms of cell wall degradation involving reactive oxygen species have been recently postulated (Fry et al. 2002) and could potentially also play a role in cell plasticity regulation in the wood.

XET might regulate differential expansion between vessel elements and fibers

Cell-type specific plasticity regulation is important for differential expansion of vessel elements and fibers and might even play a role in determining cell fate. XET activity and PttXET16A protein show differential distribution in fibers and vessel elements (Bourquin et al. 2002). The enzyme appears to be more expressed in the developing fibers than vessel elements. We speculated that higher XET activity in developing fibers might lead to more microfibril cross-linking by xyloglucan and more resistance to turgor pressure preventing radial expansion. In support of this idea, CCRC-M1 monoclonal antibodies detected more xyloglucan in the fiber wall than in vessel element wall (Bourquin et al. 2002). Differential xyloglucan cross-linking might be therefore the basis of differential cell expansion of fibers and vessel elements. In contrast, the major expansin of developing wood (*PttExp1*) seems to be equally expressed in vessel elements and fibers, at least as observed at the mRNA level (Gray-Mitsumune et al. 2004).

Cell-region specific mechanisms acting in wood: transcript and protein targeting, and pH

It is important to realize that transcriptional regulation alone cannot explain growth phenomena that are cell-region specific as it is the case of developing wood. For example, intrusive tip growth in fibers requires the transport of wall building and modifying enzymes as well as wall carbohydrates to the growing tip. In root hairs, tip growth depends on intact filamentous actin network (Baluska et al. 2000; Ringli et al. 2002). Growing root hair tips accumulate longitudinally arranged actin filaments that are thought to deliver wall building materials precisely where it is needed. Developing wood fibers have a longitudinal arrangement of actin filaments (Chaffey et al. 2000) that could be involved in tip growth.

An intriguing possibility is the targeting of mRNA to the growing tips (Im et al. 2000). Such targeted mRNA would be translated close to the tip ensuring a high yield of the ready protein for the delivery to growing wall. We found that mRNA of expansin *PttEXP1* along with ribosomal RNA were accumulated at the tips of elongating fibers in poplar suggesting a role for RNA targeting and localized translation during fiber tip growth (Gray-Mitsumune et al. 2004).

Localized growth occurs also in radially expanding vessel elements and fibers whose radial (R) walls expand while tangential (T) walls remain unexpanded (Figure 2B). The differences between R and T walls arise already in the cambium during wall deposition and maturation. This can be explained as follows. Cambial cells divide periclinally, i.e. by depositing a new T wall inside the cell. T walls are therefore developmentally younger than R walls and their properties might simply reflect their maturation status. Wall homogalacturonans are deposited in a methylated form and are gradually demethylated *in muro* by wall-residing PMEs. It might be therefore expected that the older R walls would contain more demethylated, acidic pectin compared to younger T walls. TEM microscopy combined with differential carbohydrate extraction indeed demonstrated more EDTA extractable material, most likely acidic and calcium-bound pectins in R walls (Catesson et al. 1994; Baïer et al. 1994). Acidic pectins were also demonstrated in R walls by immunolocalization using JIM5 monoclonal antibody (Guglielmino et al. 1997). Acidic pectin might contribute either to the sequestration of Ca^{2+} ions and wall stiffening, or to the local decreasing of wall pH, which in turn might activate expansins and other wall residing proteins causing the differential expansion of R walls. The hypothesis might be put forward that in the cambium, the action of PME would remove some methyl groups from pectin, causing wall acidification, but the remaining methyl groups would prevent efficient binding of Ca^{2+} ions. At this stage the walls would be most extensible. However, as cells differentiate, the prolonged PME action would eventually remove remaining methyl groups, thus creating long stretches of acidic pectin with a very high affinity for Ca^{2+} ion binding. Ca^{2+} binding and formation of egg-box-like structures with a layer of Ca^{2+} ions sandwiched between layers of acidic pectin would lead to wall stiffening and would prevent further expansion.

Intrusive growth: specialty of the house

The intrusive growth, in contrast to other forms of cell growth, involves cell separation and wall yielding during the penetration of a growing cell between other cells, and it is particularly prevalent in developing wood. Cambial fusiform initials elongate via intrusive tip growth. In addition, in angiosperm species, xylem fiber cells use also this mechanism to elongate to their ultimate size. Vessel element expansion can also involve intrusive growth between fibers. In this case, the intrusion occurs sideways rather than longitudinally.

Previous studies of intrusive growth have focused mostly on anatomical aspects and hormonal regulation (Wenham and Cusick 1975, Ridoutt and Pharis 1996). We speculated that the intrusive growth might be regulated by the properties of middle lamella and in particular, by the presence of stiff egg-box-like structures. To explore this possibility, we cloned, in collaboration with Dr. Luc Richard from the Institut Jacques Monod in France, a PME encoding gene *PttPME1* from poplar developing wood and made transgenic poplars using sense and antisense constructs (EJ Mellerowicz, A Siedlecka, M-A Péronne, F Micheli, J Lesniewska, L Richard, and B Sundberg, unpublished). Down regulation of *PttPME1* resulted in a lower amount of pectins, increased methylesterification levels, the appearance of a novel homogalacturonan epitope recognized by the monoclonal antibody LM7, and in longer fibers. Conversely, the overexpression of *PttPME1* caused a decrease in methylesterification, a substantial increase in the non-methylated homogalacturonan recognized by the monoclonal antibody Pam1 and in shorter fibers. However, the changes induced in the fiber length albeit significant, were rather modest, between -12% and +8% of the wild type level. This indicates that *PttPME1* is one of several enzymes that limit fiber intrusive growth.

Where to go from here?

Cell wall plasticity in developing wood cells can be conceived as regulated by wall modifying enzymes, both, plasticizing and strengthening, and by wall biosynthetic enzymes, many of which are still unknown. Our work has initiated the identification a few agents regulating cell size and shape in developing wood. The genes so far characterized were only representatives of large enzymatic families involved in wall extension during wood differentiation. This analysis needs to be completed by a functional characterization of remaining members expressed during wood development. Novel wall carbohydrate synthesizing and modulating enzymes are being discovered and their role in wood development should be evaluated. Wall protein network might be of special importance for wood development, especially the mysterious arabinogalactan proteins that are very abundant in this tissue. Recent findings (Fry et al. 2002) prompt to evaluate agents regulating reactive oxygen species in wood development.

The resources for identifying new agents for xylem cell growth in woody species are being developed at a high speed. Poplar model system is particularly attractive because of availability of EST databases, genomic data, transgenics and expression databases. The microarray analysis is becoming possible for very small amount of tissue approaching the cell-type resolution and opening a possibility of high-resolution high throughput expression analysis. Exploration of *Arabidopsis* model for wood developmental studies (Chaffey et al. 2002) will increase undoubtedly our understanding of wood cell expansion in forest trees. Insertional mutant analysis and QTL analysis are still not fully explored for wood development. With the advantages of *Arabidopsis* model for functional genomic studies, the function of these unknown genes will eventually going to be determined and the nearest future will undoubtedly be a very exciting period for wood cell wall expansion and cell wall research in general.

Acknowledgements

I thank Deborah Delmer for providing inspiration for me as well as for many generations of scientists working on plant cell walls. I also thank Björn Sundberg for stimulating discussions on wood development and his appreciation for plant anatomy. Support from the Wood Ultrastructure Research Centre (WURC) and the Foundation for Strategic Research, the European project Eden QLK5-CT-2001-00443 is acknowledged.

References

Baïer M, Goldberg R, Catesson A-M, Liberman M, Bouchemal N, Michon V, Hervé du Penhoat

C (1994) Pectin changes in samples containing poplar cambium and inner bark in relation to the seasonal cycle. Planta 193: 446-454.

Baluška F, Salaj J, Mathur J, Braun M, Jasper F, Šamaj J, Chua NH, Barlow PW, Volkmann D (2000) Root hair formation: F-actin-dependent tip growth is initiated by local assembly of profilin-supported F-actin meshworks accumulated within expansin-enriched bulges. Dev Biol 227: 618-632.

Bourquin V, Nishikubo N, Abe H, Brumer H, Denman S, Eklund M, Christiernin M, Teeri TT, Sundberg B, Mellerowicz, EJ (2002) Xyloglucan endotransglycosylases have a function during the formation of secondary cell walls of vascular tissues. Plant Cell 14: 3073-3088.

Catesson A-M, Funada R, Robertbaby D, Quinetszely M, Chuba J, Goldberg R (1994) Biochemical and cytochemical cell-wall changes across the cambial zone. IAWA J 15: 91-101.

Chaffey N, Barlow P, Barnett J (2000) A cytoskeletal basis for wood formation in angiosperm trees: the involvement of microfilaments. Planta 210: 890-896.

Chaffey N, Barlow P, Sundberg B (2002) Understanding the role of the cytoskeleton in wood formation in angiosperm trees: hybrid aspen (*Populus tremula* x *P. tremuloides*) as the model species. Tree Physiol 22: 239-249.

Fry SC, Miller JG, Dumville JC (2002) A proposed role for copper ions in cell wall loosening. Plant Soil 247: 57-67.

Funada R (2000) Control of wood structure. In: Plant microtubules. P Nick (ed), Springer Verlag, pp: 51-82.

Gray-Mitsumune M, Mellerowicz EJ, Abe H, McQueen-Mason S, Winzéll A, Sterky F, Blomqvist K, Schrader J, Teeri TT, Sundberg B (2004) Expansins abundant in secondary xylem belong to Subgroup A of the α-expansin gene family. Plant Physiol 135: 1552-1564.

Guglielmino N, Liberman M, Jauneau A, Vian B, Catesson A-M, Goldberg R (1997) Pectin immunolocalization and calcium visualization in differentiating derivatives from poplar cambium. Protoplasma 199: 151-160.

Hertzberg M, Aspeborg H, Schrader J, Andersson A, Erlandsson R, Blomqvist K, Bhalerao R, Uhlen M, Teeri TT, Lundeberg J, Sundberg B, Nilsson P, Sandberg G (2001) A transcriptional roadmap to wood formation. Proc Natl Acad Sci USA 98: 14732-14737.

Im K-H, Cosgrove DJ, Jones AM (2000) Subcellular localization of expansin mRNA in xylem cells. Plant Physiol. 123: 463-470.

Kroll RE, Ritter DC, Gertjejansen RO, Au KC (1992) Anatomical and physical-properties of balsam poplar (*Populus balsamifera* L) in Minnesota. Wood Fiber Sci. 24: 13-24.

Langer K, Ache P, Geiger D, Stinzing A, Arend M, Wind C, Regan S, Fromm J, Hedrich R (2002) Poplar potassium transporters capable of controlling K^+ homeostasis and K^+-dependent xylogenesis. Plant J 32: 997-1009.

Larson PR (1994) The Vascular Cambium. Springer Verlag, Berlin.

Mellerowicz EJ, Baucher M, Sundberg B, Bojeran W (2001) Unraveling cell wall formation in the woody dicot stem. Plant Mol Biol 47: 239-274.

Micheli F, Sundberg B, Goldberg R, Richard L (2000) Radial distribution pattern of pectin methylesterases across the cambial region of hybrid aspen at activity and dormancy. Plant Physiol. 124: 191-199.

Ridoutt BG, Pharis RP, Sands R (1996) Fibre length, gibberellins A(1) and A(20) are decreased in Eucalyptus globulus by acylcyclohexanedione injected into the stem. Physiol Plant 96: 559-566.

Ringli C, Baumberger N, Diet A, Frey B, Keller B (2002) ACTIN2 is essential for bulge site selection and tip growth during root hair development of Arabidopsis. Plant Physiol 129: 1464-1472.

Ryden P, Sugimoto-Shirasu K, Smith AC, Findlay K, Reiter WD, McCann MC (2003) Tensile properties of Arabidopsis cell walls depend on both a xyloglucan cross-linked microfibrillar network and rhamnogalacturonan II-borate complexes. Plant Physiol 132: 1033-1040.

Shi HZ, Kim Y, Guo Y, Stevenson B, Zhu JK (2003) The Arabidopsis SOS5 locus encodes a putative cell surface adhesion protein and is required for normal cell expansion. Plant Cell 15: 19-32.

Tyree MT, Zimmermann MH (2002) Xylem structure and ascent of sap. Springer Verlag, Berlin.

Wenham MW, Cusick F (1975) The growth of secondary fibers. New Phytol 74: 247-261.

Cellulose Synthase Genes in Conifers: What We Know and What We Need to Learn

Anita Sherrie Klein and Josquin Tibbits

Comparative genomics have provided a means to begin to study cellulose synthesis in economically important pines. In this paper we describe the cloning of a full-length cellulose synthase gene from mRNA in radiata pine. Bioinformatic approaches allow us to predict that there are ten and fifteen different *CesA* genes in pines, and that the genes are highly conserved among species that diverged as long as 80 million years ago. Preliminary analysis of expression patterns suggests that three cellulose synthase genes are highly transcribed during earlywood formation and these are orthologs of secondary cell wall *CesA* genes from angiosperms.

How I came back to cellulose research

With T. Hayashi's kind request to contribute to this book, I have returned to a research community that I left more than 24 years ago. In the intervening years I pursued interests in molecular genetics and eventually turned to study of population and evolutionary genetics of conifers. As a graduate student with Debby Delmer, I used a protoplast cell wall regeneration system to study cellulose, the most abundant biopolymer (Klein et al. 1981; Willison and Klein 1982). With the cloning of the first higher plant cellulose synthase genes (Pear et al. 1996) and accumulating observations that the *CesA* gene family is highly conserved among various angiosperms (Delmer 1999), I realized it might now be possible to study cellulose biosynthesis in conifers. I compared the sequence of the cotton *CesA1* gene to the 'expression sequence tagged' (EST) clones from the pine Megagenome project (http://pine.ccgb.umn.edu/); a large number of these EST clones showed high degrees of sequence similarity to *GhCesA1*. Based on these bioinformatics I wrote a grant proposal to fund a sabbatical leave (2000-2001) project on cellulose synthase genes in economically important pines.

Why wood?

Wood is a primary resource for fuel and fiber and a very important carbon sink in natural ecosystems. With the need to produce more wood comes the need to breed for faster growth and better quality wood (Fengel and Wegener 1983). Worldwide conifers including pines and spruces are an important source of wood for construction, paper pulping, and biomass production. Wood conducts photosynthate to the roots (phloem) and water and nutrients to the crown of the tree (xylem). The production of thick cell walls in the xylem tracheids also provides strength and rigidity (Bailey 1952). Cellulose constitutes 40-60% of the mass of the secondary cell wall and these secondary cell walls represent the bulk of wood biomass (Fengel and Wegener 1983). Cellulose synthase (*CesA*) genes, the catalytic subunit of the Cellulose Synthase Complex (CSC) have clear roles in wood formation.

Wood consists of the secondary vascular system, where the lateral meristematic tissue, the vascular cambium, annually undergoes an ordered process of cell division, expansion, and differentiation referred to as xylogenesis (Bailey 1952). The annual growth ring in temperate conifers differentiates from the cambial ray and fusiform initials (Higuchi 1997). The fusiform initials divide in a longitudinal fashion to produce phloem and xylem mother cells. The xylem mother cells may divide two to three more times producing radial files of tracheids (Harada and Côté 1985) (Figure 1A). These cells first expand in diameter and length (primary cell wall deposition) and then the cell wall thickens (secondary cell wall deposition). Mature tracheids undergo apoptosis, programmed cell death, during the final stages of cell wall synthesis; the dead cells conduct water from the roots to the needles. The ray initials divide to form ray parenchyma and ray tracheids that elongate radially (Figure 1B). The xylem also contains epithelial cells, which surround radial resin canals. Early and latewood are an annual cycle in conifer wood formation where there is a progression towards slower cell division, an increase in wall thickness, and an decrease in cellulose to lignin ratio (more lignin, less cellulose).

As axial tracheids develop, they lay down walls processively forming several layers. The primary cell wall underlays the middle lamella and is the thinnest layer. When cell expansion ceases, the several layers of secondary cell wall are synthesized. The S1 layer is 0.2-0.3 μm in width, with 3-4 lamella, and the cellulose microfibrils circle with a gentle helical slope to the right or left against the long axis of

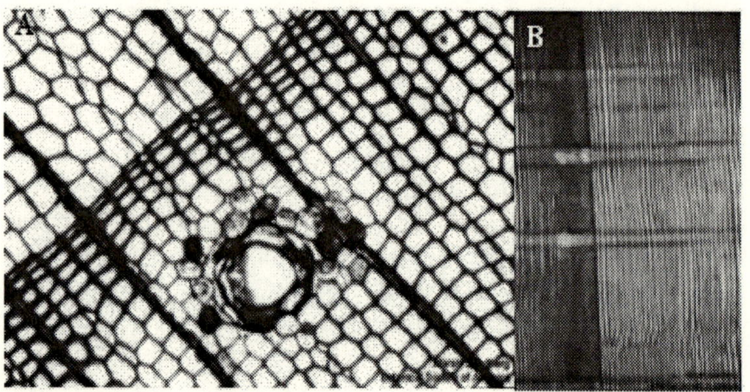

Fig. 1 Anatomy of pine stem
A: In cross section the cambium divides to form phloem and xylem initials. The xylem initials divide to form radial files of tracheid cells, which after differentiation undergo apoptosis. A resin canal is seen in between files of tracheids. The tracheids are separate by ray parenchyma. B: Radial section illustrating theradial elongations of tracheids. (reprinted with permission, Botanical Society of America http://www.botany.org/plantimages/PlantAnatomy.php).

the cell. The S2 layer, with up to 150 lamella, is the thickest layer in the tracheid cell wall. The microfibrils of the S2 layer run at steep angle approximately parallel to the long axis of the wall (Harada and Côté 1985); variation in the angle and packing results in irregular lamellation of S2. The S2 varies from 1 μm in early wood to 5 μm in latewood. The tensile strength of wood results from the structure of the S2 layer and is correlated with cellulose content (Jones and Corson 1996). The innermost layer is the relatively thin, warty tertiary (T) layer in conifers (also called the S3 layer) (Fengel and Wegener 1983). The microfibrils of the S3 layer run with a gentle slope but not necessarily in parallel order. Microfibril angle also changes between earlywood and latewood.

The highest lignin concentration is in the middle lamella between tracheids but since the S2 layer accounts for the largest volume of the cell walls (80-90% of wood mass), 25% of the lignin is deposited in the S2 (Fengel and Wegener 1983). Special conditions for wood formation exist when trees are bent, either naturally, where branches form, or when trees are mechanically bent (compression wood). In compression wood tracheids have rounded cell walls with intercellular spaces; the lignin content is much higher than in normal wood and wood and the cellulose content is lower (Fengel and Wegener 1983). Le Provost et al. (2003) showed that many of the genes involved in lignin biosynthesis are up regulated, and those in the cellulose pathway are down regulated in compression wood forming tissues compared to normal wood.

With this detailed description of the developmental processes of conifer xylogenesis in mind, what do we need to know about the role of cellulose synthase genes in wood formation? From elegant studies in Arabidopsis, we know that that at least three distinct CES proteins are required to produce an active Cellulose Synthase Complex (CSC) in secondary cell walls (Taylor et al. 2003). It appears that three other CES proteins are required for primary cell wall cellulose synthesis. A variable segment of the gene referred to as the class specific region (CSR) (Vergara and Carpita 2001) appears to distinguish subclasses of *CesA* genes. There are ten *CesA* genes in Arabidopsis (Richmond and Somerville 2000); the roles of the remaining four *AtCesA* genes are unclear. How many genes are required to form the cellulose synthase complex during wood formation in pine? Are different sets of pine *CesA* genes involved in the synthesis of the primary cell wall during xylogenesis than in secondary cell walls? Do expression levels of individual *CesA* genes contribute to the variation of cellulose synthase content seen in provenance progeny trials (Tibbits unpublished observations) as hypothesized by (Dhugga 2001)? Is it possible to alter the timing and level of *CesA* gene expression, either by molecular breeding or genetic engineering, to improve the wood quality and quantity?

Cloning *PrCesA1* from radiata pine

While on sabbatical with Tony Bacic and Steve Read at the University of Melbourne, I constructed cDNA libraries from early developing vascular cambium: one from radiata pine and one from eastern white pine. This tissue contained both developing phloem and xylem layers, and was approximately 1 cm thick in *P. radiata* and only 2-3 mm thick in *P. strobus*. RNAs were isolated by the method of (Chang et al 1993). PolyA-enriched RNA (mRNA) was isolated and converted to copy DNA (cDNA) using reverse transcriptase. The cDNAs were converted phage libraries, arrays of all the different expressed gene sequences.

Primary libraries were screened using probes amplified from two putative loblolly pine *CesA* cDNAs (GenBank # BE187012 & AW290811). These ESTs correspond to amino-acids 627-775, spanning the class specific region and glycosyl transferase domain B as defined by Liang and Joshi 2004. Putative radiata and white pine *CesA* cDNA clones, present at a frequency of 0.01-0.02% of the each library, were converted to plasmids, and those with larger inserts (>2000 bp) were selected for DNA sequencing. The 5'-end of one long cDNA (*PrCesA1*) was isolated by 5'Rapid Amplification of cDNA ends (RACE).

A radiata pine *CesA* gene: The full length *PrCesA1* cDNA (GenBank #AY639654) is 3881 bp in length. The predicted protein structure is shown in Figure 2. *PrCesA*1 is 67% identical to *AtCesA7* (*IRX*3), one of three secondary cell wall *CesA* genes in Arabidopsis. The 5' UTR of *PrCesA1* is 169 bp while the 3' UTR is 456 bp. This gene is allelic but not identical to GenBank #BD236020, an anonymous gene patented (JP 2002527056-A 42) as part of "materials and methods for modification of plant cell wall polysaccharides". The two alleles differ at the sequence level by 0.4%, primarily in the 5'and 3' untranslated regions, but also across the coding region.

PrCesA1 is highly conserved across pines: it is 99% identical at the DNA level to a 1711 bp *Ces*A contig 7803 from loblolly pine (http://pine.ccgb.umn.edu/). Another very similar EST (99.1%) has been reported in maritime pine (BX250234). This degree of sequence

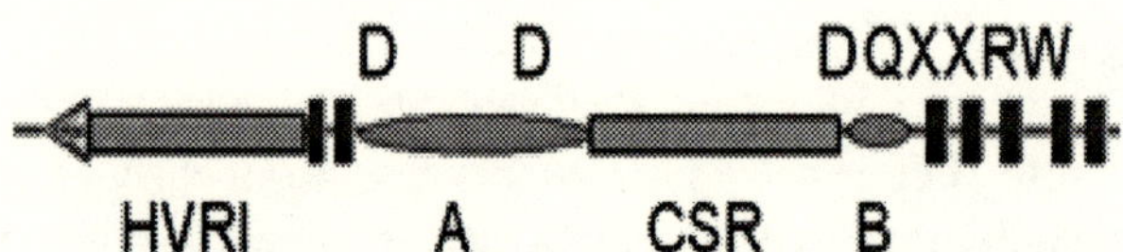

Fig. 2 Predicted structure of *PrCes*A1. Structural domain analysis was performed using SMART algorithms http://smart.embl-heidelberg.de. Orange triangle: Ring finger domain. Blue: Transmembrane domain. Turquoise: HVRI (hypervariable domain I) and Class Specific Region (CSR) of CESA Proteins as predicted by Delmer 1993 and Liang and Joshi 2004. Green: conserved domains of CES proteins. Amino acid abbreviations: D: aspartate, Q: glutamine, X: any amino acid; R: arginine, W tryptophan

conservation is quite striking as the three species in the subgenus *Pinus* (hard pines) probably diverged 10-80 million years ago (Price et al 1998). These observations suggest that while pines have inherently high levels of nucleotide variation, much of that variation arose before the radiation of extant species. Another explanation to explain this level of sequence conservation is that strong selection acts on these genes (Tibbits unpublished observation).

PrCesA1 maps to linkage group 11 of Loblolly pine (Garth Brown, David Neale Personal Comm.) and to LG6 of *Pinus pinaster* (Tibbits and Pot unpublished). In one mapping experiment the interval associated with the *CesA* gene on linkage group 11 co-located to a quantitative trait locus (QTL) for percent volume of latewood and one QTL for latewood chemistry (Brown et al 2003); however these QTLs were not observed in a second mapping population. Brown and coworkers emphasized the need for much larger mapping populations to confirmed candidate gene associations. However these results are exciting in that they suggest that *CesA* genes may be useful for marker-aided selection, a molecular plant-breeding tool.

How complex is the pine *CesA* gene family?

By the time I started my sabbatical leave in July 2000, Todd Richmond had set up a datamining web site (http://cellwall.stanford.edu/) that extracts new *CesA* and *Csl* sequences as new entries are deposited in GenBank. A Blast search (Altschul et al. 1997) of full length Arabidopsis and other plant *CesA* genes is used to sort the new sequences into different family groups. Once the *CesA* sequences from a given species are sorted into 'bins' representing different orthologs of the *CesA* gene family, sequences in each bin are assembled into 'contigs' – overlapping fragments of that gene. The largest EST sequence database for a pine species is that of loblolly pine, *Pinus tadea.* Richmond's datamining predicts there are as many as fourteen different *CesA* genes in loblolly pine. The longest contig is only 1962 bp, representing less than 2/3's of the predicated open reading frame for the cellulose synthase protein. Most of the predicted contigs are considerably shorter (290 bp –1710 bp).

Sequences for six different pine xylem libraries have been independently analyzed at the Center for Computational Genomics and Bioinformatics ("The Genomics of Wood Formation in Loblolly

Pine." Supported by the National Science Foundation Plant Genome Research Program under NSF award DBI 9975806; http://pine.ccgb.umn.edu/). Contigs were built from 59,447 ESTs using the algorithm Phred/Phrap/Consed for sequence assembly; the ESTs collapsed into 8046 contigs and 12, 437 single sequences (singletons). Each contig has been annotated based on a Blast sequence comparison to GenBank. In the CCGB database at least thirteen contigs and two singletons from http://pine.ccgb.umn.edu/ have been tentatively identified as *PtCesA* genes.

I have compared the two predicted *PtCesA* gene families: there are differences between the two sets of predicted genes and in some instances the predicted contigs overlap. Some of the discrepancies result from the inherent high levels of sequence variation in pines, and the fact that various cDNA libraries were constructed with different genetic stocks (Ross Whetten, personal communication). After matching the predicted *CesA* contigs from both data sets there are up to fifteen different *CesA* gene in the pine *CesA* gene family (Table 1). It is possible to tentatively match the longer contigs to Arabidopsis orthologs that have defined functions in primary and secondary cell wall synthesis.

Table 1 *Ces*A genes in Loblolly Pine

	Contig ID or singleton NSF Pine MegaGenome Directory 20 http://pine.ccgb.umn.edu/	Length bp	Plant Cell Wall contig http://cellwall.stanford.edu/	Length bp	Potential Arabidopsis Ortholog
1	7803	1710	*PtCesA3*	1510	*AtCes*A7 (*IRX3*)
2	7881	1962	*PtCesA1*	1785	*AtCes*A8 (*IRX1*)
3	6485, 7766	627, 1763	*PtCes*A2	1764	*AtCes*A4 (*IRX5*)
4	5828	767	*PtCesA6*; *PtCes*A5	660, 1321	
5	1432	797			
6	5607, 3607	607, 801	*PtCes*A9; *PtCes*A7	1017,367	
7	1567	415			
8	2633	330	*PtCes*A4	290	
9	12653	449			
10			*PtCes*A11	429	
11			*PtCes*A12	980	
12	6864	719	*PtCes*A7, *PtCes*A10	1210	
13	993	260	*PtCes*A14	215	
14	5258	878			
15	10078	501	*PtCes*A8	522	

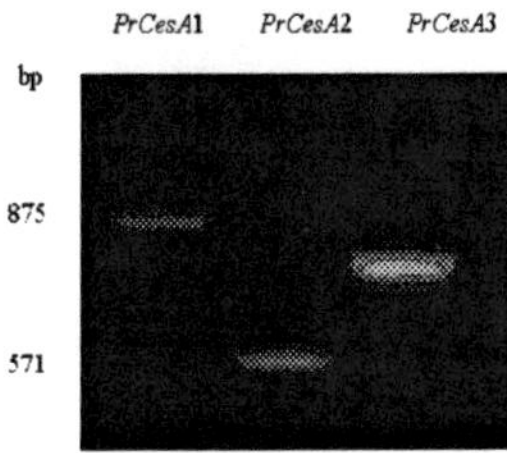

Fig. 3 PCR amplification of cDNAs prepared from mRNA isolated from radiata pine cambial peels made during the spring. **Primers were designed to selectively amplify *PrCes*A1 *PrCes*A2 and *PrCes*A3, corresponding to the Arabidopsis secondary cell wall *Ces*A genes (*AtCes*A7, *AtCes*A8 and *AtCes*A4 respectively.)**

What are the roles of different pine *CesA* genes in wood formation? For a preliminary look at which pine *CesA* genes that are expressed during early wood formation, we isolated mRNA from radiata cambial peels, used reverse transcriptase to produce cDNA. Then we used primers for the Polymerase Chain Reaction (PCR^{Tm}) designed to selectively amplify the cDNAs from eight of pine *CesA* genes to see which and how many genes were transcribed during early xylogenesis. Three distinct *CesA* genes were detected corresponding to orthologs of the Arabidopsis secondary cell wall *AtCesA7*, *AtCesA8*, and *AtCesA4* (Figure 3). Two of these three pine genes corresponding to contigs 7881 and 7766 are most highly represented among the loblolly *CesA* ESTs available in public data bases (about 50% of *CesA* ESTs; http://cellwall.stanford.edu/). The third class of pine *Ces*A genes (*PrCesA1* or loblolly contig 7803) is fainter in the RT-PCR products and is proportionally represented in half as many ESTs. These three genes are likely to be the most abundantly expressed *CesA*s during earlywood formation; more sensitive detection methods such as *in situ* hybridization are likely to be needed detect the expression of the other members of the pine *CesA* family for example in parenchyma or epithelial cells.

In summary, comparative genomic methods provide new tools to study cellulose synthesis during wood formation in conifers. Early results suggest that roles of individual *CesA* genes are highly conserved from angiosperms to conifers as are many other gene functions (Kirst et al. 2003).

Acknowledgements

This work was supported by NCGRIP USDA Grant "Isolation of Cellulose synthase genes from economically important conifers." Paul St. Onge, Leah Gaumont, and David Schrier assisted with DNA sequencing. A Klein thanks Professors Tony Bacic, Steve Read and Pauline Ladiges, of the School of Botany, Melbourne University, for their hospitality and assistance during her 2000-2001 sabbatical.

References

Altschul SF, Madden TL, Schäffer AA, Zhang J, Zhang Z, Miller W, Lipman DJ (1997) Gapped BLAST and PSI-BLAST: a new generation of protein database search programs. Nucleic Acids Res 25: 3389-3402.

Bailey IW (1952) Biological processes in the formation of wood. Science 115: 255-259.

Brown GR, Bassoni DL, Gill GP, Fontana JR, Wheeler NC, Megraw RA, Davis MF, Sewell MM, Tuskan GA, Neale DB (2003) Identification of Quantitative Trait Loci Influencing Wood Property Traits in Loblolly Pine (*Pinus taeda* L.). III. QTL Verification and Candidate Gene Mapping. Genetics 164: 1537-1546.

Chang S, Puryear J, Cairney J (1993) A simple and efficient method for isolating RNA from pine trees. Plant Mol Bio Report 11: 113-116.

Delmer DP (1999) Cellulose biosynthesis: exciting times for a difficult field of study. Ann Rev Plant Physiol Plant Mol Biol 50: 245-276.

Dhugga KS (2001) Building the wall: genes and enzyme complexes for polysaccharide synthases. Curr Opin Plant Biol 4: 488-493.

Fengel D, Wegener G (1983) Wood: Chemistry, Ultrastructure, Reactions. Walter de Gruyter & Co., Berlin.

Harada H, Côté WFJr (1985) Structure of Wood. In Higuchi T (Ed) Biosynthesis and biodegradation of wood components. Academic Press Orlando pp 1-42.

Higuchi T (1997) Biochemistry and Molecular Biology of Wood. Springer, Heidelberg Germany.

Jones TG, Corson SR (1996) Relationships of TMP and wood properties of radiata pine trees. Appita Annual General Conference Proceedings 2: 553-560.

Kirst M, Johnson AF, Baucom C, Ulrich E, Hubbard K, Staggs R, Paule C, Retzel E, Whetten R, Sederoff R (2003) Apparent homology of expressed genes from wood-forming tissues of loblolly pine (*Pinus taeda* L.) with *Arabidopsis thaliana*. Proc Natl Acad Sci USA 100: 7383-7388.

Klein AS, Montezinos D, Delmer DP (1981) Cellulose and 1,3 glucan synthesis during the early stages of wall regeneration in soybean protoplasts. Planta 152: 105-114.

Le Provost G, Paiva J, Pot D, Brack J, Plomion C (2003) Seasonal variation in transcript accumulation in wood-forming tissues of loblolly pine (*Pinus tadea* L.) with *Arabidopsis thaliana*. Proc Natl Acad Sci USA 100: 7383-7388.

Liang X, Joshi CP (2004) Molecular cloning of ten distinct hypervariable regions from cellulose synthase gene superfamily in aspen trees. Tree Physiol 24: 543-550.

Pear JR, Kawagoe Y, Schreckengost WE, Delmer DP, Stalker DM (1996) Higher plants contain homologs of the bacterial *celA* genes encoding the catalytic subunit of cellulose synthase. Proc Natl Acad Sci USA 93: 12637-12642.

Price RA, Liston A, Strauss SH (1998) Phylogeny and systematics of *Pinus*. In: Richardson DM (Ed) Ecology and Biogeography of *Pinus*. Cambridge University Press, pp 49-68.

Richmond T, Somerville C (2000) The cellulose synthase superfamily. Plant Physiol 124: 495-498.

Taylor NG, Howells RM, Huttly AK, Vickers K, Turner SR (2003) Interactions among three distinct CesA proteins essential for cellulose synthesis. Proc Natl Acad Sci USA 100: 1450-1455.

Vergara CE, Carpita NC (2001) Beta-D-glycan synthases and the CesA gene family: lessons to be learned from the mixed-linkage (1→3),(1→4)-beta-D-glucan synthase. Plant Mol Biol 47: 145-160.

Willison JHM, Klein AS (1982) Cell-wall regeneration by protoplasts isolated from higher plants. In: Brown RMJr (Ed.), Cellulose and other natural polymer systems. Plenum, New York, pp 61-85.

What Makes a Good Monolignol Substitute?

John Ralph

Lignin and lignification

Among the more intriguing components in the plant cell wall are the lignin polymers. Their formation, occurring after the polysaccharides are laid down, provides structural integrity to those lignified cell walls, facilitates water transport, and provides defensive functions. But what makes the polymers enigmatic is their mode of formation. Unlike the polysaccharides and proteins, no exact chemical "sequence" of units is dictated by the cell. Although there is considerable control by the cell over aspects of the structure by the supply of the various lignin monomers to the wall and the supply and control of oxidants (in the peroxidase-H_2O_2 system), the assembly of the polymer is a combinatorial process, under the auspices of simple chemical control, i.e. governed by normal chemical concerns such as the concentrations of reactants, their natural coupling and cross-coupling propensities, the matrix, and the physical conditions during the polymerization (Boerjan et al. 2003; Ralph et al. 2004). The process is not under the control of enzymes, for example, so results in racemic polymers with astronomical numbers of possible isomers; no two lignin molecules need have the same structure. This theory has been recently challenged (Gang et al. 1999), and even a text book has pronounced that lignification is a process under absolute structural control (Croteau et al. 2000), but this challenge is wholly without merit and should now be dismissed (Ralph et al. 2004).

In addition to providing a flexible system for the plant to respond to various challenges and stresses, such a polymerization mode provides unparalleled opportunities to re-engineer this component of the wall. Nature herself has explored many variations on the theme already. Certain plants use monomers that most texts would not consider to be lignin monomers; it is widely assumed that lignins derive from only three monolignols, *p*-coumaryl, coniferyl, and sinapyl alcohols but, as reviewed recently (Boerjan et al. 2003; Ralph et al. 2004), other lignin monomers must be recognized. And nature of course experiments with genetic engineering. Mutations sporadically appear in which "crucial" lignin-biosynthetic-pathway genes are knocked out. Some such plants have made substitutions for the monolignols whose biosynthesis has been negatively impacted.

The process of lignification has been well implemented. In what remains best described as a process of radical coupling of phenols (Harkin 1967), the monolignols are rather ideal monomers. Like most other chemical polymerizations, the primary polymerization reaction involves the coupling of a monomer with the growing polymer (Figure 1a); monomer-monomer reactions certainly occur but are less prominent in the overall scheme. Polymer-polymer (or oligomer-oligomer) coupling reactions play a significant role in lignification, creating branching points in the polymer. The radical coupling process is what allows the polymerization to be somewhat combinatorial. The monolignols are potentially capable of coupling at various sites (β-, 4-*O*-, 1-carbons for coniferyl and sinapyl alcohols, and additionally the 5-carbon for coniferyl alcohol, but almost overwhelmingly react via their β-positions with phenolic ends on the growing polymer, Figure 1a. This conveys a certain linearity to the polymer. The phenolic end can couple at its 4-*O*- and, less frequently, 1-positions, and/or (in the case of guaiacyl units only) the 5-position. In many ways it is the favored β-*O*-4-coupling of a monolignol with the growing polymer that defines lignins, but it is the other reactions including β-5- and β-1-coupling, as well as 5–5- and 4-*O*-5-coupling between oligomers, that characterizes them.

Non-monolignol lignin monomers

So what kinds of monomers can substitute for the primary monolignols? From studies to date, those that are capable of undergoing analogous β-*O*-4-coupling with the growing polymer seem to be the most successful. This may be the major requirement, although it still provides considerable variability, as will be seen below. Plants which have available, and apparently attempt to utilize, monomers that can not undergo this type of coupling reaction fare more poorly, presumably because their lignin substitutes do not adequately exhibit the properties required of them.

Acylated monolignols

A class of successful lignin precursors are the variously acylated monolignols implicated in an assortment of plants, Figure 1b (Ralph et al. 2004). Sinapyl acetate has been demonstrated to be a monomer

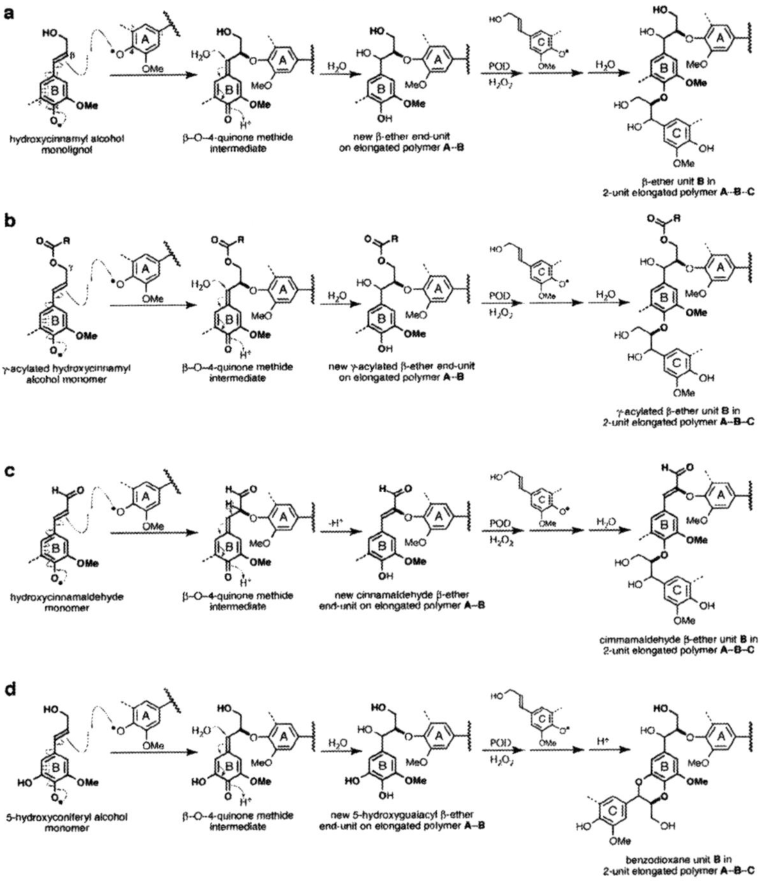

Fig. 1 Differences in cross-coupling and post-coupling reactions for various well-suited "monomers" incorporated into lignification. Illustration is for the major β–*O*–4-coupling only. a) Normal hydroxycinnamyl alcohol radicals B cross-couple with the phenolic end of the growing polymer A, mainly by β–*O*–4-coupling, to produce an intermediate quinone methide which rearomatizes by nucleophilic water addition to produce the elongated lignin chain A–B. The subsequent chain elongation via a further monolignol radical C etherifies the unit created by the prior monomer B addition, producing the 2-unit-elongated polymer unit A–B–C. b) Various γ-acylated monolignols (*p*-coumarate, *p*-hydroxybenzoate, and acetate) cross-couple equally well producing analogous products but with the β-ether unit B γ-acylated in the lignin polymer unit A–B–C. c) Hydroxycinnamaldehydes B may also cross-couple with the phenolic end of the growing polymer A, again mainly by β–*O*–4-coupling, to produce an intermediate quinone methide again, but one which rearomatizes by loss of the acidic β-proton, producing an unsaturated cinnamaldehyde-β–*O*–4-linked B end-unit. Incorporation further into the polymer by etherification is analogous to a). The unsaturated aldehyde units B give rise to unique

thioacidolysis markers. d) 5-Hydroxy-coniferyl alcohol monomer A also cross-couples with the phenolic end of the growing polymer A, mainly by β–*O*–4–coupling, to produce an intermediate quinone methide as usual which rearomatizes normally by nucleophilic water addition to produce the elongated lignin chain A–B bearing a novel 5-hydroxyguaiacyl phenolic end-unit. The subsequent chain elongation via a further monolignol radical C coupling β–*O*–4 to the new phenolic end of A–B, but this time the rearomatization of the quinone methide (not shown) is via internal attack of the 5-OH producing novel benzodioxane units B–C in the 2-unit-elongated polymer unit A–B–C. 5-Hydroxy-coniferyl alcohol incorporation produces a lignin with a structure that deviates significantly from the "normal" lignin. The bolded bonds are the ones formed in the radical coupling steps.

in kenaf bast fiber and palm lignification, and at low levels in other hardwoods. Sinapyl *p*-hydroxybenzoate is similarly a monomer in palm, aspen, poplar and willow lignification. And sinapyl *p*-coumarate (as well as lower levels of coniferyl *p*-coumarate) analogously contribute to lignification in grasses. What benefit the plant receives from lignin acylation is little understood. *p*-Hydroxybenzoate and *p*-coumarate are excellent substrates for many of the peroxidases that only slowly oxidize sinapyl alcohol. And, since they form less stable radicals, they readily undergo radical transfer with sinapyl alcohol, generating the sinapyl alcohol radical required for its incorporation into the polymer (Takahama et al. 1996). Their roles as radical transfer agents make them a sort of catalyst for lignification, especially with respect to sinapyl alcohol monomers. These phenolic esters then may have a role in facilitating the oxidation of sinapyl alcohol and oligomer substrates. Despite being phenolic, they themselves do not undergo coupling reactions and are found adorning lignins as pendant free-phenolic entities. In model reactions, it is not until other phenolics in the system are depleted that these components couple (Ralph et al. 2004). Their presence in free-phenolic form in lignins provides evidence that the cell limits radical concentrations. The role of the over 50% levels of acetylated sinapyl alcohol in kenaf is even less understood. Presumably the resultant polymer is more hydrophobic than normal lignins, so the acetylation may be associated with drought tolerance.

Whatever the reasons for these acylated components, the three types of acylated monolignols (acetate, *p*-hydroxybenzoate, and *p*-coumarate) are now well implicated as authentic lignin precursors in their respective plants. It is just becoming apparent by *in vitro* studies that the acylation does not significantly affect the course of the coupling reactions (Figure 1b). That means that the acylated monolignols also behave nicely as lignin monomers, most importantly undergoing the β-O-4-coupling reactions that incorporate them into the chain of the polymer. Acylated monolignols alter the structure of the lignins by more than just adorning them with pendant groups, however. This is because the γ-OH group on a monolignol, Figure 1a, functions in some post-coupling reactions, internally trapping the quinone methide following β–β-coupling, for example (not shown). With the γ-OH group acylated, Figure. 1b, such internal reactions are no longer possible and the quinone methide must be rearomatized by trapping an external nucleophile, usually water, and forming quite different products in the lignin as a result (Lu and Ralph 2002). Additionally, the stereochemistry of water attack on the quinone methide intermediate following β-*O*-4-coupling is altered. The lignins therefore differ from "normal" lignins in both substantial and subtle ways.

Hydroxycinnamaldehydes and dihydroconiferyl alcohol in CAD-deficient plants

A pertinent example of poor monomer substitution is in a viable but unhealthy CAD-deficient mutant pine (MacKay et al. 1997; Ralph et al. 1997). CAD is the last enzyme on the monolignol biosynthetic pathway, reducing conifer-aldehyde to the monolignol coniferyl alcohol in softwoods. Coniferaldehyde might be anticipated to build up in this plant, and evidence is that it does. On paper, coniferaldehyde appears to be a good candidate for a lignin monomer, Figure 1c. It doesn't have the alcohol group, but it has the requisite conjugated double bond of the cinnamyl unit, and it was early on shown that it undergoes the same range of monomer-monomer coupling reactions as coniferyl alcohol, forming its own versions of β-*O*-4-, β–5-, and β–β-dehydrodimers (Connors et al. 1970). It could even make a dehydrogenation polymer. The drawback, as was not revealed until later studies on angiosperms (Ralph et al. 2001; Kim et al. 2003), is that coniferaldehyde simply will not chemically β-*O*-4-cross-couple with phenolic guaiacyl units in the polymer. It is purely a chemistry problem; it will undergo such coupling with suitably unsaturated units, but with a normal guaiacyl unit, the reactants are not sufficiently chemically compatible to react either *in vitro* or, apparently, *in vivo*.

Since the CAD-deficient pine still makes some coniferyl alcohol, the polymer needs to incorporate monomers compatible with a normal guaiacyl lignin. Coniferaldehyde simply does not comply. It therefore becomes relegated to homo-coupling reactions and cross-coupling reactions solely with the monolignol coniferyl alcohol, and is therefore left adorning only the periphery of the lignin molecule (as end-groups) and not significantly incorporating into the polymer chain. Interestingly, this pine also produces, for unknown reasons, high levels of dihydroconiferyl alcohol (Ralph et al. 1997; Lapierre et al. 2000). It also finds itself in the lignin. Obviously dihydroconiferyl alcohol cannot couple at its β-position — there is no way to get the required single electron density to the β-position without the presence of the double bond. It therefore is also limited to a set of reactions that do not incorporate it into the lignin chain, again relegating it to the periphery of the structure. Nevertheless, it is found in the lignin at striking levels due to its ability to still 5- and 4-*O*-couple with guaiacyl units, as well as with the monolignol, coniferyl alcohol. This pine then appears to have attempted to augment its polymer by substituting available coniferaldehyde and dihydroconiferyl alcohol monomers for some of the deficient coniferyl alcohol. It remains viable, but not vigorous.

The above example would be unfulfilling if it were not for the fact that coniferaldehyde is a well-behaved monomer in angiosperms. How can this be? Again, it is simple chemistry. Angiosperms have guaiacyl/syringyl lignins (deriving from both coniferyl and sinapyl alcohols). Coniferaldehyde, which does not β-*O*-4-cross-couple with guaiacyl units, readily undergoes β-*O*-4-cross-coupling reactions with syringyl end-groups. As it turns out, sinapaldehyde readily undergoes cross-coupling with either guaiacyl or syringyl end-groups. As a result, both coniferaldehyde and sinapaldehyde couple in a similar way as the monolignols do, and can therefore incorporate integrally into the body of the polymer in CAD-deficient angiosperms. These details are revealed by NMR studies on the lignins (Kim et al. 2003), and by the release of thioacidolysis marker compounds specifically from these hydroxy—cinnamaldehyde-β-*O*-4-linked units (see Figure 1c) (Kim et al. 2002). The plants appear to grow essentially normally, but the hydroxy-cinnamaldehydes may not be quite perfect monomers. It appears that coupling to the new type of conjugated β-*O*-4-phenolic end-units that these create, the next step in the polymerization, becomes difficult. The result is that many of the incorporated coniferaldehyde/ sinapaldehyde units remain as free-phenolic end-groups (Lapierre, unpublished), limiting the degree of polymerization of the lignin. In addition to the property changes caused by the structural changes, these lower molecular weight lignins are presumably less ideal for the plant. An interesting side-benefit however is that the lignins are much more easily broken down and removed in chemical pulping (Pilate et al. 2002), so plants with limited CAD-deficiency are being pursued for their enhanced pulping potential.

5-Hydroxyconiferyl alcohol in COMT-deficient plants

Beyond the (partial) substitution of the hydroxycinnamaldehydes for their hydroxy-cinnamyl alcohol monolignol analogs in CAD-deficient angiosperms is a particularly successful substitution in the case of COMT-deficiency (Ralph et al. 2001). COMT is a methyl transferase enzyme necessary for the biosynthesis of sinapyl alcohol and ultimately syringyl groups in lignins. Knock-out mutants are essentially or totally devoid of syringyl components, and COMT down-regulation will reduce the syringyl content. As CAD-deficient angiosperms incorporated the immediate CAD precursor (the hydroxy-cinnamaldehydes), COMT-deficient plants must deal with the un-methylated 5-hydroxy-coniferaldehyde precursor. Apparently CAD is able to reduce this aldehyde as it is 5-hydroxy-coniferyl alcohol, not the aldehyde, that is exported to the wall and incorporated into lignins. 5-Hydroxy-coniferyl alcohol has all the makings of an ideal monolignol substitute (Figure 1d). It beautifully β-*O*-4-couples with guaiacyl, syringyl, or new 5-hydroxy-guaiacyl phenolic endgroups, integrating into the polymer as would a primary monolignol. The lignin structure becomes strikingly different however. The presence of the extra phenolic-OH, the 5-OH, drastically affects the post-coupling reactions. Novel benzodioxane units are formed in the polymer as a result of incorporating 5-hydroxy-coniferyl alcohol, at striking levels (essentially replacing the syringyl units in the control plants). The plant does not seem to mind; COMT-deficient plants appear to grow normally. In this case, the severe structural changes are a serious detriment to chemical pulping. Despite still being β-ethers, these units will not efficiently cleave under pulping conditions, as do the syringyl units which they displace. However, COMT-deficient plants appear to be more digestible (Guo et al. 2001). A possible reason is that, by providing a rapid alternative internal pathway for rearomatizing the quinone methide intermediate, these units cannot cross-link with polysaccharides in the wall (via addition to quinone methides). Lignin-polysaccharide cross-linking has been shown to have a detrimental effect on cell wall digestibility.

Conclusions

A small range of monomers are now known to substitute for the conventional monolignols in various natural and transgenic plants. Monolignol substitution appears to be most successful when the novel monomer behaves, in its chemical radical coupling and cross-coupling reactions, like a normal monolignol. Most important is the β-*O*-4-coupling reaction with the phenolic end of the growing polymer to extend the polymer chain, as shown in Figure 1. The post-coupling reactions that may be altered by the different functionality on the monomer seem to have less effect. Thus massive changes in the lignin structure occur when 5-hydroxyconiferyl alcohol substitutes for sinapyl alcohol, for example — the coupling reactions are analogous, but post-coupling steps produce novel benzodioxane structures that drastically change the lignin. Observations that plants with monolignol substitution and profoundly altered lignin structure can fare well supports the heretical tongue-in-cheek idea expressed at a conference some time back that the exact structure of lignins is not that important to the functioning of the plant (Ralph 1997). The plant requires certain properties and functionality of its lignins, but does not expend resources dictating those properties by exactly stipulating lignin primary structure. Such biosynthetic malleability functions well for the plant, and also provides significant opportunities for engineering the polymer. Already it has been demonstrated that natural and industrial processes ranging from ruminant digestibility to chemical pulping can be both positively and negatively impacted by alterations to lignin composition and structure. It is also apparent that phenolic components from beyond the monolignol pathway itself (such as the acylated monolignols) may be incorporated into lignins if they have compatible reaction chemistry and are transportable to the wall. Future work should reveal opportunities beyond the interesting deviations achieved by up- and down-regulating genes on the monolignol pathway to date.

Acknowledgements
The author is indebted to many people with whom he did not clear this article! These include students, postdocs, coworkers, and many collaborators. Partial funding of the more solid studies upon which this commentary derives by the USDA-National Research Initiatives and the USDOE's Energy Biosciences program is gratefully acknowledged.

References

Boerjan W, Ralph J, Baucher M (2003) Lignin biosynthesis. Ann Rev Plant Biol 54: 519-549.

Connors WJ, Chen C-L, Pew JC (1970) Enzymic dehydrogenation of the lignin model coniferaldehyde. J Org Chem 35: 1920-1924.

Croteau R, Kutchan TM, Lewis NG (2000) Chapter 24. Natural products (secondary metabolites), In Buchanan B, Gruissem W, Jones R (eds) Biochemistry and molecular biology of plants. Am Soc Plant Physiologists Rockville, MD, pp 1250-1318, specifically pp 1294-1299.

Gang DR, Costa MA, Fujita M, Dinkova-Kostova AT, Wang HB, Burlat V, Martin W, Sarkanen S, Davin LB, Lewis NG (1999) Regiochemical control of monolignol radical coupling: A new paradigm for lignin and lignan biosynthesis. Chem Biol 6: 143-151.

Guo DG, Chen F, Wheeler J, Winder J, Selman S, Peterson M, Dixon RA (2001) Improvement of in-rumen digestibility of alfalfa forage by genetic manipulation of lignin *o*-methyltransferases. Transgenic Res 10: 457-464.

Harkin JM (1967) Lignin - a natural polymeric product of phenol oxidation, In Taylor WI, Battersby AR (eds) Oxidative coupling of phenols. Marcel Dekker, New York, pp 243-321.

Kim H, Ralph J, Lu F, Pilate G, Leplé JC, Pollet B, Lapierre C (2002) Identification of the structure and origin of thioacidolysis marker compounds for cinnamyl alcohol dehydrogenase deficiency in angiosperms. J Biol Chem 277: 47412-47419.

Kim H, Ralph J, Lu F, Ralph SA, Boudet A-M, MacKay JJ, Sederoff RR, Ito T, Kawai S, Ohashi H, Higuchi T (2003) NMR analysis of lignins in cad-deficient plants. Part 1. Incorporation of hydroxycinnamaldehydes and hydroxybenzaldehydes into lignins. Org Biomol Chem 1: 158-281.

Lapierre C, Pollet B, MacKay JJ, Sederoff RR (2000) Lignin structure in a mutant pine deficient in cinnamyl alcohol dehydrogenase. J Agric Food Chem 48: 2326-2331.

Lu F, Ralph J (2002) Preliminary evidence for sinapyl acetate as a lignin monomer in kenaf. J Chem Soc, Chem Commun 90-91.

MacKay JJ, O'Malley DM, Presnell T, Booker FL, Campbell MM, Whetten RW, Sederoff RR (1997) Inheritance, gene expression, and lignin characterization in a mutant pine deficient in cinnamyl alcohol dehydrogenase. Proc Natl Acad Sci USA 94: 8255-8260.

Pilate G, Guiney E, Holt K, Petit-Conil M, Lapierre C, Leple JC, Pollet B, Mila I, Webster EA, Marstorp HG, Hopkins DW, Jouanin L, Boerjan W, Schuch W, Cornu D, Halpin C (2002) Field and pulping performances of transgenic trees with altered lignification. Nature Biotechnol 20: 607-612.

Ralph J (1997) Recent advances in characterizing 'non-traditional' lignins and lignin-polysaccharide cross-linking. 9th International Symposium on Wood and Pulping Chemistry, Montreal, Quebec, Vol 1: pp 1-7, paper PL2.

Ralph J, Lapierre C, Marita J, Kim H, Lu F, Hatfield RD, Ralph SA, Chapple C, Franke R, Hemm MR, Van Doorsselaere J, Sederoff RR, O'Malley DM, Scott JT, MacKay JJ, Yahiaoui N, Boudet A-M, Pean M, Pilate G, Jouanin L, Boerjan W (2001) Elucidation of new structures in lignins of cad- and comt-deficient plants by NMR. Phytochem 57: 993-1003.

Ralph J, Lundquist K, Brunow G, Lu F, Kim H, Schatz PF, Marita JM, Hatfield RD, Ralph SA, Christensen JH, Boerjan W (2004) Lignins: Natural polymers from oxidative coupling of 4-hydroxyphenylpropanoids. Phytochem Reviews 3: 29-6.

Ralph J, MacKay JJ, Hatfield RD, O'Malley DM, Whetten RW, Sederoff RR (1997) Abnormal lignin in a loblolly pine mutant. Science 277: 235-239.

Takahama U, Oniki T, Shimokawa H (1996) A possible mechanism for the oxidation of sinapyl alcohol by peroxidase-dependent reactions in the apoplast: Enhancement of the oxidation by hydroxycinnamic acids and components of the apoplast. Plant Cell Physiol 37: 499-504.

New Insights on the Occurrence and Role of Lignin in the Secondary Cell Wall Assembly

Jean-Paul Joseleau and Katia Ruel

Structural studies of the plant cell wall are often restricted to the primary wall. However, considering the importance of lignocellulosic materials in many industrial uses, it is important to better understand the lignified secondary wall that constitutes more than 90% of the Wood biomass. The purpose of this article is to review a portion of our work regarding the ultrastructural distribution of lignin in relation to xylem fiber secondary wall formation. The immunocytochemical identification of the first lignin epitopes synthesized in the space between loose cellulose microfibrils in the innermost layer of the incipient thickening wall underscores the functional role of non-condensed structural motifs in the cohesion between microfibrils. Another level of organization of the fiber secondary walls is thought to exist in lamellar association of microfibrils. On the basis of transmission electron microscopy examination of various materials, consisting of immature fiber walls, wood degraded by ligninolytic fungi, lignin altered transgenic plants, and wood pulp fibers, we suggest that lignin plays a specific role in the assembly of the lamellar microstructure of the S2 layer.

For more than a century, lignin has been regarded as an amorphous material deposited with hemicelluloses in the spaces left between cellulose microfibrils during secondary cell wall thickening. Although the polyphenol nature of lignin was first recognized by Payen more than a century ago, its macromolecular characteristics are still mostly unknown. The chemistry and physicochemistry of lignin have been investigated primarily in relation to the wood and pulping industries. Investigations of lignin in the context of cell wall formation and physiology have been carried out by biochemical (Higuchi 1990; Anterola and Lewis 2002; Boerjan et al. 2003) and cytochemical (Terashima et al. 1993; Joseleau and Ruel 1997) approaches. However, the analytical data have allowed only partial characterization of the structure of lignin because of the inability of the available methods to completely extract and analyze lignins. Analytical characterization by NMR is severely limited by the low solubility of lignin, resulting in the description of only a small fraction of the polymer, whereas the chemical degradative method by thioacidolysis rarely exceeds 50 % in yield (Rolando et al. 1992), leaving a large part of the protolignin uncharacterized. Our interest in plant cell wall organization at the ultrastructural level, concerning the mode of distribution of the wall macromolecules and their function in wall assembly, necessitated the non-invasiveness and high resolution of transmission electron microscopy (TEM). We have also used other methods and techniques allowing the specific visualization and identification of different wall polymers. In particular, immunocytochemical probes directed against precise structural epitopes have proven to be invaluable tools for the study of primary wall polysaccharides (Knox 1997; Ruel 2003). However, antibodies specifically recognizing lignin were initially lacking. We therefore initiated a research program to develop the immunological characterization of lignins *in planta,* with the objective to better describe the ultrastructural organization of lignified plant cell walls, and to better understand the role of lignins and lignin substructures in the assembly of secondary walls.

Preparation of immunological probes against lignin model compounds of various structures

To overcome the carbohydrate contamination that is unavoidable in extracted lignin preparations, we used synthetic lignin dehydrogenative polymers (DHPs) as antigens for the production of antibodies. Thus, by varying the mixtures of monolignols subjected to peroxidase-catalyzed polymerization in the presence of hydrogen peroxide, several dehydrogenative polymers of various compositions were obtained (Joseleau and Ruel 1997). Another important feature of the dehydrogenative polymers is that by controlling the conditions of polymerization, the inter-unit linkages could be preferentially favored. The result was a variety of polyclonal antibodies with defined specificities, for both monomer epitopes and inter-unit linkages.

Lignin is active in cellulose microfibril cohesion

Cellulose microfibrils (CMFs) are the basic architectural unit of all plant cell walls, both primary and secondary, at the ultrastructural level. Observed by TEM, the walls of mature cells appear compact, indicating that the macromolecular constituents are maintained by tight interactions. In the internal part of developing walls in young, growing cells of *Arabidopsis thaliana* stems, the newly deposited cellulose microfibrils exhibit a loose disposition and irregular orientation (Ruel et al. 2002). As secondary thickening occurs, however, cellulose microfibrils become tightly associated and take a regular orientation, while the newly formed cellulose microfibrils again show the loose pattern (Figure 1a and b). This might be explained by the recent demonstration by Tokoh et al. (2002) that cellulose microfibrils bundles that form in a pure glucuronoxylan medium are loose. Immunolabeling with the antibodies directed against model lignin polymers demonstrated that the first type of deposited lignin on the unassociated, individualized cellulose microfibrils was of the condensed guaiacyl type (Figure 1c). On the other hand, wherever cohesion between cellulose microfibrils occurred, both condensed guaiacyl and non-condensed guaiacyl–syringyl lignin were present (Figure 1d and e). This suggests that non-condensed lignin may play a role in the cohesion between secondary wall cellulose microfibrils. Similar conclusions were obtained by immunolabeling the fiber walls from normal and genetically engineered plants. Tobacco and *A. thaliana* plants with antisense-downregulation of the gene encoding for cinnamoyl CoA reductase (CCR), an enzyme of the monolignol biosynthesis pathway (Goujon et al. 2003), had about a 50% reduction of their lignin content and showed specific alteration of their secondary walls, consisting of a lack of cohesion of the internal part of S2 (Figure 2a and b). The significant loosening, as observed by immuno-TEM, coincided with a particularly weak concentration of the non-condensed lignin epitopes, as compared to the normal plant (Figure 2c), and was confirmed by semi-quantitative measurement of the gold particles (Ruel et al. 2001). Conversely, the condensed epitopes were well distributed in the loose inner part of S2 (Figure 2d). Again, the non-condensed substructures of lignins appear to be important in providing cohesion between cellulose microfibrils. It is clear that in the case of cinnamoyl CoA reductase downregulation, the plant was not able to overcome the defect in lignin synthesis by a compensatory mechanism consisting of the enhanced synthesis of another wall polymer, as was shown by Delmer and co-workers (Shedletzky et al. 1990) for the defect in cellulose synthesis induced by the herbicide dichlorobenzonitrile. On the other hand, our results indicate that lignification of the secondary wall proceeds as soon as the first cellulose microfibrils are laid down, and well before the polysaccharide matrix of the S2 layer is completed; this is in contrast to Donaldson's report (2001). Our ultrastructural evidence supports Terashima's view of guaiacyl lignin as being the first lignin deposited (Terashima and Fukushima 1989). In addition, it demonstrates that although lignin forms deposits in a pre-formed polysaccharide environment, the nature of lignin contributes to the wall organization. Its deposition follows the progressive thickening of the secondary wall in a co-regulated formation that corresponds to the pattern of assembly described as a "self-perpetuating cascade" by Taylor and Haigler (1993).

The lamellar sub-layer structure of fiber secondary wall

The lamellar sub-structure, as an element of fiber wall architecture, was shown by TEM observations on diverse materials such as developing immature fibers, genetically modified plants, and wood

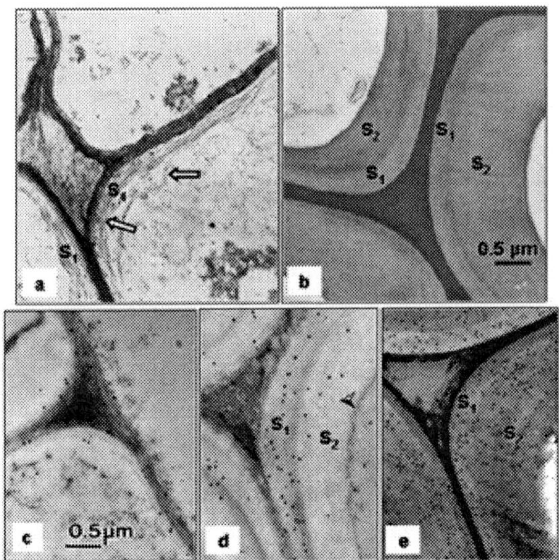

Fig. 1 Fiber walls from *A. thaliana*; a, the young, developing secondary walls consist of loose microfibrils (arrows). The mature cells have a compact appearance, with S1 and S2 well differentiated. b, the mature cells have compact secondary walls with well differentiated S1 and S2 layers. c, early deposition of condensed lignin epitopes in the incipient S1 and middle lamella. d, non-condensed lignin epitopes are concentrated in the compact S1, and are almost absent in the loose network (arrowhead) of the immature S2. e, in the mature and fully compact secondary wall, the non-condensed lignin epitopes are evenly distributed throughout S1 and S2.

cell walls degraded by fungi or selectively attacked with enzymes, as well as pulp fibers at various stages of industrial processes. In all these materials, the TEM images constitute a body of evidence in favor of a degree of organization of the fiber secondary wall corresponding to fundamental lamellar elements of about 55 to 70 nm in thickness, existing in association with cellulose microfibrils. This lamellar sub-layering is distinct from the early concept of lamellar structure for the tracheid secondary wall, as proposed by Kerr and Goring (1975), which was demonstrated at the ultrastructural level for several land plant species and consisted of interlamellar spaces of 7.1 to 8.6 nm between cellulose microfibrils (Ruel et al. 1978).

The examination of an incipient secondary wall of poplar (a woody plant relative to *A. thaliana*) seedlings shows that the first steps of secondary thickening involve the formation of microfibrillar material

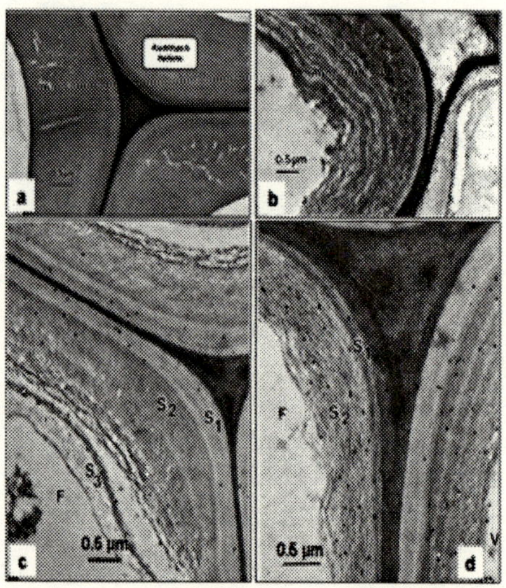

Fig. 2 Cinnamoyl CoA reductase downregulation of *A. thaliana* plants. **a, fiber and vessel walls in xylem of normal plant, exhibiting compact and cohesive walls. b, loose S2 layer of cinnamoyl CoA reductase - downregulated plant, showing the incapacity to associate cellulose microfibrils lamellae. c, labeling of non-condensed epitopes shows that they are essentially present in the compact S1 and the compact outer part of S2, but less visible in the loose, inner part of S2. d, labeling of the condensed epitopes shows that they are formed equally well in compact S1 and loose S2.**

laid down as lamella against the already completed S1 (Figure 3). As secondary thickening proceeds, we observe that cellulose microfibrils tend to associate to form a new lamella (Figure 3b) that, in a later stage, will merge into an apparently homogeneous and compact S2 (Figure 3c and d). A similar pattern was observed in the developing fibers of *A. thaliana* (data not shown). This suggests, therefore, that an intermediary step in secondary thickening consists of the aggregation of cellulose microfibrils into lamellae. It is also notable that the junctions between lamellae constitute mechanically weak points in the secondary wall architecture. This concept of lamellar sub-structure as an integral part of S2 organization was further supported by various other experimental evidences. First, note the way a fiber subjected to chemo-mechanical pulping is delaminated into fibrillar lamellae (Figure 4a and b). Individualized lamellae also separate from S2 during lignin removal by chemical pulping and refining, or during recycling processes (Figure 4c), giving rise to the delamination process. Interestingly, similar microfibrillar lamellae were released after enzymatic treatment of Kraft fibers with manganese peroxidase (MnP) (data not shown), a lignin degrading enzyme (Ruel et al. 1998). The specificity of this enzyme implicates

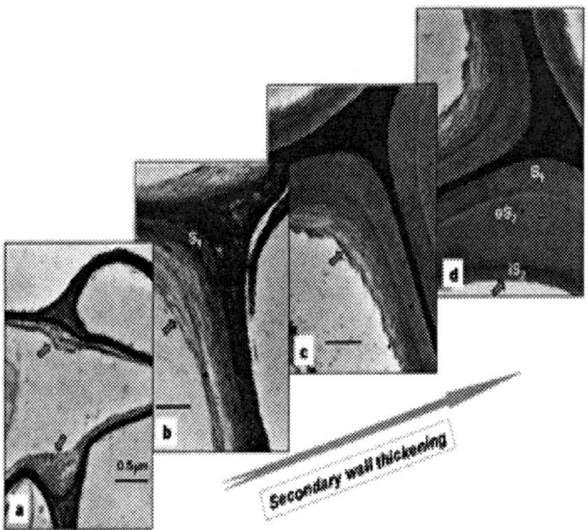

Fig. 3 Intermediary steps of secondary wall formation in poplar wood. From incipient S1 formation (a) to fully mature lignified stage (d), intermediary stages show the loose lamellae (b) beginning to associate (c).

lignin in lamellae assembly during secondary wall formation. The function of lignin in the aggregation of lamellae is also observable in the pattern of disorganization exhibited by the fiber secondary wall of plants after cinnamoyl CoA reductase downregulation (Goujon et al. 2003) and immunocytochemical identification of lignin epitopes at the site of the altered zones. Again, the results show that the altered zones of the walls of the transformed plants were markedly depleted in the non-condensed type of lignin, whereas the condensed forms were normally synthesized, thus supporting the function of the non-condensed forms of lignin in association between lamellae.

Our TEM approach, associated with the preparation of new immunocytochemical probes, allowed the visualization and characterization of typical *in planta* lignin epitopes and the monitoring of their deposition during fiber wall edification. Our results provide evidence of two distinct levels of ultrastructural organization in secondary walls: one at the level of the cellulose microfibrils, and the second at the level of cellulose microfibrils lamellae. It is noteworthy that all examined materials showed a concentric organization of the lamellae of cellulose microfibrils, contrary to the concept of radial

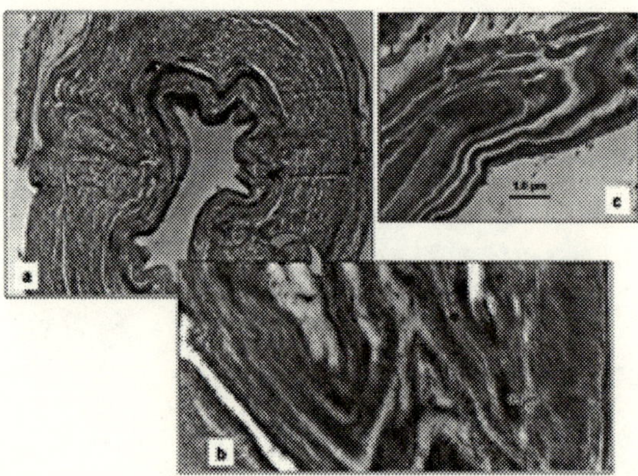

Fig. 4 Evidence of sub-lamellar organization of the secondary walls: fibers after chemo-mechanical pulping. **The secondary wall has been deeply delaminated by the process (a) revealing the concentric sub-lamellar organization. Higher magnification of lamellae (b) shows that they consist of individualized cellulose microfibrils; c, recycled fiber: the process opens the S2 layer revealing an organization of sub-lamellae formed of cellulose microfibrils.**

organization of the lignified fiber wall (Singh et al. 1998). In both elementary structures, lignin is implicated in assembly, between either microfibrils or lamellae. Such a role for lignin, and more particularly for non-condensed lignin sub-structures, is in agreement with the idea of highly-regulated lignin deposition during formation of the composite secondary wall (Campbell and Sederoff 1996), and supports the influence of the natural microenvironment at the site of lignin polymerization (Roussel and Lim 1995). The fractal dimensionality of lignin macromolecules (Radotic 1994) could explain its ability to participate in regular assemblies.

The exact sequence of events that leads to the formation of various lignin types in the different layers of a cell wall, as well as the different patterns of lignification observable in the different cell types of the same plant, are still far from being fully understood. However, the economical importance of lignified cell walls constituting the majority of trees and plant biomass justifies great effort in the development of novel approaches and tools to allow better understanding of the formation, assembly, and mechanical properties of lignified secondary walls.

Acknowledgments

This work was supported in part by the European Program "COPOL" (QLK5-2000-1493), and by the French Program AGRICE (AIP P00080).

References

Anterola AM, Lewis NG (2002) Trends in lignin modification: a comprehensive analysis of the effects of genetic manipulation/mutations on lignification and vascular integrity. Phytochemistry 61: 221-294.

Boerjan W, Ralph J, Baucher M (2003) Lignin biosynthesis. Ann Rev Plant Biol 54: 519-546.

Campbell MM, Sederoff RR (1996) Variation in lignin content and composition Plant Physiol 110: 3-13.

Donaldson LA (2001) Lignification and topochemistry – an ultrastructural view. Phytochemistry 57: 859-873.

Goujon T, Ferret V, Mila I, Pollet B, Ruel K, Burlat V, Joseleau J-P, Barriere Y, Lapierre C, Jouanin L (2003) Down-regulation of *AtCCR1* gene in *Arabidopsis thaliana*: effects on phenotype, lignins and cell wall degradability. Planta 217: 218-228.

Higuchi T (1990) Lignin biochemistry: biosynthesis and biodegradation. Wood Sci Technol 24: 23-63.

Joseleau J-P, Ruel K (1997) Study of lignification by non-invasive techniques in growing maize internodes. An investigation by FTIR, 13C-NMR spectroscopy and immunocytochemical transmission electron microscopy. Plant Physiol 114: 1123-1133.

Kerr AJ, Goring DAI (1975) The ultrastructural arrangement of the wood cell wall. Cellulose Chem Technol 9: 563-573.

Knox JP (1997) The use of antibodies to study the architecture and developmental regulation of plant cell walls. International Review of Cytology (Series: International review of cytology – A survey of Cell Biology) 171: 79-120.

Radotic K, Simic-Krstic J, Jeremic M, Trifunovic M (1994) A study of lignin formation at the molecular level by scanning tunneling microscopy. Biophys J 66: 1763-1767.

Rolando C, Monties B, Lapierre C (1992) Thioacidolysis. In : Lin SY, Dence CW (eds), Methods in Lignin Chemistry, Springer-Verlag, Berlin, pp 334-349.

Roussel MR, Lim C (1995) Dynamic model of lignin growing in restricted spaces. Macromolecules 28: 370-376.

Ruel K (2003) Immunochemical Probes for Microscopy Study of the Plant Cell Walls. In: A. Mendez-Vilas (Ed.), Science, Technology and Education of Microscopy: an Overview; FORMATEX (Spain) Microscopy Series n°1, vol. 2, pp 445-454.

Ruel K, Barnoud F, Goring DAI (1978) Lamellation in the S2 layer of softwood tracheids as demonstrated by Scanning Transmission Electron Microscopy. Wood Sci Technol 12: 287-291.

Ruel K, Burlat V, Comtat J, Moukha S, Sigoillot JC, Asther M, Joseleau JP (1998) Visualization of the Action of ligninolytic enzymes on high-yield pulp fibers. In: Eriksson K-El and Cavaco-Paulo A (eds) Am Chem Soc Symp Series 687, Washington DC, pp 116-132.

Ruel K, Chabannes M, Boudet A-M, Legrand M, Joseleau J-P (2001) Reassessment of qualitative changes in lignification of transgenic tobacco plants and their impact on cell wall assembly. Phytochemistry 57: 875-882.

Ruel K, Montiel M-D, Goujon T, Jouanin L, Burlat V, Joseleau J-P (2002) Interrelation between lignin deposition and polysaccharide matrices during the assembly of plant cell walls. Plant Biol 4: 2-8.

Shedletzky E, Shmuel M, Delmer D, Lamport DTA (1990) Adaptation and growth of tomato cells on the herbicide 2,6-dichlorobenzonitrile leads to production of unique cell walls virtually lacking a cellulose-xyloglucan network. Plant Physiol 94: 980-987.

Singh AP, Sell J, Schmitt U, Zimmermann T, Dawson B (1998) Radial striations of the S2 layer in mild compression wood tracheids of *Pinus radiata.* Holzforschung 52: 563-566.

Taylor JG, Haigler CH (1993) Patterned secondary cell-wall assembly in tracheary elements occurs in a self perpetuating cascade. Acta Bot Neerland 42: 153-163.

Terashima N, Fukushima K (1989) Biogenesis and structure of macromolecular lignin in the cell wall of tree xylem a studied by micro-autoradiography. In: Lewis NG, Paice MG (Eds) Plant Cell Wall polymers, Biogenesis and Biodegradation. Am Chem Soc Symposium series 339, Washington DC, pp 160-168.

Terashima N, Fukushima K, He LF, Takabe K (1993) Comprehensive model of the lignified plant cell wall. In: Jung HG, Buxton DR, Hatfield RD, Ralph J (eds), Mdison: ASA-CSSA-SSSA, Forage cell wall structure and digestibility, pp247-270.

Tokoh C, Takabe K, Sugyiama J, Fujita M (2002) Cellulose synthesized by Acetobacter xylinum in the presence of plant cell wall polysaccharides. Cellulose 9: 65-74.

Hydroxyproline-Rich Glycoproteins in Plant-Microorganism Interactions

Marie-Thérèse Esquerré-Tugayé

Despite their long lasting interest in plant cell walls, scientists considered them for a long time as a rigid pecto-cellulosic envelope, composed of structural polysaccharides and polymers. After the work of Lamport and Northcote (1960) the unequivocal presence of a hydroxyproline-containing protein in the cell wall was admitted. Lamport called it "extensin" (Lamport 1965) assuming its involvement in cell wall extension. From that time, the concept of a dynamic structure containing complex polysaccharides and various proteins emerged progressively. It is within this context that I started working in the early 1970's on the role of plant cell walls in plant – microorganism interactions.

Hydroxyproline as a marker of plant cell walls during pathogenesis
Owing to their strategic location, plant cell walls are the first barrier encountered by most phytopathogens. They must be degraded in order to allow penetration, colonisation, and nutrient supply. The genus *Colletotrichum* contains several pathogen species. *Colletrotrichum lagenarium*, the causal agent of anthracnose on melon (*Cucumis melo*) plants, displays a strong proteolytic activity when grown in vitro on melon cell walls as the sole source of carbon and nitrogen (Pladys and Esquerré-Tugayé 1972). The proteolytic activity within diseased tissues also increased to a high level soon after inoculation. Upon purification and characterization, we found that the proteolytic activity of the fungus corresponds to serine proteases. Other microbial pathogens, notably *Colletotrichum lindemuthianum*, were also shown to secrete proteases, suggesting that these enzymes are pathogenic traits involved in host colonisation. Hence we hypothesized that cell wall proteins might be their first target. A simple experiment, aimed at measuring the protein content and analysing the amino acid pattern of cell walls from diseased melon tissues was conducted. Two major results came out from the first attempt : 1) the amounts of protein and amino acids were increased in response to infection and 2) among amino-acids, the proportion of hydroxyproline (Hyp) was highly

enhanced. Since *Colletotrichum* cell walls do not contain Hyp, the inescapable conclusion was that Hyp-containing compounds were increased in infected melon plants (Esquerré-Tugayé 1973).

This prompted us to chemically characterize the Hyp compound that accumulated and its effect on the outcome of the host-pathogen interaction. Meanwhile Lamport demonstrated that extensin is made of repetitive motifs containing the Ser-Hyp_4 pentapeptide glycosylated by 1 to 4 L-arabinose residues attached to Hyp, and 1 galactose residue attached to serine (Lamport 1967; Lamport 1973). It became clear that the Hyp that accumulated in infected plants had the same glycosylation pattern (Esquerré-Tugayé and Mazau 1974) and the same Hyp-containing pentapeptides as extensin (Esquerré-Tugayé and Lamport 1979).

In order to understand the biological significance of this response, melon plants with higher or lower extensin levels than control plants were obtained by treating them with ethylene, or growing them in the presence of L-trans Hyp, an inhibitor of peptidyl prolyl hydroxylase. Upon inoculation, an inverse relationship was found between the amount of the glycoprotein and the extent of fungal colonisation. Altogether, this was the first demonstration that the Hyp-rich glycoprotein extensin accumulates in the cell wall of a host plant in response to pathogen infection, and that its accumulation is a defence response (Esquerré-Tugayé et al. 1979). To avoid confusion with wall extension, it was thereafter called HRGP.

That it is a defence reaction was then confirmed through the analysis of host-pathogen systems involving various host plants and diverse pathogens including fungi, bacteria, and viruses (Esquerré-Tugayé et al. 1999). Interestingly, HRGP increased earlier in resistant than in susceptible cultivars, consistent with a role in disease resistance. Our research provided a novel molecular marker for studying plant defence, an emerging field still mainly focused on phytoalexins in the late 1970's.

HRGPs and plant defense

How HRGP contributes to plant defence was then investigated. In all likelihood, it acts by strengthening the cell wall, owing to its structural characteristics, but this might not be its unique property. Indeed, new features arose from the isolation of an intact HRGP, and the cloning of HRGP-cDNAs, both accomplished for the first time in Pr. J. Varner's laboratory (Stuart and Varner 1980 ; Chen and Varner 1985) from carrot, then from other plants by other groups. The knowledge gained at the molecular level indicated that the wall HRGP is encoded by a multigene family. Several distinct HRGPs are present in plant cell walls, albeit at a generally low level. Due to its relatively high proportion of lysine, the major HRGP which accumulates in diseased melons might behave as a polycation, enabling it to interact with negatively charged molecules and cell surfaces. This property allowed similar HRGPs from tobacco and potato to agglutinate two microbial pathogens of these host plants. It is believed that the agglutinating properties of HRGPs participate in the immobilization of bacteria and fungi observed during incompatibility (Leach et al. 1990).

In order to study HRGP responses at the cellular level, we prepared antibodies against the glycoprotein, and against its peptide portion (HRP). The use of ultrastructural cytochemistry and gold labeling techniques showed that HRGPs accumulate particularly in defence structures such as papillae and paramural deposits (O'Connell et al. 1990). In the incompatible interaction between *Phaseolus vulgaris* cv kievitsboon koekoek and race β of *Colletotrichum lindemuthianum*, the fungal hyphae appear encased in the papillae and are highly altered (O'Connell et al. 1990). HRGPs are clearly localized inside and at the periphery of the papillae (Figure 1). Together with other newly synthesized compounds, notably wall polysaccharides, they contribute to the formation of a matrix whose protective role as a barrier against colonization has been proposed. In the same way, HRGPs may contribute to the limitation of symbiont spread since they also accumulate strongly at the host-symbiont interface in legume nodules as well as in plant mycorhizae (Benhamou et al. 1991; Bonfante-Fasolo et al. 1991). Thus, the enrichment of plant cell walls in HRGP appeared as a general response to microbial ingress.

Signaling leading to HRGP biosysnthesis

In order to study the recognition phenomena underlying the HRGP response, we began working on elicitors and ethylene. Our approach proceeded from separate reports showing: 1) the ability of microbial antigens to elicit plant defence (phytoalexins), and 2) the increased levels of cell wall Hyp upon ethylene treatment. An elicitor preparation consisting of glycopeptides was obtained from *C. lagenarium*. This led us show that both the synthesis of ethylene and of HRGP are highly induced in melon plants upon elicitor treatment. The role of ethylene in signaling was deduced from the ability of effectors of the ethylene pathway to similarly modulate the synthesis of the hormone and of HRGP (Toppan et al. 1982). Since then, the induction of HRGPs by various elicitors has been demonstrated, notably those originating from *Colletotrichum lindemuthianum* and *Phytophthora parasitica* var. *nicotianae* (Showalter et al. 1985 ; Villalba-Mateos et al. 1997). Several transcripts ranging from 1.5 to 6.0 kb are induced in *Phaseolus vulgaris* by an elicitor of C. *lindemuthianum* (Showalter et al. 1985).

The notion of elicitor, itself, has been extended to plant endogenous compounds, and more particularly to oligogalacturonides (OGAs) derived from plant cell walls. The endopolygalacturonase (endoPG) of *C. lindemuthianum* releases pectic fragments from the cell walls of its host plant (*Phaseolus vulgaris*), both in vitro and in planta. During pathogenesis, the enzyme diffuses freely in the host cell wall,

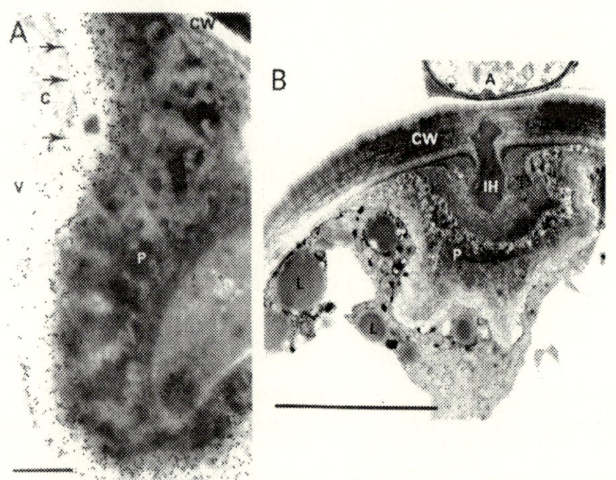

Fig. 1 Incompatible interaction between *Phaseolus vulgaris* cv. Kievitsboon koekkoek and race β of *Colletotrichum lindemuthianum*. A. Immunogold labeling of HRGPs between the newly expanded plasmalemma (arrows) and the cell wall (cw) within a papilla; C: cytoplasm, V: vacuole, Bar = 0.5 µm. B. Papilla reaction: a small infection hypha (IH) is encased by a multilayered papilla (P); A: appressorium, L: lipid body, Bar = 5 µm.

causing drastic degradation of pectic polysaccharides, as shown by ultrastructural cytochemical analysis of infected tissus labelled with the host polygalacturonase inhibitor protein (PGIP) which specifically binds to endopolygalacturonase.

The pectic fragments recovered through the action of endopolygalacturonase differ as regard to their chemical composition and elicitor effects, the fragments released from a resistant line being less acidic and more active than the ones obtained from a susceptible line (Boudart et al. 1998). Besides elicitors, the cell walls of *P. vulgaris* contain complex oligogalacturonides that suppress HRGP gene expression, most particularly the 2.5 kb transcript. Thus, the cell wall has the potential to modulate the level of its own components. The role of endopolygalacturonase in signaling was recently confirmed by using the *Agrobacterium tumefaciens* transient delivery system to express active and mutagenised inactive versions of the enzyme. The data demonstrated that a functional catalytic site and cell wall localization are required for the elicitor activity of endopolygalacturonase (Boudart et al. 2003).

More recently, the role of plant cell walls in signaling leading to HRGP gene expression and defence was further illustrated by the effect of Cellulose Binding Elicitor Lectin from *Phytophthora* (Villalba-Mateos et al. 1997) whose elicitor activity depends on the ability of its two CBD domains to bind crystalline cellulose. Since Cellulose Binding Elicitor Lectin has no hydrolytic activity, it is unlikey that it acts through the release of cell wall oligosaccharides. From the work on the cellulose synthase *cev1* mutant (Ellis et al. 2002) it has been proposed that cellulose might participate in a mechanosensitive signal transduction pathway that responds to perturbations of cell wall structural integrity. The use of mutants of *Arabidopsis thaliana* impaired in one of the three main transduction pathways that involve ethylene, salicylic acid, and jasmonic acid led us to show that the elicitation of HRGP by Cellulose Binding Elicitor Lectin is regulated by salicylic acid in this system (Khatib et al. 2004).

Conclusion

Thirty years after the finding that the level of cell wall Hyp is increased in diseased plants, the accumulation of HRGPs has become one of the most general and best characterized cell wall responses to pathogen attack. This has allowed us to study the role of cell surfaces in signaling leading to defence. Gene expression profiling experiments conducted on *Medicago truncatula* have recently confirmed the strict association of HRGPs to plant resistance (Torregrosa et al. 2004). Yet genetic evidence supporting such a role is still lacking. In the future HRGP mutants should help increase understanding of the functions attributed to these glycoproteins in plant development and in plant defence.

Acknowledgement

I thank Dr Richard O'Connell for kindly providing Figure 1, and reading this manuscript.

References

Benhamou N, Lafontaine PJ, Mazau D, Esquerré-Tugayé MT (1991) Differential accumulation of hydroxyproline-rich glycoproteins in bean root nodule cells infected with a wild-type strain or a C_4-dicarboxylic acid mutant of *Rhizobium leguminosarum* bv. *phaseoli*. Planta 184: 457-467.

Bonfante-Fasolo P, Tamagnone L, Peretto R, Esquerré-Tugayé MT, Mazau D, Mosiniak M, Vian B (1991) Immunocytochemical location of hydroxyproline rich glycoproteins at the interface between a mycorrhizal fungus and its host plant. Protoplasma 165:127-138.

Boudart G, Lafitte C, Barthe JP, Frasez D, Esquerré-Tugayé MT (1998) Differential elicitation of defense responses by pectic fragments in bean seedlings. Planta 206: 86-94.

Boudart G, Charpentier M, Lafitte C, Martinez Y, Jauneau A, Gaulin E, Esquerré-Tugayé MT, Dumas B (2003) Elicitor activity of a fungal endopolygalacturonase in tobacco requires a functional catalytic site and cell wall localization. Plant Physiol 131: 93-101.

Chen J, Varner JE (1985) An extracellular matrix protein in plants: characterization of a genomic clone for carrot extensin. EMBO J 4: 2145-2151.

Ellis C, Karafyllidis I, Wasternack C, Turner JG (2002) The Arabidopsis mutant cev1 links cell wall signalling to jasmonate and ethylene responses. Plant Cell 14: 1557-1566.

Esquerré-Tugayé MT (1973) Influence d'une maladie parasitaire sur la teneur en hydroxyproline des parois cellulaires d'épicotyles et de petioles de plantes de melon. C R Acad Sci 276: 525-528.

Esquerré-Tugayé MT, Mazau D (1974) Effect of a fungal disease on extensin, the plant cell wall glycoprotein. J Exp Bot 25: 509-513.

Esquerré-Tugayé MT, Lamport DTA (1979) Cell surfaces in plant-microorganism interactions. I. A structural investigation of cell wall hydroxyproline-rich glycoproteins which accumulate in fungus-infected plants. Plant Physiol 64: 314-319.

Esquerré-Tugayé MT, Lafitte C, Mazau D, Toppan A, Touzé A (1979) Cell surfaces in plant-microorganism interactions. II. Evidence for the accumulation of hydroxyproline-rich glycoproteins in the cell wall of diseased plants as a defense mechanism. Plant Physiol 64: 320-326.

Esquerré-Tugayé MT, Campargue C, Mazau D (1999) The response of plant cell wall hydroxyproline – rich glycoproteins to microbial pathogens and their elicitors, In: Datta SK, Muthukrishnan S (eds) Pathogenesis – Related Proteins in Plants. CRC Press, Boca Raton, pp 157-170.

Khatib M, Lafitte C, Esquerré-Tugayé MT, Bottin A, Rickauer M (2004) The CBEL elicitor of Phytophthora parasitica var. nicotianae activates defence in Arabidopsis thaliana via three different signalling pathways. New Phytol 162: 501-510.

Lamport DTA, Northcote DH (1960) Hydroxyproline in primary cell walls of higher plants. Nature 188: 665-666.

Lamport DTA (1965) The protein component of primary cell walls. Adv Bot Res 2: 151-218.

Lamport DTA (1967) Hydroxyproline-O-glycosidic linkage in the plant cell wall glycoprotein extensin. Nature 216: 1322-1324.

Lamport DTA (1973) The glycopeptide linkage of extensin: O-D-galactosyl serine and O-L-arabinosyl hydroxyproline, In: Loewus L (ed) Biogenesis of Plant Cell Wall Polysaccharides. Academic Press, New-York, London, pp 149-163.

Leach JE, Cantrell MA, Sequeira L (1982) Hydroxyproline-rich bacterial agglutinin from potato. Plant Physiol 70: 1353-1358.

O'Connell RJ, Brown IR, Mansfield JW, Bailey JA, Mazau D, Rumeau D, Esquerré-Tugayé MT (1990) Immunocytochemical localization of hydroxyproline-rich glycoprotein accumulating in melon and bean at sites of resistance to bacteria and fungi. Mol Plant-Microbe Interact 3: 33-40.

Pladys D, Esquerré-Tugayé MT (1972) Activité protéolytique et pathogenèse dans l'anthracnose du melon. Ann Phytopathol 4 : 277-284.

Showalter AM, Bell JN, Cramer CL, Bailey JA, Varner JE, Lamb CJ (1985) Accumulation of hydroxyproline-rich glycoprotein mRNAs in response to fungal elicitor and infection. Proc Natl Acad Sci USA 82:6551-6555.

Stuart DA, Varner JE (1980) Purification and characterization of a salt-extractable hydroxyproline-rich glycoprotein from aerated carrot discs. Plant Physiol 66: 787-792.

Toppan A, Roby D, Esquerré-Tugayé MT (1982) Cell surfaces in plant-microorganism interactions. III. In vivo effect of ethylene on hydroxyproline-rich glycoprotein accumulation in the cell wall of diseased plants. Plant Physiol 70: 82-86.

Torregrosa C, Cluzet S, Fournier J, Huguet T, Gamas P, Prosperi JM, Esquerré-Tugayé MT, Dumas B, Jacquet C (2004) Cytological, genetic and molecular analysis to characterize compatible and incompatible interactions between Medicago truncatula and Colletotrichum trifolii. Mol Plant-Microbe Interact 17:909-920.

Villalba Mateos F, Rickauer M, Esquerré-Tugayé MT (1997) Cloning and characterization of a cDNA encoding an elicitor of *Phytophthora parasitica* var *nicotianae* that shows cellulose binding and lectin-like activities. Mol Plant- Microbe Interact 10:1045-1053.

Cell Wall-Associated Hydroxyproline-Rich Glycopeptide Defense Signals

Clarence A. Ryan, Gregory Pearce and Javier Narvaez-Vasquez

A family of defense signaling peptides of 15-20 amino acids in length, called systemins, are functionally and structurally similar, but not necessarily homologous. Tomato systemin, an 18 amino acid peptide, was the first peptide hormone isolated from plants. It is released at sites of pest attacks and is transported thoughout the plants where it activates the synthesis of proteinase inhibitors and polyphenolase, which negatively affect the digestive systems of attacking pests. A search for other defense-related peptide signals in plants led to the discovery of a family of small hydroxyproline-rich glycopeptides, called HypSys peptides, associated with the plant cell wall that are also defense signals. The characterization of the cell wall associated peptides led to the inclusion of these peptides in the systemin family of defense signals (Ryan and Pearce 2003). The properties of the known systemin family peptides are presented in this chapter, as is a brief description of the research that led to the discovery of the cell wall associated systemins. This research demonstrates that the cell wall harbors protein-derived peptide signals in addition to the well-known polysaccharide- and cutin-derived signals that activate defensive genes in response to pathogen and herbivore attacks.

The identification of tomato systemin as a defense signal evolved from the discovery of proteinase inhibitors in potatoes nearly 40 years ago (Ryan and Balls 1962). Studies of the biochemistry and physiology of these proteinase inhibitor proteins led to the discovery that herbivore attacks on potato and tomato leaves cause signals to be released at the wound sites that travel throughout the plants to activate the synthesis of the inhibitor proteins in leaves throughout the plants (Green and Ryan 1972). Many years were devoted to the isolation of the signal(s), and their mechanism of signal transduction and activation of defense genes, and in 1991 the initial wound signal was isolated and identified as an 18 amino acid peptide hormone that was named tomato systemin (Pearce et al. 1991). In 1992 the systemin precursor of 200 amino acids in length, called prosystemin, was identified

(McGurl et al 1992) and the role of systemin was established as a signal for a lipid based signaling pathway that activates defensive genes (Farmer and Ryan 1992).

Prosystemin is synthesized constitutively at low levels within the cytoplasm of the phloem parenchyma cells of the vascular bundles of leaves, petioles and stems (Jacinto et al. 1997; Narvaez-Vasquez and Ryan 2004). Wounding causes the release of systemin from vascular bundle cells and initiates a signaling cascade (Pearce et al. 1991; Jacinto et al. 1997; Narvaez-Vasquez and Ryan 2004) that releases linolenic acid from cell membranes and its conversion to jasmonic acid through the octadecanoid pathway (Farmer and Ryan 1992). Evidence suggests that both jasmonic acid and systemin are transported within vascular bundles to distal cells of the plant (Pearce et al. 1991; Narvaez-Vasquez and Ryan 2004; Li et al. 2000). The tomato systemin receptor, SR 160, was isolated from the cell surface membranes of tomato cells and shown to be an LRR receptor kinase (Scheer et al. 2002). SR 160 is identical to the brassinolide receptor, BRI1, a complex organic hormone that is involved with growth and development (Scheer et al. 2003; Montoya et al. 2002). The brassinolide receptor is ubiquitious in the plant kingdom (McCarty and Chory 2000), which suggests that its dual role in defense and development may occur in other plant families. The HypSys peptide receptors have not yet been identified, but resolving their identities will provide new insights into the functional relationships among the systemin family of peptides.

Neither tomato systemin nor its precursor protein contains posttranslational modifications (McGurl et al. 1992). Prosystemin gene transcripts have been identified in tomato, potato, nightshade and pepper plants (Constabel et al. 1998), which are members of the subfamily Solanoideae of the Solanaceae family. A prosystemin transcript could not be found in leaves of tobacco, which is a members of the Cestroideae subfamily of the Solanaceae, but tobacco exhibits a systemic wound response (Pearce et al. 1993), although the response is not as strong as is found in tomato plants.

A search for the signal(s) for the systemic wound response of tobacco plants was initiated using a novel assay, called "the alkalinization assay" that was devised to detect peptide signals in leaf extracts

(Pearce et al. 2001). The assay is based on the ability of receptor-mediated peptide ligands to cause the pH of the medium of suspension-cultured tobacco cells to rapidly increase, due to the blockage of a proton pump in the cell membranes (Meindl et al. 1998; Pearce et al. 2001). As fractions of leaf extracts elute from HPLC columns, small aliquots (~10 μL) are added to 1mL of suspension-cultured cells and the pH is monitored. If a receptor mediated peptide hormone is present in a fraction, it is identified within minutes as the pH of the cell medium increases. With the assay, two peptides were isolated from tobacco leaves and three from tomato leaves that exhibited powerful defense gene inducing activities (Pearce et al. 2001). The newly isolated peptides are distinct from the original systemin peptide, which does not contain hydroxylated prolines nor attached carbohydrate residues (Pearce and Ryan 2003). All five of the newly isolated peptides are composed of between 15 and 20 amino acids and each contains regions of 5 to 11 continuous amino acid sequences consisting of varying numbers of proline, hydroxyproline, serine or threonine residues flanked by charged amino acids. Because of the presence of tandem prolines and/or hydroxyprolines in all six of the defense signaling peptides, and their powerful defense signaling properties, the name systemin is now used to include all of the peptides in a functionally and structurally similar family of defense signals (Ryan and Pearce 2003). The newly identified tobacco and tomato peptides were named TobHypSys I and II and TomHypSys I, II and III (Ryan and Pearce 2003), and tomato systemin has retained its original name. The amino acid sequences of the five HypSys peptides are shown in Figure 1 compared to the sequence of tomato systemin (TomSys). Hydroxyproline residues (O) are the result of post-translational hydroxylations of proline residues (P) that occur along with glycosylations in the ER and Golgi. The multiple proline and/or hydroxyproline residues found in the peptides, together with carbohydrate decorations, are common components of proteins associated with plant cell walls (Kieliszewski 2001). Peptides with tandem proline and hydroxyproline residues form polyproline II (PP II) conformations, that tend to orient the polyproline region in a left-handed helix (Reiersen and Rees 2001). A PP II structure was identified in systemin (Tomagdje and Johnson 1995), and similar PP II conformations are predicted to be present in other systemin peptides

		CARBOHYDRATE Pentoses/peptide
TobHypSys I	[1] RGANLPOOSOASSOOSKE [18]	9
TobHypSys II	[1] NRKPLSOOSOKPADGQRP [18]	6
TomHypSys I	[1] RTOYKTOOOOTSSSOTHQ [18]	8-17
TomHypSys II	[1] GRHDYVASOOOOKPQDEQRQ [20]	12-16
TomHypSys III	[1] GRHDSVLPOOSOKTD [15]	10
TomSys	[1] AVQSKPPSKRDPPKMQTD [18]	0

Fig. 1 Amino acid sequences of systemin peptides. **The linkages of the carbohydrate residues are not known. The amino acids, proline (P), hydroxyproline (O), threonine (T), and serine (S), are shown in red for comparison between peptides. These amino acids are important to the conformations of the peptides and their recognition by receptors.**

containing tandem hydroxyproline or proline residues.

A cDNA from tobacco leaves codes for a precursor protein of 165 amino acids that contains the sequences for both TobHypSys I and II peptides (Pearce et al. 2001), while a cDNA from tomato leaves codes for a protein of 146 amino acids that contained all three TomHypSys peptides. Box diagrams of the deduced HypSys precursor proteins from tobacco and tomato, compared to tomato prosystemin are shown in Figure 2. Both the pre-proTobHypSys and the pre-proTomHypSys have a short N-terminal leader or transit sequence for directing the nascent protein to the secretory pathway. By transforming tomato plants with the *preproTomHypSys* gene fused to the *green fluorescent protein (GFP)* gene the newly synthesized protein was found in walls of leaf epidermal cells (Figure 3). The nucleotide and amino acid

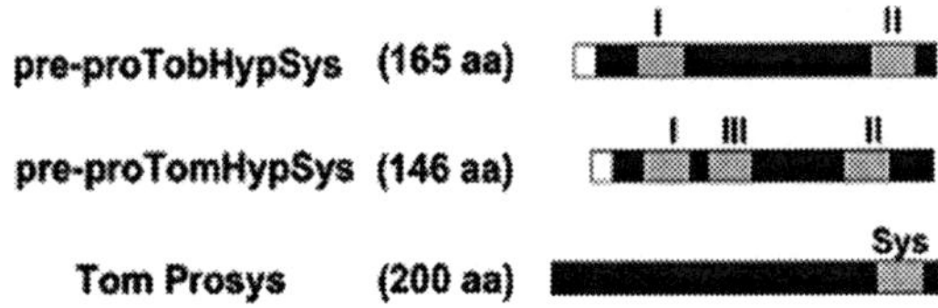

Fig. 2 Box diagrams of systemin precursors. **The open boxes represent transit sequences and the hatched boxes represent the systemin peptides that are cleaved from the precursors in response to herbivore attacks or other mechanical wounding (Pearce and Ryan 2003).**

sequences of the two *pre-proHypSys* genes display weak homologies, indicating that both of the *pre-proHypSys* genes were derived from a common ancestral gene. Neither gene exhibits any homology to the prosystemin gene, except for the tandem proline residues. The two HypSys pre-proproteins are the first examples in plants in which

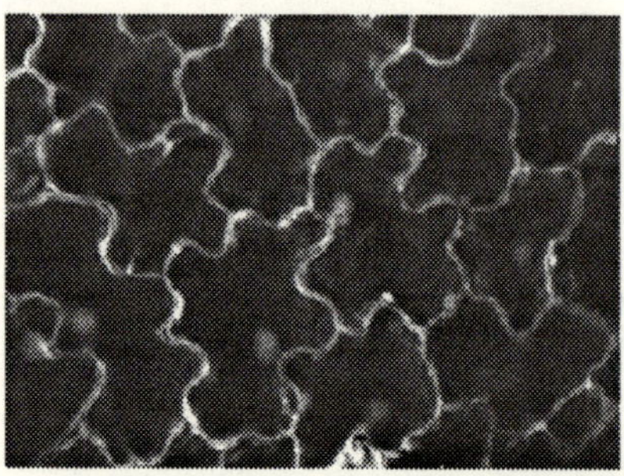

Fig. 3 Confocal image showing the cell wall localization of a tomato hydroxyproline-rich systemin precursor fused with a green fluorescent protein (GFP) in leaf cells of transgenic tomato plants. **(J Narvaez Vasquez, G Pearce, and CA Ryan (2005) Proc Natl Acad Sci USA 102: 12974-12977).**

multiple peptide hormones are derived from single precursors, a scenario commonly found among animal pre-prohormones.

Tomato and other plants from the subfamily Solanoideae of the Solanaceae family(Figure 4) express both prosystemin and the *pre-proHypSys* genes, which may generate a synergistic response by the various systemins to initiate and amplify a signaling cascade in vascular bundles that increases levels of jasmonic acid, which can amplify the levels of prosystemin and pre-proHypSys and their processing enzymes, as well as other signaling components including polygalacturonase, active oxygen species and signal transduction pathway genes along the vascular bundles to amplify the defense response. The cascade continually enhances both signals and signaling pathway components in a wave that moves through the plants to activate defensive genes such as proteinase inhibitors and polyphenol oxidase. In tobacco and petunia of the subfamily Cestdroideae (Figure 4), where prosystemin transcripts have not been found, the wound signaling cascade appears to involve only HypSys peptides in the

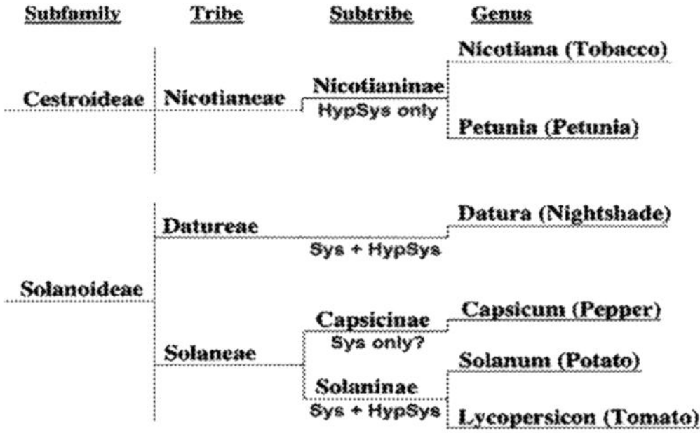

Fig. 4 Phylogenetic tree showing the subfamilies, tribes, and subtribes of the Solanaceae family in which genes coding for systemin family precursor genes have been isolated.

cascade, which may explain the weaker systemic signal found in tobacco plants (Pearce et al. 1993). Tomato plants transformed with an antisense *HypSys precursor* gene, produced a phenotype that retained a systemic wound response, indicating that HypSys peptides are not an absolute requirement for distal signaling and may be localized wound signals. Although their presence has not yet been investigated in species of other plant families, it is now known that wounding of leaves of many plant species by herbivores or other mechanical damage results in the activation of defense responses (Karban and Baldwin 1997). The possible role of peptides as common signals for inducible defense responses in plants is being investigated.

Summary

The isolation and characterization of the hydroxyproline-rich glycopeptides of the systemin family has provided new insights into the role of the plant cell wall in defense signaling. The HypSys peptides join other cell wall-derived defense signals, including oligosaccharides and cutin monomers that are produced from larger cell wall-associated macromolecules in response to attacking predators to activate defensive genes.

Acknowledgement

The research reported herein was supported by grants from the National Science Foundation.

References

Constabel CP, Bergey DR, Ryan CA (1998) Prosystemin from potato, black nightshade, and bell pepper: Primary structure and biological activity. Plant Mol Biol 36: 55-62.

Farmer EE, Ryan CA (1992) Octadecanoid jasmonate precursors activate the synthesis of wound-inducible proteinase inhibitors. Plant Cell 4: 129-134.

Green TR, Ryan CA (1972) Wound-induced proteinase inhibitor in plant leaves: A possible defense mechanism against insects. Science 175: 776-777.

Jacinto T, McGurl B, Franceschi V, Delano-Freier J, Ryan CA (1997) Tomato prosystemin confers wound-inducible, vascular bundle-specific expression of the β-glucuronidase gene in transgenic tomato plants. Planta 203: 406-411.

Karban R, Baldwin IT (1997) Induced responses to Herbivory. University of Chicago Press, Chicago, IL.

Kieliszewski MJ (2001) The latest hype on Hyp-o-glycosylation codes. Phytochemistry 57: 319-323.

Li L, Li C, Lee GI, Howe GA (2000) Distinct roles for jasmonic acid synthesis and action in the systemic wound respnse of toato. Proc Natl Acad Sci USA 99:6416-6421.

McCarty DR, Chory J (2000) Conservation and innovation in plant signaling pathways. Cell 103: 201-209.

Montoya T, Nomura T, Farrar K, Kaneta T, Yokata T, Bishop GJ (2002) Cloning the tomato curl3 gene highlights the putative dual role of the leucine-rich receptor kinase tBRI1/S160 in plant steroid hormone and peptide hormone signaling. Plant Cell 14: 3163-3176.

McGurl B, Pearce G, Orozco-Cardenas M, Ryan CA (1992) Structure, expression and antisense inhibition of the systemin precursor gene. Science 255: 1570-1573.

Meindl T, Boller T, Felix G (1998) The plant wound hormone systemin binds with the N-terminal part to its receptor but needs the C-terminal part to activate it. Plant Cell 10: 1561-1570.

Narvaez-Vasquez J, Ryan CA (2004) The cellular localization of prosystemin: A functional role for phloem parenchyma in systemin signaling. Planta 218: 360-369.

Pearce G, Moura DS, Stratmann J, Ryan CA (2001) Production of multiple plant hormones from a single polyprotein precursor. Nature 411: 817-820.

Pearce G, Ryan CA (2003) Systemic signaling in tomato plants for defense against herbivores: Isolation and characterization of three novel defense-signaling glycopeptide hormones coded in a single precursor gene. J Biol Chem 278: 30044-30050.

Pearce G, Strydom D, Johnson S, Ryan CA (1991) A polypeptide from tomato leaves activates the expression of proteinase inhibitor genes. Science 253: 895-897.

Pearce G, Johnson S, Ryan CA (1993) Purification and characterization from tobacco leaves of six small, wound inducible, proteinase iso-inhibitors of the potato inhibitor II family. Plant Physiol 102: 639-644.

Ryan CA (1987) Oligosaccharide signalling in plants. Ann Rev Cell Biol 3: 295-317.

Ryan CA, Balls, AK (1962) An inhibitor of chymotrypsin from *Solanum tuberosum* and its behavior toward trypsin. Proc Natl Acad Sci USA 48: 1839-1844.

Ryan CA, Pearce G (2003) Systemins – A functionally defined family of peptide signals that regulate defensive genes in Solanaceae species. Proc Natl Acad Sci USA 100: 14573-14577.

Scheer J, Pearce G, Ryan CA (2003) Generation of systemin signaling in tobacco by transformation with the tomato systemin receptor-kinase gene. Proc Natl Acad Sci USA 100: 10114-10117.

Scheer JM, Ryan CA (2002) The systemin receptor SR160 from *Lycopersicon peruvianum* is a member of the LRR receptor kinase family. Proc Natl Acad Sci USA 99: 9585-9590.

Tomagdje A, Johnson WC Jr. (1995) Systemin has the characteristics of a poly(L-proline) II type helix. J Am Chem Soc 117: 7023-7024.

Reiersen H, Rees AR (2001) The hunchback and its neighbors: proline as an environmental modulator. Trends in Biochem Sci 26: 679-684.

Cell Wall Proteins as a Tool for Plant Cell Modification

Ziv Shani and Oded Shoseyov

From seedless citrus fruit to fast-growing plants

It all started with a very practical project aiming to reduce the number of seeds in citrus fruits...

The idea was to apply recombinant cellulose binding domain (CBD) to the stigmata of citrus flowers, hoping that the CBD would bind to pollen tubes and thereby inhibit their elongation and the consequent fertilization and seed development.

A graduate student in the laboratory, E. Shpigel, performed the following experiment: Pollen was geminated overnight *in vitro* in the presence or absence of CBD. We anticipated that CBD would inhibit pollen tube growth, but to our surprise, pollen tubes that were grown in the presence of CBD were significantly longer compared to the controls. So we gave up the seedless citrus idea and entered the fascinating world of plant cell growth.

Plant cell walls are important structures specifically designed for a variety of apparently opposing functions. On one hand, cell walls are responsible for tensile strength, cell shape, and resistance to pathogen invasion. On the other, they must maintain reasonable flexibility against breaking forces and just enough permeability to allow building blocks and signaling molecules to enter the living cells. Therefore, modifications of living cell walls require a sensitive, highly synchronized system of signals, enzymes, and building blocks.

It is now well established that living organisms contain complex systems for the management of cellulose-containing materials. One of the pivotal players in these systems is endo-1,4-ß-glucanase, which hydrolyzes polysaccharides possessing 1,4-ß-glucan. Another player, which appears to be present primarily in systems devoted to cellulose degradation rather than cellulose synthesis, is the cellulose-binding domain (CBD). This important, naturally occurring protein entity is contained within many EGases and other enzymes. It plays an essential role in cellulose degradation and has the potential to modify cellulose-containing materials.

Both gene families, endo-1,4-ß-glucanase and CBD have potential technological applications in their ability to modify polysaccharides and in shaping plant cell development and wall properties. This paper summarizes progress made in recent years in the understanding and use of these genes as tools for plant improvement in modern agricultural and forest plantations.

The role of endo-ß-glucanase proteins in plant cell elongation

The interaction between cellulose microfibrils and the hemicellulose-xyloglucan network is believed to represent the major load-bearing structure in the primary cell wall. In physical terms, cell shape and size are governed by the mechanics of the cell wall. Cell expansion occurs via strictly regulated reorientation of the wall's components (Carpita and Gibeaut 1993). Loosening of the cellulose-xyloglucan network is proposed to be the rate-limiting factor for this process. Several mechanisms for this relaxation have been suggested, including the activities of endo-1,4-ß-glucanase (Verma *et al.* 1975; Shani *et al.* 1997), endo-xyloglucan transferase (Fry *et al.* 1992; Nishitani and Tominaga 1992), and expansins (McQueen-Mason *et al.* 1992). During cell elongation, native endo-1,4-ß-glucanases in the plant hydrolyze cellulose-xyloglucan linkages, allowing cellulose chains to move freely relative to one another. Cleavage of xyloglucan cross-links by endo-1,4-ß-glucanase would present a rapid and simple means of achieving wall loosening and cause the cells to elongate more freely and rapidly, resulting in faster growth and development.

Endo-1,4-ß-glucanase are encoded by multi-gene families of plant enzymes, which hydrolyze polysaccharides with a 1,4-β-glucan backbone to non-substituted glucose residues. Plant endo-1,4-ß-glucanases are unable to hydrolyze crystalline cellulose but can hydrolyze cellulose chains at para-crystalline areas (Hayashi 1989; Rose & Bennet 1999). The ability of poplar endo-1,4-ß-glucanase to act on cellulose, and the capability of anti-poplar cellulose antibodies to prevent the solubilization of xyloglucan in a suspension of cultured poplar cell walls, has been demonstrated (Ohmiya *et al.* 2000). Recently, a protein inducing extension of the plant cell wall was isolated from snails (*Helix pomatia*). Interestingly enough, N-terminal sequence analysis revealed that this protein belongs to the family 9 endoglucanases (Tekada and Cosgrove 2004), which also includes

*Arabidopsis thaliana cel*1. *cel*1 is the endo-1,4-ß-glucanase gene from *A. thaliana* (Shani *et al.* 1997). The intensive accumulation of *cel*1 mRNA in young, expanding tissues of *A. thaliana* (Shani *et al.* 1997) and specific *cel*1 promoter expression in developing-stage young leaves of transgenic poplar plants (Shani *et al.* 2000) suggest its role in cell elongation. Furthermore, we have shown by *in situ* localization of CEL1 protein that it accumulates in the cell wall of *A. thaliana* plants and is located primarily in the developing xylem cells, where growth and cellulose deposition is most significant (Shani 2000). Transgenic *A. thaliana* plants expressing antisense endo-1,4-ß-glucanase (*cel*1) exhibited a dwarf phenotype and abnormal, "wrinkled" cell walls (Tsabary *et al.* 2002), further demonstrating the obligatory role of this enzyme in normal plant development.

Expression of *cel*1 and other endo-1,4-ß-glucanase proteins in the cell walls of transgenic plants

The loosening of xyloglucan in the cell walls of transgenic plants was shown to promote plant growth in several independent studies (Shani *et al.* 1999a; Park *et al.* 2003; Park *et al.* 2004; Shani *et al.* 2004).

Poplar (*Populus tremula*) plants transformed with and over-expressing *Arabidopsis thaliana cel*1 cDNA exhibited significant phenotypic alterations, including taller plants, larger leaves, increased stem diameter, higher wood volume index, larger dry weight, and a higher percentage of cellulose and hemicellulose, compared to wild type plants (Shani *et al.* 1999a; Shani *et al.* 2004).

Transgenic *A. thaliana* plants over-expressing poplar endo-1,4-ß-glucanase were reported to have enhanced growth rate, larger leaves, larger cells, increased biomass, and increased cellulose content per plant (Park *et al.* 2003). Recently, constitutive expression of *Aspergillus* xyloglucanase in poplar (*Populus alba*) increased stem length. The increased stem growth and biomass were accompanied by a decrease in Young's elastic modulus in the growth zone, but an increased elasticity in mature tissues. The increase in internode length was found to correspond to an increase in cellulose content as well as specific gravity, indicating that the removal of xyloglucan might cause an increase in cellulose density in the secondary xylem (Park *et al.* 2004).

CBD genes

Cellulose binding domains (CBDs) are naturally found as discrete domains in proteins such as cellulases (Gilkes *et al.* 1991), as well as in proteins with no hydrolytic activity (Shoseyov *et al.* 1992). A distinct Cellulose binding domains family, called expansins, was identified in plants (McQueen-Mason *et al.* 1992). The *in vitro* effect of expansins on plant cell walls was found to be similar to that of bacterial and fungal Cellulose binding domains, including swollenin, which is a Family II carbohydrate binding module with sequence similarity to plant expansins (Saloheimo *et al.* 2002). A phylogenetic tree was built from different genes sharing significant homology by the BLAST search tool (Figure 1). The results indicate that the genetic distance between expansins and swollenin is the same as between cellulases and putative endoglucanases. A three-dimensional model of the cellulose-binding domain of the rye-grass pollen allergen Lol pI (β-expensin) was built by homology modeling. Barre and Rouge (2002) found a groove and an extended strip of aromatic and polar residues that strikingly resemble the 3D structure of family III cellulose-binding domain. Cellulose binding domains proteins play an essential role in cellulose degradation and have the potential to modify cellulose-containing materials (to learn more about the different Cellulose binding domains families, see Gilkes *et al.* (1991) and Levy *et al.* (2002).

The effects of CBD on cellulose synthesis in a model system

Shpigel *et al.* (1998) demonstrated that cellulose binding domains can modulate cellulose biosynthesis. cellulose binding domains increased the rate of cellulose synthesis activity in *Acetobactor xylinum* up to five-fold over the control. Electron microscopy of cellulose synthesized in the presence of cellulose binding domains revealed that the newly formed fibrils are spread into a splayed ribbon, instead of the uniform, thin, packed ribbon of the control fibers (Figure 2). The mechanism by which cellulose binding domains affects cell wall metabolism is unknown, but we postulate a physico-mechanical mechanism whereby cellulose binding domains slides between cellulose fibers and separates them in a wedge-like action (Levy *et al.* 2002).

The synthesis of cellulose can be divided into an initial polymerization step and a second step in which the individual glucan chains associate

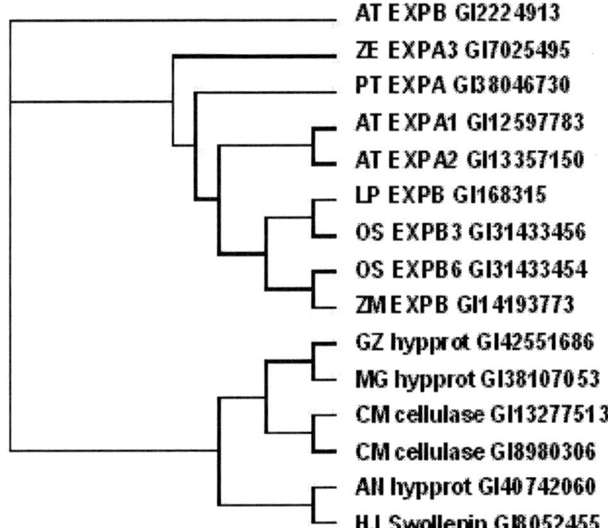

Fig. 1 A phylogenetic tree was built from different genes sharing significant homology with *A. thaliana* α-expansin I, using the BLAST search tool (www.ncbi.nlm.nih.gov).
AT EXPA1 GI12597783 and AT EXPA2 GI13357158, *A. thaliana* α-expansins; PT EXPA GI38046730, poplar α-expansin; ZE EXPA3 GI7025495, *Z. elegans* α-expansin; LP EXPB GI168315, ryegrass β-expansins; OS EXPB3 GI31433456 and OS EXPB6 GI31433454, rice β-expansins; ZM EXPB GI14193773, corn β-expansin; AT EXPB GI2224913, *A. thaliana* β-expansin; GZ hypprot GI42551686, *Gibberella zeae* putative EGase; AN hypprot GI40742060, *Aspergillus nidulans* putative EGase; CM cellulase GI13277513, *Clavibacter michiganensis* Cellulase; HJ Swollenin GI8052455, *Hypocrea jecorina* swollenin; MG hypprot GI38107053, *Magnaporthe grisea* putative EGase; CM Cellulase GI8980306, *Clavibacter michiganensis* cellulase.

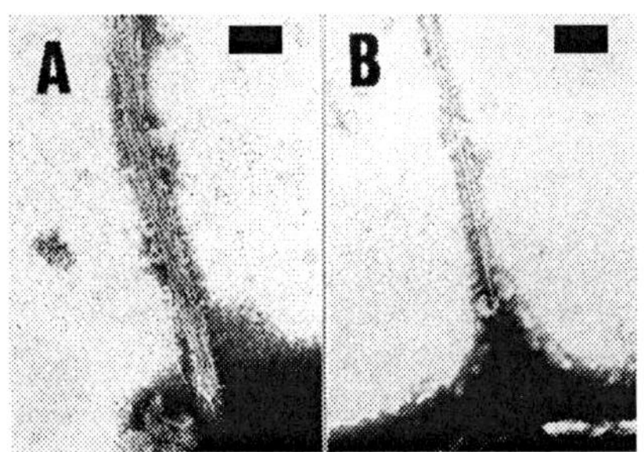

Fig. 2 The effect of Cellulose binding domains on cellulose ribbon produced by *A. xylinum* (A), or a control without cellulose binding domains (B). Bar = 0.2 μm.

to form crystalline cellulose (Brown *et al.* 1996). By introducing the Cellulose binding domain genes into plants, we are able to express Cellulose binding domain proteins within the cell wall of plant tissue. This physico-mechanical interference uncouples the cellulose-biosynthetic polymerization step from the crystallization step. This results in an increased rate of synthesis of the cellulose polymer, improved polymer qualities, and enhanced biomass (Figure 3).

This postulate was further supported by additional *in vitro* experiments in which application of recombinant cellulose binding domains significantly reduced the wet tensile strength of cellulose paper when tested in an Instron Universal Testing Machine. The observed results resembled the *in vivo* effects of expansin. Furthermore, petunia cell suspensions treated with increasing concentrations of cellulose binding domains displayed abnormal shedding of cell wall layers, indicating that cellulose binding domains can cause non-hydrolytic cell wall disruption *in vivo* (unpublished data).

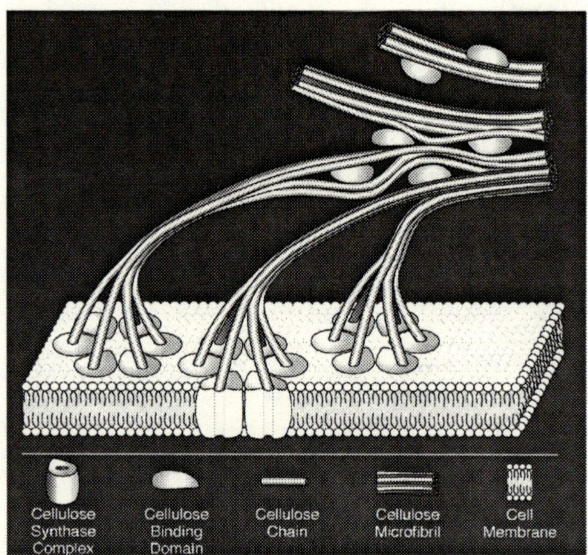

Fig. 3 A model describing the interactions between cellulose binding domain (CBD) and cellulose fibers. **Cellulose binding domains may interact with the cellulose microfibrils at different stages. When the interaction occurs during the initial stages of crystallization, the result is an increased rate of synthesis and splayed fibrils. A post-synthesis interaction results in non-hydrolytic fiber disruption.**

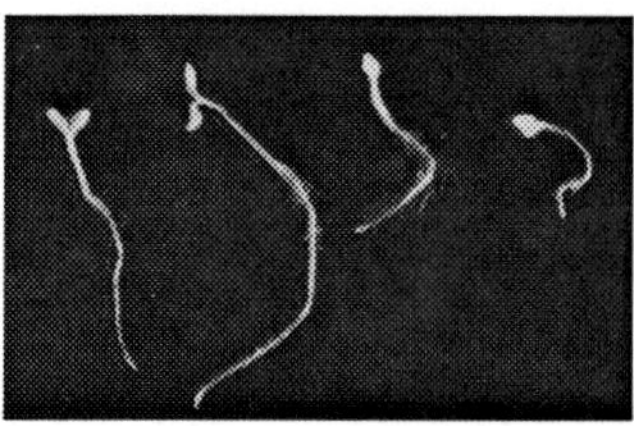

Fig. 4 The effect of cellulose binding domains on *Arabidopsis* seedlings. **Representative seedlings from left to right: control (no cellulose binding domains), 10^{-2}, 100, and 500 µg mL^{-1} cellulose binding domains .**

The effects of Cellulose binding domains on cell elongation in a model system

In peach (*Prunus persica* L.) pollen tubes and in *A. thaliana* seedlings, cellulose binding domains can modulate cell elongation. At low concentrations, cellulose binding domains enhanced the elongation of pollen tubes and roots, whereas at high concentrations, cellulose binding domains inhibited root and pollen elongation in a dose-dependent manner (Shpigel *et al.* 1998). The effects of different Cellulose binding domains concentrations on the development of *Arabidopsis* roots are shown in Figure 4.

Expression of cellulose binding domains protein in the cell wall of transgenic plants

In transgenic plants, expression of bacterial cellulose binding domains resulted in accelerated plant growth, as demonstrated in tobacco (Shani *et al.* 1999), poplar (Levy *et al.* 2002; Shoseyov *et al.* 2002), and potato plants (Safra *et al.* 2000). A similar effect was also observed with plant Cellulose binding domains (expansin) in transgenic *A. thaliana* (Cho and Cosgrove 2000) and in transgenic poplar (Mellerowicz *et al.* 2004) plants. Introduction of the Cellulose binding domains gene under the control of the elongation-specific *cel*1 promoter into transgenic poplar plants led to significant increases in biomass production in selected clones, as compared to wild type control plants (Levy *et al.* 2002). Analysis of wood characteristics from transgenic poplar trees showed significant increases in fiber cell length in the average degree of cellulose polymerization. In addition, a significant decrease in the microfibril angle was observed

(Shoseyov *et al.* 2002). Our results coincided with increased burst, tear, and tensile indices of paper prepared from these transgenic wood fibers.

Concluding remarks

It is now evident that both endo-1,4-ß-glucanases and cellulose binding domains possess immense potential as tools to modify forest trees in various aspects, including increased growth rate, biomass accumulation, and fiber improvement. The recent breakthrough in Eucalyptus and pine transformation prepares the way for the comprehensive development of new, improved transgenic clones for commercial plantations. The benefits from these fast-growing forests are not limited to the pulp and paper industries, but will also include improvements in the global environment. These benefits will be apparent in decreased pressure to cut natural forests and increased CO_2 sequestration from the atmosphere, in accordance with the Kyoto protocol.

References

Barre A and Rouge P (2002) Homology modeling of the cellulose-binding domain of a pollen allergen from rye grass: structural basis for the cellulose recognition and associated allergenic properties. Biochem Biophys Res Comm 296: 1346-1351.

Brown RMJ, Saxena IM, Kudlica K (1996) Cellulose biosythesis in higher plants. Trends Plant Sci 1: 149-155.

Carpita NC, Gibeaut DM (1993) Structural models of primary cell walls in flowering plants: consistency of molecular structure with the physical properties of the walls during growth. Plant J 3: 1-30.

Cho HT, Cosgrove DJ (2000) Altered expression of expansin modulates leaf growth and pedicel abscission in *Arabidopsis thaliana.* Proc Natl Acad Sci USA 15: 9783-9788

Fry SC, Smith RC, Renwick KF, Martin DJ, Hodge SK, Matthews KJ (1992) Xyloglucan endotransglycosylase, a new wall-loosening enzyme activity from plants. Biochem J 282: 821-828.

Gilkes NR, Henrissat B, Kilburn DG, Miller RJ, Warren RA (1991) Domains in microbial beta-1,4-glycanases: sequence conservation, function, and enzyme families. Microbiol Rev 55: 303-315.

Hayashi T (1989) Xyloglucans in the primary cell wall. Annu. Rev Plant Physiol Plant Mol Biol 40: 139-168.

McQueen-Mason SJ, Durachko DM, Cosgrove DJ (1992) Two endogenous proteins that induce cell wall extension in plants. Plant Cell 4: 1425-1433.

Mellerowicz E, Nishikubo N, Gray-Mitsumune M, Siedlecka A, Sundberg B (2004) cell expansion in developing wood. X cell wall meeting, Sorrento, Italy.

Levy I, Shani Z, Shoseyov O (2002) Modification of polysaccharides and plant cell wall by endo-1,4-β-glucanase and cellulose-binding domains. Biomol Eng 19: 17-30.

Nishitani K, Tominaga R (1992) Endo-xyloglucan transferase, a novel class of glycosyltransferase that catalyzes transfer of a segment of xyloglucan molecule to another xyloglucan molecule. J Biol Chem 267: 21058-21064.

Ohmiya Y, Samejima M, Shiroishi M, Amano Y, Kanda T, Sakai F, Hayashi T (2000) Evidence that endo-1,4-β-glucanase act on cellulose in suspension-cultured poplar cells. Plant J 24: 147-158.

Park YW, Tominaga R, Sugiyama J, Furuta Y, Tanimoto E, Samejima M, Sakai F, Hayashi T (2003) Enhancement of growth by expression of poplar cellulase in *Arabidopsis thaliana*. Plant J 33:1099-106.

Park YW, Baba K, Furuta Y, Iida I, Sameshima K, Arai M, Hayashi T (2004) Enhancement of growth and cellulose accumulation by overexpression ofxyloglucanase in poplar. FEBS Lett 564: 183-187.

Rose JKC, Bennett AB (1999) Cooperative disassembly of the cellulose-xyloglucan network of plant cell walls: parallels between cell expansion an fruit ripening. Trends Plant Sci 4: 176-183.

Safra L, Shani Z, Shoseyov O, Wolf S (2000) Growth modulation of transgenic potato plants by cellulose-binding domain (CBD). 6TH International Congress of Plant Molecular Biology, Quebec, Canada.

Saloheimo M, Paloheimo M, Hakola S, Pere J, Swanson B, Nyyssonen E, Bhatia A, Ward M, Penttila M (2002) Swollenin, a *Trichoderma reesei* protein with sequence similarity to the plant expansins, exhibits disruption activity on cellulosic materials. Eur J Biochem 269: 4202-4211.

Shani Z (2000) The effect of endo-1,4-β-glucanase and a cellulose binding protein on plant cell elongation. Ph D Thesis, The Hebrew University of Jerusalem.

Shani Z, Dekel M, Sig Jensen C, Tzfira,T, Goren R, Altman A, Shoseyov O (2000) *Arabidopsis thaliana* endo-1,4-β-glucanase (*cel*1) promoter mediates uidA expression in elongating tissues of aspen (*Populus tremula*). J Plant Physiol 156: 118-120.

Shani Z, Dekel M, Tzbary G, Goren R, Shoseyov O (2004) Growth enhancement of transgenic poplar plants by overexpression of *Arabidopsis thaliana* endo-1,4-β-glucanase (*cel*1). Molecular Breeding 14: 321–330.

Shani Z, Dekel M, Tsabary G, Shoseyov O (1997) Cloning and characterization of elongation specific endo-1,4-β-glucanase (*cel*1) from *Arabidopsis thaliana*. Plant Mol Biol 34: 837-842.

Shani Z, Dekel M, Tzbary G, Jensen CS, Tzfira T, Goren R, Altman A, Shoseyov O (1999a) Expression of *Arabidopsis thaliana* endo-1,4-β-glucanase (*cel*1) in transgenic plants, In: Altman A, Ziv M, Izhar S (eds) Plant Biotechnology and In Vitro Biology in the 21st Century. Kluwer Academic Publishers. pp. 209-212.

Shani Z, Shpigel E, Roiz L, Goren R, Vinocur B, Tzfira T, Altman A, Shoseyov O (1999) Cellulose binding domain, increases cellulose synthase activity in *Acetobacter xylinum*, and biomass of transgenic plants, In: Altman A, Ziv M, Izhar S (eds) Plant Biotechnology and In Vitro Biology in the 21st Century. Kluwer Academic Publishers. pp. 213-218.

Shpigel E, Roiz L, Goren R, Shoseyov O (1998) Bacterial cellulose-binding domain modulates in-vitro elongation of different plant cells. Plant Physiol 117: 1185-1194.

Shoseyov O, Levy I, Shani Z, Mansfield S (2002) Modulation of wood fibers and paper by cellulose binding domains (CBDs). Proceedings of *The 223th American Chemical Society National Meeting*. Orlando, Florida, USA.

Shoseyov O, Takagi M, Goldstein MA, Doi RH (1992) Primary sequence analysis of *Clostridium cellulovorans* cellulose binding protein A. Proc Natl Acad Sci USA 89: 3483-3487.

Takeda T. and Cosgrove D (2004) A snail protein inducing extension of plant cell wall. X Cell Wall Meeting, Sorrento, Italy.

Tsabary G, Shani Z, Roiz L, Levi I, Riov J, Shoseyov O (2002) Abnormal "wrinkled" cell walls and retarded development of transgenic *Arabidopsis thaliana* plants expressing endo-1,4-β-glucanase (*cel*1) antisense. Plant Mol Biol 51: 213-224.

Verma DPS, Maclachlan GA, Byrne H, Ewings D (1975) Regulation and *in vitro* translation of messenger ribonucleic acid for cellulase from auxin-treated pea epicotyls. J Biol Chem 250: 1019-1026.

The Biotechnology of Plant Cell Walls – Established Technology and New Trends

Tuula T. Teeri

The cell walls of plant fibers constitute an important raw material for many industries. The use of enzymes instead of chemicals offers environmentally friendly tools for fiber processing. In addition to microbes, plants produce enzymes that can be used to modify fiber morphology and chemistry. Genomic approaches offer shortcuts for the identification of plant enzymes involved in fiber formation. A functional genomics program on hybrid aspen has already identified hundreds of genes that are specifically activated during xylogenesis. Moreover, studies of the cell wall organization and properties have led to the development of new biomimetic approaches for fiber modification.

Plant biomass and the microbial enzyme technology

Plant biomass represents an important raw material for the manufacture of paper, textiles and construction materials as well as a sustainable source of future fuels and chemicals. Recent scientific and technological advances promise to expand the use of plant fibers towards novel types of products such as natural fiber reinforced biocomposites (Brumer et al. 2004; Zhou et al. 2005). All plant cells, including the fiber cells used for the manufacture of paper and textiles, are encapsulated by a cell wall that constitutes the major component of biomass. Cellulose microfibrils represent the load-bearing component of the cell walls, while hemicelluloses contribute to the dynamics of the walls by providing flexible cross-links between cellulose and lignin. Lignin is an aromatic polymer, which is abundantly available in the secondary cell walls of trees. It has an important role in plant defense and wood properties. However, the presence of lignin also causes yellowing of paper upon storage, and it is therefore usually removed during most pulping processes.

Chemical technology has long dominated the processing of biomass for human exploitation. However, increasing environmental concerns and the need for sustainable development have led to the development of biotechnological means to replace and complement previous chemical processes. In nature, many different fungi and bacteria contribute to the degradation of the different components of biomass. Cellulose and hemicelluloses are degraded by a wide range of different glycosyl hydrolases while lignin is oxidized by various oxidases and peroxidases. Because of the insoluble nature of the substrate these enzymes are necessarily secreted to the culture medium of the microbes for extracellular digestion and uptake of the solubilized nutrients. Since extracellular enzymes are relatively easy to produce and purify in a large scale, microbial enzymes dominate the present market of industrial enzymes. The current biotechnological processing of lignocellulosic substances relies on limited hydrolysis of the undesired components of the substrate. Thus, pectinases and laccases are used for the clarification of fruit juices and endolytic cellulases provide useful tools for surface treatment of cotton garments to achieve a silk-like appearance. Cellulases are also added in modern laundry detergents to remove fuss and pills for better maintenance of color. Xylanases have been used to facilitate non-chlorine bleaching of pulps to remove the residual lignin trapped on the pulp surfaces by reprecipitated xylan. Most of the present day industrial enzymes have been obtained by activity-based screening of microbial enzymes secreted to the culture media, followed by enzyme engineering programs to optimize the protein properties for industrial use. However, in spite of their excellent specificity and environmental advantages, the use of enzymes in large-scale operations is often limited by the high cost of the treatment.

Plant genomics – identification of enzymes involved in fiber biosynthesis

In addition to microbes, plants produce enzymes that could be used to modify the structure and properties of lignocellulosic materials for industrial purposes. In contrast to microbial enzymes, most fiber-active enzymes in plants are engaged in the synthesis rather than degradation of the cell wall components. The advantage of synthetic enzymes is that they can be used to alter the raw material for both improved yield and fiber performance. The disadvantage of plant enzymes is that they have evolved to operate in moderate temperatures and protected by the plant cell walls. They are therefore more difficult to adapt for industrial use than the more robust microbial enzymes. Glycosyltransferases that are involved in carbohydrate biosynthesis are particularly challenging since they are often membrane-bound, and thus difficult to isolation and characterize (Scheible and Pauly 2004). Therefore, the enzymology of the plant cell wall biosynthesis and remodeling has remained poorly understood and the use of synthetic enzymes from plants to upgrade the fiber properties is still in its infancy.

The advent of genomics has offered much needed shortcuts for detailed studies of the biochemical pathways in a wide range of living organisms. The genome of a tiny weed, *Arabidopsis thaliana*, was the first plant genome that was completely sequenced (*Arabidopsis* genome initiative, 2000). Other plants of commercial interest such as maize and rise are close followers (Lunde et al. 2003, Cyranoski 2003), and the genome sequence of the first tree, *Populus trichocarpa*, is nearly completed (Brunner et al. 2004, http://genome.jgi-psf.org/Poptr1/Poptr1.home.html). The annotation of these plant genomes has led to the identification of many cell wall specific enzymes and proteins (Coutinho et al. 2003). However, a genome sequence is only the starting point for the functional exploration of the metabolic processes of an organism. In most genomic projects a large proportion of the discovered genes cannot be assigned a function, and further studies are required for determining their general roles and molecular function. The tools commonly used for the downstream investigations of gene function include the generation and phenotypic analysis of gene knock-outs, gene-expression profiling, *in situ* gene expression analysis and immunolocalization studies as well as genetic mapping of quantitative trait loci (QTLs) (Steinmetz and Davis 2004). In the early stages, gene expression profiling using microarrays provides powerful means for high-throughput screening of genes active in defined tissues or during different developmental or physiological processes. Once the key genes of a given metabolic pathway, and their putative substrates have been identified, the corresponding proteins can be expressed in heterologous hosts for detailed biochemical studies.

Functional genomics of wood formation in hybrid aspen

The hybrid aspen, *Populus tremula* x *tremuloides*, has recently emerged as a convenient model system to obtain a comprehensive view of the secondary cell wall biogenesis in trees (Bhalerao et al. 2003). The Swedish *Populus* Genome Project has so far established a database of well over 100,000 EST-sequences from 19 different tissue specific cDNA libraries from the European aspen (*Populus tremula*), the hybrid aspen (*Populus tremula x tremuloides* T89), and the black cottonwood (*Populus trichocarpa*) (http://www.populus.db.umu.se) (Sterky et al. 1998, 2004). Since lignin biosynthesis has been extensively studies by others (Boerjan et al. 2003), the gene discovery in this program has been mainly focused on the cell wall carbohydrate biosynthesis and modification. The present dataset of ESTs contains many known cell wall associated enzymes including 10 isoenzymes of the cellulose synthase (CESA) (Djerbi et al, 2004), two different plant cellulases (Cel9A, Cel9B) (Master et al. 2004), as well as several expansins (Gray-Mitsumune et al. 2004) and xyloglucan endotransglycosylases/hydrolases (XTH16) (Sterky et al. 1997). All of these enzymes and proteins target different carbohydrates in the cell wall and thus represent potential tools for fiber modification.

Further enzymes and proteins specifically involved in xylogenesis have since been sought by using expression profiling on microarrays containing 2,995 or 13,490 unique ESTs from different hybrid aspen libraries (Hertzberg et al. 2001; Djerbi et al. 2004; Aspeborg et al. 2005). Both microarrays were probed with cDNAs prepared by tangential cryosectioning of narrow tissue sections from the different stages of developing wood (Figure 1) (Hertzberg et al. 2001; Aspeborg et al. 2005). Several hundred genes were found to be particularly highly expressed in the sections corresponding to secondary cell wall synthesis, and thus probably represent the major xylem associated genes in poplar. Figure 1 shows some examples of genes encoding enzymes of known function that were found to be abundant during cell expansion, including the primary cell wall synthesis (Figure 1), and during early xylogenesis (Figure 1). A subset of the actively transcribed genes encoded enzymes of previously unknown function as judged by routine bioinformatic analyses. These genes were then subjected to full-length sequencing and bioinformatic analysis relying on tools routinely used to assign proteins to the Carbohydrate-active enzyme database (CAZy: http://afmb.cnrs-mrs.fr/CAZY) (Coutinho and Henrissat 1999). In this way, it was possible to identify many new enzymes in 8 families of glycosyl transferases and 5 families of glycosyl hydrolases (Aspeborg et al. 2005). The family classification

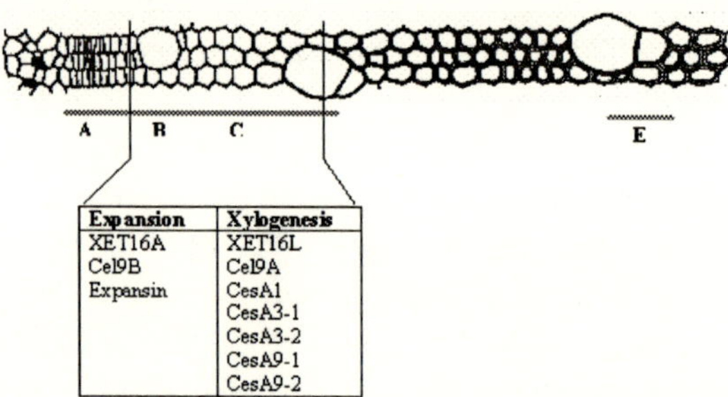

Fig. 1 A microscopic image of the wood-forming tissues in poplar. **Zone A, cell division; zone B, early expansion; zone C, late expansion; zone D, xylogenesis; zone E, dell death. A) Examples of genes highly expressed during cell expansion (zones B and C) (data from Sterky et al, 2001; Gray-Mitsumune et al, 2004). B) Examples of genes highly expressed during xylogenesis (zone D) (data from Sterky et al. 2001; Aspeborg et al. 2005; Djerbi et al. 2004).**

is diagnostic of the general protein fold and the chemical reaction mechanism, which are shared by the family members, but it cannot necessarily predict substrate specificity. A comparison of the carbohydrate active enzymes known today reveals that a similar overall protein fold can accommodate active site structures suitable for many different substrates. For example, the glycosyl hydrolase Family GH16 groups together prokaryotic and eukaryotic enzymes of widely differing substrate specificities, including (1→3;1→4)-β-glucanase (lichenase, EC 3.2.1.73), xyloglucan: xyloglucosyltransferase (EC 2.4.1.207), keratan-sulfate endo-(1→4)-β-galactosidase (EC 3.2.1.103), glucan endo-(1→3)-β-glucosidase (EC 3.2.1.39), endo-[1→3(4)]-β-glucanase (EC 3.2.1.6), agarase (EC 3.2.1.81) and κ-carragenase (EC 3.2.1.83). However, although the substrate specificities remain to be determined for many of the new xylem specific enzymes in poplar, their classification into known enzyme families will help to focus the functional studies to a limited number of potential substrates.

A novel biomimetic principle for fiber modification

Identification of the genes encoding interesting cell wall active enzymes opens the way for their detailed biochemical characterization and exploitation by way of heterologous expression in microbial hosts. Among the hybrid aspen cell wall enzymes, the xyloglucan endotransglycosylase, PttXET16A was the first one successfully expressed in the methylotrophic yeast, *Pichia pastoris* (Kallas et al. 2005). XETs are interesting enzymes that act during cell expansion to catalyze a transient weakening and religation of the xyloglucan-cellulose cross-links in the primary cell walls (Rose et al. 2002). Although XETs belong to the glycosyl hydrolase family GH16, the catalyze transglycosylation reactions by effectively excluding water as an acceptor substrate in favor of carbohydrate acceptors. Characterization of the recombinant enzyme revealed that PttXET16A is indeed a strict transglycosylase devoid of any detectable hydrolytic activity (Kallas et al. 2005). The high yield of the recombinant protein obtained in yeast permitted structure determination of the PttXET16A (Johansson et al. 2004), which opens the way for detailed studies of the molecular basis of the transglycosylating *versus* hydrolytic activity of XETs.

The discovery and efficient production of PttXET16A also permitted the development of a new biotechnological approach for cellulose surface modification (Brumer et al. 2004). Many current applications of plant fibers would benefit from chemical engineering of cellulose surfaces. However, direct chemical modification of cellulose usually leads to the weakening of the microfibrillar structure, which depends of the same hydroxyl groups that are necessarily the subject of chemical modification. Xyloglucan binds spontaneously to cellulose with high affinity in aqueous solutions in room temperature, and it has been shown to improve paper formation and wet strength (Christiernin et al. 2001). Tamarind seed powder is a rich source of xyloglucan, which is commonly used as a sizing agent for textiles and a gelling agent in foodstuffs. Because of its unique transglycosylating activity, XET can join together two xyloglucan molecules. In addition to high-molecular weight xyloglucan polymers, XET can also use short xyloglucan oligosaccharides (XGOs) as the acceptor substrate, even when they are extensively chemically modified at their reducing end. Thus, the xyloglucan oligosaccharides can be chemically modified to desired functionality and offered as substrate for XET for joining to polymeric xyloglucan, which in turn binds to cellulose surfaces and acts as an anchor for the new chemistry (Figure 2). Such a modification can be carried out with no impact on the fiber strength and it offers a route for a versatile chemical modification of cellulose surfaces (Brumer et al. 2004). In particular, the technology can be used to improve the interfacial stability between the naturally hydrophilic plant fibers and the hydrophobic matrix polymers in fiber reinforced biocomposite materials (Zhou et al. 2005). This will improve the strength properties and permit increased fiber loading thus paving the way for light-weight composites of high performance.

Towards fiber engineering *in vivo*

Identification of the major cell wall specific enzymes and the recent rapid progress in plant genetic engineering are paving the way for fiber engineering *in vivo*. When modifying fiber properties by metabolic engineering of the plant, the problem of the enzyme cost suddenly becomes irrelevant since the enzymatic treatment occurs already during the fiber production. The approach also overcomes the difficulties of extracting the membrane-bound plant enzymes for industrial use. While microbial enzymes can be used to modify the quality parameters of wood fibers, *in vivo* engineering has the potential to influence both the quality and the quantity of the wood components. This opens completely new possibilities for innovative fiber engineering for future needs. Pioneering work on the lignin biosynthetic pathways has already led to transgenic poplar lines with a different composition of lignin that is consequently easier to remove

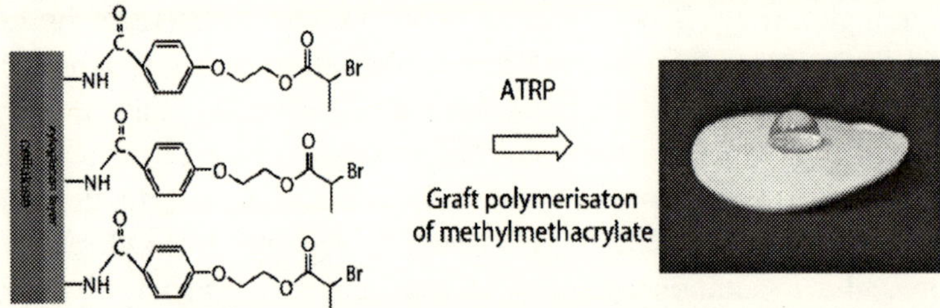

Fig. 2 Principle of cellulose surface modification using chemo-enzymatically modified xyloglucan. **Small xyloglucan oligosaccharides (XGO) can be chemically modified to carry reactive amino groups, which can be further chemically modified with e.g. initiators of atom radical polymerization (ATRP). The modified oligosaccharides can be coupled to xyloglucan by using the XET enzyme. The modified xyloglucan binds spontaneously and irreversibly to cellulose. Graft polymerization of methyl methacrylate from the xyloglucan-bound initiators generates a highly hydrophobic surface on filter paper discs (Data from Brumer et al. 2004; Zhou et al. 2005). Figure courtesy of Qi Zhou.**

during pulp processing (Boudet et al. 2003). In a similar fashion, other types of changes on the fiber structure and properties can be introduced by selective regulation of the expression or activities of carbohydrate-active enzymes during the cell wall biosynthesis. The present dataset of xylem specific genes in poplar serves as a basis for an improved understanding of the fiber biosynthesis and as a rich source of potential targets for future fiber engineering.

References

Arabidopsis genome initiative (2000) Analysis of the genome sequence of the flowering plant *Arabidopsis thaliana*. Nature 408: 796-815.

Aspeborg H, Schrader J, Coutinho PM, Stam M, Nilsson P, Denman S, Master E, Djerbi S, Sandberg G, Mellerowicz E, Sundberg B, Henrissat B, Teeri TT (2005) Carbohydrate-active enzymes involved in the secondary cell wall biogenesis in hybrid aspen. Plant Physiol 137: 983-997.

Bhalerao R, Nilsson O, Sandberg G (2003) Out of the woods: forest biotechnology enters the genomic era. Curr Opin Biotechnol. 14: 206-213.

Boudet AM, Kajita S, Grima-Pettenati J, Goffner D (2003) Lignins and lignocellulosics: a better control of synthesis for new and improved uses. Trends Plant Sci 8: 576-81.

Boerjan W, Ralph J, Baucher M (2003) Lignin biosynthesis. Annu Rev Plant Physiol Plant Mol Biol 54: 519-546.

Brumer H 3rd, Zhou Q, Baumann MJ, Carlsson K, Teeri TT (2004) Activation of crystalline cellulose surfaces through the chemoenzymatic modification of xyloglucan. J Am Chem Soc 126: 5715-21.

Brunner AM, Busov VB, Strauss SH (2004) Poplar genome sequence: functional genomics in an ecologically dominant plant species. Trends Plant Sci 9: 49-56.

Christiernin M, Laine J, Brumer H, Henriksson G, Lindström M, Teeri TT, Lindström T (2003) The effects of xyloglucan on the properties of paper made from bleached kraft pulp. Nord Pulp Pap Res J 18: 182-187.

Coutinho PM, Henrissat B (1999) Carbohydrate-active enzymes: an integrated database approach. In HJ Gibert, G Davies, B Henrissat, B Svensson (Eds) Recent Advances in Carbohydrate Bioengineering. The Royal Society of Chemistry, Cambridge, pp 3-12.

Coutinho PM, Stam M, Blanc E, Henrissat B (2003) Why are there so many carbohydrate-active enzyme-related genes in plants? Trends Plant Sci 8: 563-565.

Cyranoski D (2003) Rice genome: A recipe for revolution? Nature 422: 796-798.

Djerbi S, Aspeborg H, Nilsson P, Mellerowicz E, Sundberg B, Blomqvist K, Teeri TT (2004) Identification and Expression Analysis of Wood Specific Cellulose Synthase Genes in Hybrid Aspen. Cellulose 11: 301-312.

Gray-Mitsumune M, Mellerowicz EJ, Abe H, Schrader J, Winzéll A, Sterky F, Blomqvist K, McQueen-Mason S, Teeri TT, Sundberg B (2003). Secondary xylem-abundant expansins belong to the Subgroup A α-expansin gene family and exhibit conservation among different wood-forming species. Plant Physiol 135: 1552-1564.

Hertzberg M, Aspeborg H, Schrader J, Andersson A, Erlandsson R, Blomqvist K, Bhalerao R, Uhlen M, Teeri TT, Lundeberg J, Sundberg B, Nilsson P, Sandberg G (2001) A transcriptional roadmap to wood formation. Proc Natl Acad Sci USA 98: 14732-14737.

Johansson P, Brumer H 3rd, Baumann MJ, Kallas AM, Henriksson H, Denman SE, Teeri TT, Jones TA (2004) Crystal structures of a poplar xyloglucan endotransglycosylase reveal details of transglycosylation acceptor binding. Plant Cell 16: 874-86.

Kallas Å, Denman S, Piens K, Henriksson H, Johansson P, Brumer H, Teeri TT (2004) Enzymatic properties of poplar xyloglucan endotransglycosylase, PttXET16A, expressed in *Pichia pastoris* Biochem J 390: 105-113.

Lunde CF, Morrow DJ, Roy LM, Walbot V (2003) Progress in maize gene discovery: a project update. Funct Integr Genomics 3: 25-32.

Master E, Rudsander U, Denman S, Wilson D, Teeri TT (2003) Enzymatic characterization of a recombinant poplar endoglucanase, PttCel9A, expressd in *Pichia pastoris*. Biochemistry 43: 10080-10089.

Rose JK, Braam J, Fry SC, Nishitani K (2002) The XTH family of enzymes involved in xyloglucan endotransglucosylation and endohydrolysis: current perspectives and a new unifying nomenclature. Plant Cell Physiol 43: 1421-35.

Scheible WR, Pauly M (2004) Glycosyltransferases and cell wall biosynthesis: novel players and insights. Curr Opin Plant Biol 7: 285-95.

Steinmetz LM, Davis RW (2004) Maximizing the potential of functional genomics. Nat Rev Genet 5: 190-201.

Sterky F, Regan S, Karlsson J, Hertzberg M, Rohde A, Holmberg A, Amini B, Bhalerao R, Larsson M, Villarroel R, Van Montagu M, Sandberg G, Olsson O, Teeri TT, Boerjan W, Gustafsson P, Uhlen M, Sundberg B, Lundeberg J (1998) Gene discovery in the wood-forming tissues of poplar: analysis of 5, 692 expressed sequence tags. Proc Natl Acad Sci USA 95: 13330-13335.

Sterky F, Bhalerao RR, Unneberg P, Segerman B, Nilsson P, Brunner AM, Campaa L, Jonsson Lindvall J, Tandre K, Strauss SH, Sundberg B, Gustafsson P, Uhlen M, Bhalerao RP, Nilsson O, Sandberg G, Karlsson J, Lundeberg J, Jansson S (2004) A Populus EST resource for functional genomics. Proc Natl Acad Sci USA 101: 13951-13956.

Zhou Q, Greffe L, Baumann M, Malmström E, Teeri TT, Brumer HIII (2005) The use of xyloglucan as a molecular anchor for the elaboration of polymers from cellulose surfaces: a general route for the production of biomaterials. Macromolecules 38: 3547-3549.

Glossary

A genome progenitor

One of two ancestoral, closely-related species that served as a parent for the commercially important allotetraploid hybrids, *Gossypium hirsutum* and *Gossypium barbadense.*The A genome progenitor was from the Old World.

adventitious buds

Buds arising from an unusual position on differentiated vegetative tissues.

aleurone

A cell layer found in some plant seeds that secretes enzymes used to promote the breakdown of starch into sugars for use by the embryo as an energy source.

alleles

Different forms of the same gene.

allotetraploid

A plant hybrid formed from two closely-related species that contains four sets of genes, two from each parent.

angiosperms

Angiosperms are vascular plants. They have stem, roots, and leaves. Unlike gymnosperms such as conifers, angiosperm's seeds are in a flower.

anticlinal cell division

Cell division where the new wall is perpendicular to the surface.

AOPP

L-α-Aminooxy-β-phenylpropionic acid an inhibitor of the enzyme phenylalanine ammonia lyase an early step in the phenylpropanoid pathway that is required for lignin biosynthesis.

arabinogalactan proteins (AGPs)

Extracellular hydroxyproline-rich proteoglycans implicated in plant growth and development. The protein backbones of AGPs are rich in proline/hydroxylproline, serine, alanine and threonine. Carbohydrate, type II arabinogalactan, can be more than 95%.

atrichoblast

See *trichoblast*.

autofluorescence

Self-induced fluorescence.

bins

When sorting DNA sequences by degrees of similarity, a bin represents all the sequences that probably represent overlapping segments of the same gene. Sequence variation within a bin represents different alleles (forms) of the same gene.

birefringence

Birefringence, or double refraction, occurs when polarized light enters an anisotropic material and splits into two beams that travel at different speeds. The cellulose in cotton fiber secondary cell walls is highly ordered, and therefore the walls are highly birefringent.

callose

A water-insoluble, linear (1→3)-β-glucan found ubiquitously in embryophytes, especially as a component of specialized cell walls in reproductive tissues and as a transient component of the cell plate.

cambial ray

Cells elongate radially to form a distinct ray through the xylem.

cauline leaves

Leaves attached to a stalk, in some cases an inflorescence stalk.

CBD

Microbial cellulose binding domains are small proteins that are found in nature as discrete domains in proteins, such as cellulases, as well as in proteins that have no hydrolytic activity (scaffoldins). Their intrinsic cellulose binding activity serves to juxtapose linked catalytic domains to their cellulose substrate. Plant CBD-like proteins (expansins) are associated with cell wall loosening during plant development as well as in fruit softening.

cDNA

Complementary DNA, which is reverse transcribed from mRNA by reverse transcriptase, a RNA dependent DNA polymerase.

cDNA-AFLP

Complementary DNA – amplified fragment length polymorphisms. An RNA fingerprinting technique involving amplification of partial cDNA sequences from messenger RNA populations for comparison by gel electrophoresis.

CESA-GFP fusion

A hybrid proteins composed of the presumed cellulose synthase catalytic subunint (CESA) and a fluorescent reporter (GFP).

chaperonin

A protein required for the proper folding of other proteins.

CHAPS

3-[(3-cholamidopropyl) dimethylammonio]-1-propanesulfonate. A detergent used to solubilize proteins from cells.

collenchyma

A plant tissue adapted to carry loads in tension, such as the 'strings' on the outside of a celery stalk.

commelinids

A group of monocotyledonous angiosperms, including the palms, bananas, gingers, grasses and other groups, that are united by DNA sequence data as well as the general occurrence of cell-wall bound, UV-fluorescent hydroxyl-cinnamates (e.g., ferulic, *p*-coumaric, diferulic acids).

composite (material)

A material containing two phases with different mechanical properties, such as fibreglass.

condensed substructures

Lignin monomers linked essentially through C-C bonds.

cortical microtubules

Array of long polymers of tubulin arranged around the outside of the cytoplasm adjacent to the plasma membrane.

contig

An overlapping set of DNA sequences that represent an individual gene or cDNA.

critical-point drying

A process where by liquid in the sample is removed without allowing an air-liquid surface to move through the sample. This involves replacing liquid in the sample with liquid CO_2 and heating it to the critical point, where liquid and gas are in equilibrium.

cryo-ultramicrotome

An attachment for an ultramicrotome (q.v.) that allows material to be sectioned when frozen.

cytokinesis

Division of the cytoplasm and plasma membrane of a single cell into two cells.

cytorrhysis

Where both the protoplast and wall shrink and the whole cell collapses. It implies that the solute cannot penetrate the wall.

D genome progenitor

One of two ancestoral, closely-related species that served as a parent for the commercially important allotetraploid hybrid, *Gossypium hirsutum* and *Gossypium barbadense.* The D genome progenitor was from the New World.

DcPRP

Carrot genomic clone encoding p33 (Ebener et al., 2001).

degeneracy

When two different structures have the same function or one structure has more than one function.

dehydration

Replacing the water in a sample with an organic solvent, usually ethanol.

DHP

Synthetic lignin polymer obtained by peroxidase-catalyzed polymerizatin of monolignols.

dicotyledons

Flowering plants (angiosperms) that produce an embryo with paired cotyledons and net-veined leaves.

diffuse growth

Uniform growth of entire cell wall.

early wood

Xylem cells that are produced in the spring; these are large with relatively thin cell walls.

endo-1,4-β-glucanase (1,4-β-endoglucanase)

An enzyme that hydrolyzes the 1,4-β-glucan linkages between the anomeric carbon of an unsubstituted glucosyl residue and the 4-carbon of the adjacent glucosyl residue.

elastic modulus

The ratio of stress to the reversible deformation that it causes.

ELISA

Enzyme-linked-immunosorbent serologic assay that relies on an enzymatic reaction and is used to detect the presence of specific substances (such as enzymes or viruses or antibodies or bacteria, proteins, polysaccharides).

embryogenesis

Development of an embryo from a fertilized egg or zygote.

embryophytes

Flowering plants, conifers, ferns, mosses and liverworts.

endosperm

Nutritive tissue surrounding the embryo within seeds of flowering plants. The endosperm cells are triploid (having three sets of chromosomes); and contain food reserves such as starch, lipids, and proteins that are utilized by the developing seedling.

enthalpy-driven elasticity

Springiness resulting from the deformation of stiff molecules.

epithelial cell

Live cells derived from the ray parenchyma, which surround resin ducts common to conifer wood. Epithelial cells secrete resins which confer resistance to insects.

epigenetic phenomena

Heritable changes in gene function that do not result from changes in DNA sequence.

EST

Expressed sequence tag, partial DNA sequence of a complementary DNA (cDNA) clone synthesized from messenger RNA. Approximately 300-700 nucleotide one-pass sequences from the 5'-ends of the randomly picked cDNA clones are available in BLASTable databases. Repetitive sequencing from multiple tissue libraries tends to capture majority of the expressed genes.

exostosin

An animal β-glucuronyltransferase involved in the synthesis of heparan sulfate that belongs to the glycosyltransferase 47 family. Pectin β-glucuronyltransferase and xyloglucan β-galactosyltransferase are members of this family from plants.

extensin

The first recognized class of structural cell wall proteins. This protein consists of multiple repeats of hydroxyproline rich sequences in which chains of from one to four arabinose residues are attached to the hydroxyl groups of each of the hydroxproline residues.

fiber

Elongated cell with secondary wall providing support and protection.

fibre cell

A much elongated cell often with lignified walls.

field-emission scanning electron microscope

An instrument that uses field-emission type of source to generate a beam of electrons for imaging. Beams generated by such sources are far smaller and more precisely controlled than beams made by conventional electron sources for scanning electron microscopy and allow magnification in the ultrastructural range.

fixation

In cell biology, the act of treating a living sample chemically or in some other way to cross link proteins and other constituents, thus killing the object and preserving the structure (to a greater or lesser extent) for subsequent processing and imaging.

formvar

A polymer, also known as polyvinyl formal, widely used in electron microscopy as a support film for sections on grids being strong even in thin layers. Capitalized because it is a trade name.

fourier transform infrared spectroscopy (FTIR)

A modern form of infrared spectroscopy, a technology used to detect specific frequencies of molecular vibrations. In FTIR spectroscopy, all frequencies of infrared light that have passed through a sample are detected simultaneously, the complex signal being converted into an absorbance spectrum using the Fourier transform mathematical algorithm.

functional genomic

Investigation of the function of genes discovered by genomics.

fusiform initials

Elongated cells of the cambium layer, the meristem that wraps around the tree.

galactomannan

A branched hemicellulosic polymer consisting of 1,4-β-mannan chain that is decorated with α-1,6-linked galactosyl residues.

GC

Gas-liquid chromatography.

GC/MS

Gas-liquid chromatography / mass spectrometry.

gene knock-out

Specific inactivation of a given gene by using molecular techniques.

gene-expression profiling

An analytical method used to measure the activity (transcription) of thousands of genes in specified conditions.

genomics

The nucleotide sequence of the entire (haploid) DNA content of an organism.

(1→3),(1→4)-β-glucan

A β-glucan contain both 1→3- and 1→4-mixed linkages in the linear glucan chain.

glucan synthase I

A Golgi-associated synthase that makes alkali-insoluble 1,4-β-linked glucan chain, which is believed to form either cellulose by pro-cellulose synthase or the backbone of xyloglucan by the free form of xyloglucan glucosyltransferase.

glucan synthase II

A plasma membrane-localized synthase also referred to as callose synthase, which makes 1,3-β-linked glucan and is highly activated by polyamines and very low levels of calcium. Most of it is constitutively present in the plasma membrane and is activated upon mechanical wounding or pathogen attack.

glucomannan

A linear 1,4-β-linked copolymer of glucosyl and mannosyl residues, which is present at low levels (1-5%) in plant cell walls.

glucuronoarabinoxylans

Cell wall polysaccharides with a 1,4-β-linked xylose backbone to which are attached occasional β-glucuronosyl and more frequently α-L-arabino-furanosyl units some of which are further esterified with ferulic acid or *p*-coumaric acid.

graminoid plants

Grasses and grass-like plants, such as sedges.

GUS

β-Glucuronidase.

gymnosperms

The seed plants are divided into the gymnosperms and angiosperms. The gymnosperms arc a non-monophyletic group, and have naked seeds (seeds are not enclosed by a carpellart structure as in the angiosperms).

heartwood

The inner, dead part of wood in a tree trunk that cannot conduct water anymore.

hechtian strands

When a cell is plasmolysed (q.v.), these strands of cytoplasm, named after the botanist who first reported them, run between the cell wall and the protoplast, apparently reflecting point-like loci of adhesion between plasma membrane and cell wall.

helicoidal

Cell walls in which the microfibril orientation changes by a small fixed angle from each layer to the next.

homologues

Orthologs in two different species that carry out the same function.

homogalacturonan

Homogalacturonan is a linear chain of 1,4-linked α-galacturonic acid residues.

housekeeping function

Something that carried out by all cells most of the time.

HPLC

High performance liquid chromatography.

hydroxycinnamic acids

Phenylpropanoids such as ferulic and *p*-coumaric acids derived from cinnamic acid.

hydroxyproline-rich glycopeptides

Small peptides in which some or all of their prolines have been hydroxylated and glycosylated during synthesis. These modifications are often found in proteins associated with plant cell walls.

immunocytochemistry

The use of antibodies to stain cells or tissues, usually in sectioned material but sometimes also in whole mounts. Also called immunohistochemistry.

in muro

In its natural place within the cell wall.

intrusive growth

A growth between cells – opposite of symplastic growth.

in vitro

This refers to an experiment which is performed outside a living organism, in an artificial environment, for instance in a test tube.

IR

Infrared, a region of the electromagnetic spectrum and the frequency range of many molecular vibrations.

jasmonic acid

A cyclic 12 carbon lipid that is derived from linolenic acid and is a powerful signal for many process in plants.

Kyoto Protocol

An international protocol on global warming which became the Kyoto Treaty on February 16th 2005. Participants commited to reduce their emissions of carbon dioxide and five other greenhouse gases, or engage in emissions trading if they maintain or increase emissions of these gases. A total of 141 countries have ratified the agreement.

lateral root primordium

Cells derived from the pericycle that differentiate to form a structure giving rise to the lateral root meristem and mature lateral roots.

late wood

Xylem cells that are formed late in the growing season; these have relatively thick walls

lectin

A type of protein which binds specifically to particular sugar residues without modifying them. Lectins are usually multimeric proteins so can crosslink carbohydrate-bearing structures. Many lectins are hemagglutinins.

lignins

Water-insoluble, high-molecular-mass, phenylpropanoid polymers formed in walls of supporting and conducting cells of vascular plants by random, dehydrogenative, radical polymerization of 4-hydroxycinnamoyl alcohols through several different, non-hydrolysable carbon-carbon and carbon-oxygen-carbon linkages.

linolenic acid

A linear membrane-bound lipid containing 18 carbon atoms and three double bonds.

load-bearing linkage

Any chemical bond in cross-linking molecules of the cell wall, cleavage of which causes cell wall loosening or yielding. Typically, glycosidic linkages in xyloglucan, calcium and boron bridges in pectic polymers, and hydrogen bonds between cellulose and hemicellulose.

lycopodiophytes

The vascular plants have been divided into two groups.the lycopodiophytes (ferns and ferns allies) and seed plants.

M1 (plants)

A population of mutagenized seeds.

M2 (plants)

Plants that have been grown from mutated seeds for two generation to increase the number of homozygotes for a particular allele.

mannan

A 1,4-β-linked polymer of mannosyl residues. walls.

meristem

Undifferentiated, but determined tissue, the cells of which are capable of active cell division and differentiation into specialized tissues.

microarray

A glass slide containing thousands of gene sequences in well organized arrays for expression profiling; mRNA, an RNA copy of the protein coding region of a gene.

microfibrils

Threadlike components of cell walls composed of regular aggregates of cellulose molecules associated through hydrogen bonds an Van der Waals forces and visible in the electron microscope.

middle lamella

A boundary layer between cells made of pectin and cementing cells together.

monoclonal antibody

An immunoglobulin derived from a single clone of antibody-producing cells with a single defined specificity against a single antigen.

monocotyledons

Flowering plants (angiosperms) having an embryo with a single cotyledon, or seed leaf, usually having narrow leaves with parallel veins and smooth edges, and hollow or soft stems.

monolignol

Cinnamyl alcohols whose radical polymerization leads to lignin.

***N*-linked glycoproteins**

Sugars are attached to the amide nitrogen atom in the side chain of asparagine.

***N*-linked oligosaccharide**

An oligosaccharide chain linked to glycoproteins via an N-glycosidic linkage through the amide nitrogen of an asparagine.

NMR

Nuclear magnetic resonance. NMR relaxation experiments can be used to measure thermal motion, which dissipates the ordering of nuclear spins in a solid or liquid.

non-condensed substructures

Lignin monomers linked essentially through β-*O*-4 ether (C-O-C) bonds.

octadecanoid pathway

The biochemical pathway leading from linolenic acid to jasmonic acid.

***O*-linked oligosaccharide**

An oligosaccharide chain linked to a glycoprotein through an *O*-glycosidic linkage to, most often, the hydroxyl group of a serine or threonine residue. There are a few examples of *O*-linked oligosaccharides linked to hydroxyl-proline, as in extensin, or to hydroxylysine, as in collagen.

organogenesis

Creation of specific organs from meristematic tissues.

Orthologs

Referring to the relationship of members of a multigene family in two different species.

p33

A 33 kDa proline-rich cell wall protein corresponding to a wound-inducible transcript in carrot root.

parenchyma

Plant tissue consisting of relatively compactly arranged and unspecialised cells usually with walls that are not lignified.

pectate lyase

An enzyme that cleaves pectin molecules leaving an unsaturated double bond in the terminal galacturonic acid, one of the reaction products.

pectin methyl esterase

The enzyme residing in plant cell walls that converts the methylesterified homogalacturonan (pectin) to the homogalacturonan with carboxylic acid residues (pectic acid) and releases methanol.

peptide hormones

Peptides, usually derived from larger proteins, that are synthesized in one location and, in response to specific cues, are released and transported to another location to regulate physiological processes.

periclinal divisons

Cell divisions occurring parallel to the surface of a tissue.

pericycle cells

Cells lying between the endodermis and the phloem.

pericycle cylinder

A ring of pericycle cells surrounding the vascular tissue of the root.

phenylpropanoids

Compounds bearing a 3-carbon chain attached to a 6-carbon aromatic ring (C6-C3 compounds), most being formed from cinnamic or *p*-coumaric acids.

phloem

The tissue immediately underneath the bark of the tree, where sugars are mainly transported from leaves to roots.

phragmoplast

The central region of the mitotic spindle of a plant cell during cell division, in which vesicles gather and fuse to form the cell plate, apparently guided by spindle microtubules.

plasmodesmatal connection

Communicating cell-to-cell junction in plants in which a channel of cytoplasm lined by plasma membrane connects two adjacent cells through a small pore in their cell walls.

plasmolysis

The act of removing enough water from a plant cell so that the cell membrane draws away from the cell wall and the cell partially rounds up.

pod dehiscence

The spontaneous opening of a fruit by splitting along defined lines to release its contents.

polyelectrolyte

A polymer carrying positive or negative charges.

polygalacturonases

Pectin degrading enzymes that produce oligosaccharide fragments.

polylamellate

Cell walls synthesised with a number of distinct layers in which the microfibril orientation is different.

polyphenoloxidase

An enzyme that oxidizes phenolics, producing polymers that are indigestible. This class of enzymes is commonly found in plant tissues and, when released by chewing insect damage, can crosslink proteins and cause them to be difficult to digest.

polyproline II (PP II) conformations

The peptide bonds of proline and hydroxyproline have different bond angles than those found between other amino acids in proteins. Single proline residues cause structural limitations that form kinks in the proteins, and tandem prolines tend to form left handed helices, called PP II structures.

pore size

Size of the minute opening in cell wall through which solutes and fluids may pass.

positional cloning

A method for the isolation of a gene based on the chromosomal location of a mutation in the gene. Also referred to as map-based cloning.

POVYK

A pentapeptide repeat found in proline-rich proteins consisting of the sequence ProHypValTyrLys.

pre-hormones

Peptide hormone precursors that have N-terminal (pre-) extensions that are important for recognition and entry into the ER and Golgi (secretory pathway) and are cleaved from the protein during the process.

prohormones

The precursors of peptide hormones that are lacking the pre-sequence.

proteinase inhibitors

Proteins that tightly bind proteinases to form stable complexes that are enzymically inactive. Plant proteinase inhibitors can bind the digestive enzymes of herbivores and interfere with normal protein digestion.

protoplast

A plant cell without its cell wall.

pteriodophytes

Land plants are divided into two groups: non-vascular plants and vascular plants. The non-vascular plants is composed of the pteriodophytes, which is a general term used to collectively refer to hornworts, liverworts, and mosses.

QTL analysis

Identification of chromosomal loci where the genes responsible for the quantitative traits (=traits regulated by many genes) are located.

quotient Q

Dilution quotient expressed as the ratio between the detector response of original dextran probing solution (DPS) and of modified (wt or *mur1*).

ray parenchyma

Cells with produce resins in the ray.

redundancy

When there is a multiplication of a given structure and related mechanism that have the same function in the system.

RG-I

Rhamnogalacturonan I is a pectic polysaccharide that contains a backbone of the repeating disaccharide: 4)-α-Galacturonic acid-(1,2)-α-Rhamnose-(1,. Between 20 and 80% of the Rha*p* residues are substituted at C-4 with neutral and acidic oligosaccharides. The oligosaccharides contain linear and branched α- Ara*f* and β- Gal*p* residues.

RG-II

Rhamnogalacturonan II is a low molecular mass (5-10 kDa) pectic polysaccharide that is solubilized by treating a cell wall with endopolygalacturonase. RG-II contains eleven different glycosyl residues.

RGP

Reversibly glycosylated polypeptide of ~40 kD, which is specific to and ubiquitous in plants, i.e., does not occur in fungi and lower organisms. It is believed to be involved in xyloglucan formation in dicots and arabinoxylan formation in monocots.

rhamnogalacturonan (RG)

A region in pectin composed of a repeating disaccharide backbone of rhamnose and galacturonic acid with sidebranches of single galactose residues, arabinans, and galactans attached to the *O*-4 of one third to one half of the rhamnose residues.

RI (refractive index)

As a ray of light passes from air into a block of glass, the direction in which it is travelling is changed. The amount of bending that takes place depends on the amount of solutes present. It is used for detection of non-UV absorbing substances.

ruthenium red

A dye used for pectin staining, which binds to the free carboxyl groups of acidic sugars.

S1 layer, S2 layer and S3 layer

Successive layers of the secondary cell wall in conifers. The angle of the cellulose fibrils change with each different layer. The S2 layer is the thickest layer.

sapwood

The outer, living part of wood in a tree trunk that is capable of conducting water.

secondary wall

A specialized wall, rich in cellulose, lignin and hemicelluloses, that are deposited at the cell surface inside the primary (growing) wall in some cell types.

siliques

Elongated, dehiscent fruits typical of crucifers.

singletons

An EST sequence which by similarity belongs to a given gene family but does not form a contig with other EST sequences in the same family.

sitosterol-cellodextrins

Sitosterol (22,23-dihydrostigmasterol) distributed as plant sterol.

plasmolysis

Where the protoplast shrinks and pull away from the inner surface of the cell wall or the separation of plant cell cytoplasm from the cell wall as a result of water loss. It implies that the solute has been able to penetrate the wall and accumulates in the periplastic space between the wall and the protoplast.

sputter coating

A process that deposits a fine coat of metal on a sample for the purpose of making conductive and hence able to be imaged for scanning electron microscopy.

stokes' radii

The effective hydrodynamic radii of a molecules.

SuSy (sucrose synthase, EC 2.4.1.13)

UDP-glucose: fructose 2a-glucosyltransferase. An enzyme that catalyzes a reversible reaction between sucrose and UDP to form UDP-glucose and fructose. The enzyme plays a major role in the degradation of sucrose in plant sink tissues and is proposed to be a molecular switch for carbon partitioning to cellulose synthesis.

symplastic growth

A growth of cells attached to each other – opposite of intrusive growth.

syncytium

A single cell with many nuclei.

T-DNA insertion

A physical mutagen that causes mutations by inserting transfer DNA into the genome, which is expected to knockout the gene that it is inserted in.

T-DNA insertion collections

Stocks of mutant seed lines in which a piece of bacterial DNA (the T-DNA) has inserted randomly into the plant genome, potentially disrupting gene function. Amplification of the sequence borders of the T-DNA usually identifies the site of insertion.

T-DNA tagging

The insertion of transfer-DNA into a gene, which marks the gene with a known DNA sequence.

***t*-fucose**

Fucose residue present in the non-reducing end of a polysaccharide as xyloglucan.

tip growth

Localized growth at cell tip, for example in root hairs or fibers – opposite of diffuse growth.

tracheary elements

Water-conducting cells (tracheids and vessel members) of the xylem tissue.

tracheid

A capillary tube formed from a series of dead cells in the xylem, or a single such cell, where at least the middle lamella of the end walls remains intact.

***trans*-differentiation**

A phenomenon whereby a cell already specialized for its function (differentiated) becomes committed and differentiates into a different cell type, exemplified, in the zinnia system, by the change in cell fate from photosynthetic palisade cells in the leaf to become tracheary elements of the xylem.

transferase

An enzyme with the function of transferring a given sugar unit into a polysaccharide as it is being synthesized. Cellulose synthase is a kind of transferase, namely cellulose 4-β-glucosyltransferase.

transposon

DNA that is able to move from one part of the genome to another.

trichoblast (atrichoblast)

A epidermal cell type that forms a root hair, in contrast to an *atrichoblast*, which is an epidermal cell type that does not form a room hair.

turgor

The excess osmotic pressure inside plant cells that provides the driving force for their expansion during growth.

UDP-glucose

A glucose sugar unit with uridine diphosphate (UDP) moiety attached to it. UDP-glucose is itself made from the reaction of glucose-1-phosphate and the action of an enzyme called pyrophosphorylase. UDP-glucose is an intermediate in the synthesis of many biologically important compounds in plants, such as cellulose (linear chains of 1,4-β-glucose units), callose (linear chains of 1,3-β-glucose units), and the backbone of xyloglucan.

ultramicrotome

A device to cut sections of embedded material with thickness ranging from 50 nm to several microns.

UTR

Untranslated leader region. Noncoding sequences found at either end of mRNAs.

vascular cambium

Plant lateral meristem that is producing secondary xylem (wood) and secondary phloem and composed of longitudinally elongated cells (fusiform initials) and isodiametric cells (ray initials).

vessel element

Xylem conducting cell, perforated at the ends and aligned end-to-end with similar cells forming a water conducting vessel.

V_{max} and K_m

These are the two parameters that define the basic kinetic behavior of an enzyme. V_{max} is defined as the maximum velocity of a reaction under a given set of conditions. K_m is the concentration of substrate required to produce a velocity that is ½ V_{max}. Km is a measure of roughly how much substrate is required to get to full speed, and is measured as a concentration (the lower the Km for an particular substrate the higher affinity the enzyme has for that substrate).

warty tertiary layer

Inner most layer of a tracheid cell wall, with distinct pores or warts. Alternative name for S3 layer.

xylem

Tissue formed underneath the bark, which conducts water from the roots to the top of the tree. Each year a new xylem layer is produced. Successive layers form the rings of the tree.

xylogalacturonan

A region in pectin consisting of a backbone of homogalacturonan with single xylose residue sidechains on *O*-3 of many of the galacturonic acid residues.

xylogenesis

The developmental process in trees leading to wood (xylem) formation.

xyloglucan

A hemicellulose composed of a 1,4-β-glucan backbone with 1,6-α-xylosyl residues along the backbone. Dicot xyloglucan is composed of repeating units of the heptasaccharide (seven sugar oligosaccharide) XXXG, while most monocot xyloglucan is composed of repeating units of the hexasaccaride (six sugar oligosaccharide) XXGG. The repeating unit of dicot and monocot xyloglucan is further substituted with galactose, fucose, and arabinose sugar residues.

XXXG, XXLG, XLLG and XXFG

There is an unambiguous nomenclature for xyloglucan-derived oligosaccharides. This nomenclature system defines the oligosaccharide by listing the substitution pattern of glucosyl residues from the non-reducing to the reducing end of the 1,4-β-glucan backbone, where G = glucose, X = xylose (assuming the presence of G), L = galactose (assuming the presence of G and X), and F = fucose (assuming the presence of G, X and L).The main backbone of xyloglucan polysaccharides is made up of repeating units of a XXXG heptasaccharide (seven sugar).

xyloglucan endotransglucosylases/hydrolases (XTHs)

A family of enzymes that specifically use xyloglucan as a substrate and catalyze xyloglucan endotransglucosylation (XET) and/or xyloglucan endo-hydrolase activities (XTH). XTH family belongs to Glycoside Hydrolase Family 16.

Index

CPSIA information can be obtained at www.ICGtesting.com
Printed in the USA
LVOW112212261011

252281LV00001B/147/A